MOST® Work Measurement Systems

MOST® Work Measurement Systems

Fourth Edition

Kjell B. Zandin

Edited by
Therese M. Schmidt

CRC Press
Taylor & Francis Group
Boca Raton London New York

CRC Press is an imprint of the
Taylor & Francis Group, an informa business

Fourth edition published 2021
by CRC Press
6000 Broken Sound Parkway NW, Suite 300, Boca Raton, FL 33487-2742

and by CRC Press
2 Park Square, Milton Park, Abingdon, Oxon, OX14 4RN

Third edition published by CRC Press 2002

CRC Press is an imprint of Taylor & Francis Group, LLC

Library of Congress Cataloging-in-Publication Data
Names: Zandin, Kjell B., 1937- author. | Schmidt, Therese M., editor.
Title: MOST work measurement systems / Kjell B. Zandin ; edited by Therese M. Schmidt.
Other titles: MOSTr work measurement systems
Description: Fourth edition. | Boca Raton, FL : CRC Press, 2021. | On title page the registered trademark follows "MOST" in the title. | Includes bibliographical references and index.
Identifiers: LCCN 2020024141 (print) | LCCN 2020024142 (ebook) | ISBN 9780367345310 (hardback) | ISBN 9780429326424 (ebook)
Subjects: LCSH: Work measurement.
Classification: LCC T60.2 .Z36 2021 (print) | LCC T60.2 (ebook) | DDC 658.5/42—dc23
LC record available at https://lccn.loc.gov/2020024141
LC ebook record available at https://lccn.loc.gov/2020024142

ISBN: 978-0-367-34531-0 (hbk)
ISBN: 978-0-429-32642-4 (ebk)

Typeset in Times
by codeMantra

Dedication

To Kjell B. Zandin

Measure of work brings knowledge. Through this knowledge, factual decisions and improvements can be made and control exercised.

> When you can measure what you are speaking about, and express it in numbers, you know something about it; but when you cannot measure it, when you cannot express it in numbers, your knowledge is of a meagre and unsatisfactory kind.
>
> — Lord Kelvin

This text is intended to give the reader a complete description of the three systems that are included in the MOST Work Measurement Technique. It is not, however, the sole training device through which a MOST System is learned. Any attempt to apply the material in this text without proper certification training will be done at the discretion of the reader.

Chapter 3 (The BasicMOST System), Chapter 4 (The MiniMOST System) and Chapter 5 (The MaxiMOST System) provide the reader with a thorough understanding of the MOST Work Measurement Technique for application in virtually any industry.

Chapter 6 provides an overview for organizations to use MOST-based standards to effectively manage performance.

Contents

Foreword

The concept of determining the time it takes to perform a task has been a part of our everyday lives. From a simple estimation of how long it takes to perform household tasks, doing yardwork or driving from one place to another, this skill has been honed without even realizing it. For those responsible for the formal determination of times to support a company's need for solid planning and managing of workforce performance, one technique stands out for its simplicity of use and its transparency to explain it to those who need to understand it. That technique is MOST. For over 50 years, this technique has been at the core of what we, as MOST applicators, have been doing to improve workforce performance and eliminate waste in operations. '*They know MOST...*' has been a badge of honor many of us have earned throughout our careers.

Developed by a good friend and mentor of mine, the late Kjell Zandin, and brought to the United States in 1975, Kjell wanted to introduce a new approach to work measurement that would make people in charge of efficiency more efficient. He started with the end in mind and built a system that is easy to use and meets accuracy requirements needed in an industrial or business setting.

Over the years, MOST has been the secret ingredient that people rarely see to drive improved workforce performance to support an environment where people want clear and fair expectations to provide the best service they can with the best quality achievable. Many well-respected companies have used MOST to unlock value in industries such as manufacturing, retail, logistics, hospitality and public service to better serve people.

MOST is considered the greatest used and widely accepted technique for establishing engineered labor standards. The discipline of creating and maintaining engineered standards is at the heart of the effective management of workforce performance. This new edition includes a chapter on leveraging MOST-developed standards to effectively manage performance. MOST can also be used in concert with software applications to drive the effectiveness of workforce management systems worldwide.

On behalf of all those who have worked at H. B. Maynard and Company and Accenture, it was a privilege to have worked with Kjell for many years. I am sure he was glad to see what MOST has done and would be proud of what it continues to do for industry and customers alike. As Kjell would say, "Wishing You MOST Success!"

James R. Wilk
Accenture LLP
Pittsburgh, Pennsylvania

In Memory of Kjell B. Zandin

1937–2011

Often referred to as 'Mr. MOST,' Kjell Zandin was born in Sweden and became known worldwide for his outstanding accomplishments in the field of industrial engineering.

MOST was developed in the late 1960s while Kjell was working for the Swedish Division of H. B. Maynard and Company. As he would explain it, *The primary reason for developing MOST was to make life easier (through improved productivity) for work measurement analysts.*

Kjell brought MOST to the United States in 1975 and was expected to stay about one month. Thirty-six years later and more than 40,000 MOST-certified applicators worldwide, Kjell made his mark as a key figure in the field of industrial engineering.

Bopaya Bidanda, Ernest Roth Professor and Chairman of the Department of Industrial Engineering at the University of Pittsburgh and President-Elect of IISE, who collaborated with Kjell on research projects, notes that Kjell was an eternal educator and was always glad to mentor students and faculty. "He routinely presented guest lectures at the University of Pittsburgh and was always available to provide guidance and counsel to student projects." Dr. Bidanda adds, "Kjell dramatically transformed the IE profession and made a significant contribution to it with the development of MOST, a predetermined motion-time system."

In addition to his work developing MOST, Kjell was proud to have capped off his career as the editor of the fifth edition of *Maynard's Industrial Engineering Handbook*, published in 2001.

Jim Wilk, Accenture Senior Manager who worked with Kjell at H. B. Maynard and Company for more than 15 years, expressed that "Kjell always led with his calm demeanor, was a source of the truth and was always eager to provide guidance to me and many others when it came to workforce management or MOST. He always made time for me and many others, and I appreciated that very much."

Kjell may be remembered for his contributions to industrial engineering, but for many, he will be remembered as a Pittsburgh Penguins hockey fan who loved to travel, take amazing pictures, spend time with his family and yet always seemed to have time for others.

Thank you Kjell for what you have given to all. You will be remembered as a humble and gentle giant, a pioneer in industrial engineering and genius in creating MOST.

Acknowledgments

Several people assisted Kjell Zandin in the development and promotion of the MOST Work Measurement Technique including Lennart Gustavsson, the Division Head of Maynard Sweden and supporter of the creation of MOST, and William (Bill) M. Aiken, who was the President of Maynard North America in the early 1970s and who insightfully invited Kjell to the United States to introduce MOST to U.S. industries. While many others were also involved in the development and growth of MOST, Kjell was the driving force behind the system and I, along with the industrial engineering community as a whole, owe him a great deal of gratitude.

Originally developed to solve a problem in one Swedish company in 1967, MOST is now successfully being applied by industrial engineers and others worldwide in a great variety of industries and organizations. After more than 50 years in use, MOST is still the standard for work measurement applications for many companies. A special thank you to Kjell Zandin for creating an easy-to-use, practical and efficient work measurement tool that provides functionality and benefits for organizations all over the world.

Updating the manuscript for a technical book involves a great deal of planning, coordinating and teamwork. It was such a rewarding experience to work with the individuals noted below who all had a vested interest in making this the best edition on Kjell's behalf. Each person used their expertise to contribute to making this edition more comprehensive, yet easier for the reader to understand. I am so grateful and honored to have led this project with this team!

First, I would like to thank Sharyn Lovelace for her contributions to the book. Sharyn is a certified MOST Master Instructor who, in addition to supporting hundreds of clients on their MOST certification training during this process, assisted me every step of the way. Working with clients for nearly 20 years, Sharyn has great insight into the nuances of MOST and how content could be clarified or expanded to make it more applicable for the reader. Thanks to Sharyn's client-centric focus and desire to make this book full of relevant examples, the quality of the book was greatly improved.

For sharing their expertise, many thanks to Jim Wilk and Mike Kaminski. Jim Wilk, Accenture Senior Manager, provided overall guidance and support for several chapters in the book. Jim's experience with standards development and client implementations contributed to his insights, especially for Chapters 1 and 2. Mike Kaminski, Accenture Federal Services Senior Manager, offered examples and technical content in Chapter 5. Both Jim and Mike also provided significant contributions for a new chapter in the book about using standards developed with MOST to improve workforce performance.

This new edition contains updated charts from the previous editions. Thank you to Andrew Lowe for expertly creating the updated charts in an amazingly efficient manner. Andrew's involvement in the development of the certification training for MOST enabled him to see the content from a different perspective and skillfully update many charts to be more user-friendly.

Thank you to Mary Ann Broderick for sharing her expertise on the managing performance process and providing key content to a new chapter in the fourth edition that shows the vital role standards play in improved workforce performance. This chapter will provide organizations with a structured approach to increase productivity through improved workforce performance.

The fourth edition would not be complete without the expert proofing assistance provided by Terry Mylan and Jennifer Jaglowski. Thank you to Terry and Jennifer for providing additional insights into each chapter.

As MOST continues to expand globally, this edition of the book includes more photographs of common tools and measuring instruments to assist the reader in identifying the tools used. Our sincere thanks to Lowe's Companies, Inc. for providing pictures of Kobalt tools and several other common hand tools used in Chapters 3 and 5, to Mitutoyo America Corporation for providing pictures of measuring tools used in Chapters 3 and 5 and to Crown Equipment Corporation for providing pictures of powered trucks used in Chapter 5.

I genuinely appreciate the support and encouragement received from Accenture to champion the update of this book and in recognizing the importance of MOST in the field of work measurement and industrial engineering.

Thank you to all of the supporters of MOST and those that have become certified applicators of a MOST System. Your comments and recommendations over the years have been invaluable to improve this book and for all future MOST applicators.

Therese M. Schmidt
Accenture LLP

Author

With over 35 years of consulting experience in the industrial engineering field, **Kjell B. Zandin** was an Owner and Director of H. B. Maynard and Company, Inc., Pittsburgh, Pennsylvania (now part of Accenture) for more than 20 of those years.

A native of Gothenburg, Sweden, Mr. Zandin joined Maynard Sweden in 1964. In the late 1960s, he developed an exciting new concept of work measurement, MOST Systems. Mr. Zandin relocated to the United States in 1975 to introduce MOST to U.S. industry and subsequently authored the book *MOST Work Measurement Systems.*

In 1986, Mr. Zandin was the recipient of the first 'Technical Innovation in Industrial Engineering Award' presented by the Institute of Industrial Engineers for his 'significant innovative contributions to the industrial engineering profession.' In 1990, Mr. Zandin received the 'Royal Charter Award' from the Institution of Production Engineers in Great Britain.

Mr. Zandin served on the Visiting Committee for the Industrial Engineering Department at the University of Pittsburgh for many years and on the Board of Directors of the Pittsburgh Chapter of the Institute of Industrial Engineers from 1995 to 2001.

Mr. Zandin was a Senior Member and Fellow of the Institute of Industrial Engineers and a Fellow of the World Academy of Productivity Science.

Mr. Zandin held an MSc degree in mechanical engineering from Chalmers University of Technology in Gothenburg, Sweden, supplemented by education in Business Administration at the Institute of Business Management in Stockholm, Sweden.

Mr. Zandin was also the Editor-in-Chief of the Fifth Edition of *Maynard's Industrial Engineering Handbook* published by McGraw-Hill, New York, N.Y.

Editor

As a Service Management Manager with Accenture, **Therese M. Schmidt**'s experience includes more than 10 years in human resource management. She began as a human resources consultant with a Pittsburgh-based trade association and later became a MOST Certification expert with H. B. Maynard and Company (now part of Accenture), providing training and support for hundreds of organizations across all industries for the past 25 years.

Therese was the Editor of the third edition of the *MOST Work Measurement Systems* book and is a certified Master Instructor in the BasicMOST, MiniMOST, and MaxiMOST Systems.

Ms. Schmidt has delivered MOST Certification training for clients in many industries including retail, distribution, and manufacturing in both the lecture and online formats in addition to developing and delivering workforce optimization training, team skills and leadership training. Ms. Schmidt has been an integral part of the training development team acting as a subject matter expert for the MOST Certification programs through the Accenture Workforce Optimization Academy and has worked with Accenture partners across the globe to continue to expand MOST worldwide.

Ms. Schmidt holds a BS degree in business administration with a focus on human resources and accounting from Indiana University of Pennsylvania and is also a certified trainer in the Skills for an Empowered Workforce Program through Development Dimensions International.

1 The Concept of MOST—An Introduction

WORK MEASUREMENT

Work Measurement, or the analysis of a task to identify a time for the average worker to complete the task, is essential to organizations looking to improve productivity or eliminate waste.

There are many reasons for wanting to know the amount of time a task will take which mainly fall into four categories: (1) to *accomplish long-term planning*, (2) to *manage performance*, (3) to *establish costs* and (4) to *determine weekly staffing hours*. Suppose an organization wishes to manufacture a new product or rearrange a distribution center or store layout, the planning and budgeting process could be accomplished with increased confidence using an economical, predetermined motion time system (PMTS). Documenting the method and knowing the time it takes to perform a task provides an organization the ability to:

- Plan for staffing associates.
- Determine labor costs based on activity-based standards.
- Identify the type and capacity of any needed equipment.
- Plan for the supply and delivery times of materials.
- Obtain predicted costs of production or service.
- Determine the overall production or service schedule.
- Analyze the feasibility of new products and services.
- Set expectations and follow-up on production or service goals.
- Manage individual or departmental performance.
- Implement a performance-based pay system.

Knowing how much time it takes to perform certain tasks enables a manager and an organization to achieve and maintain a high utilization of staff, material and equipment. This results in an overall efficiency that will make sustainable organizational growth possible.

The original form of work measurement was simply guessing or estimating the time. It is interesting to note that the primitive guessing technique employed thousands of years ago is still in use today in many modern organizations. Today's version is a more advanced form of the original technique, however, and is known as an *educated guess*. The educated guess is unscientifically supported by intuition, individual personal experience, the importance of the estimation to be made and the inherent ability or inability of the applicator to make a confident-sounding response. This technique is neither scientific (well-documented or statistically supported) nor accurate (with any degree of confidence or consistency). The sole benefit is that it can be done quickly.

Once products began to be manufactured or work tasks completed, another source of information was available from which future times could be estimated. The *historical data* concept of work measurement evolved from estimation. Predicting time for future situations came from records of previous work that had been accomplished. Using historical data does one thing very well; it accurately represents what has already happened. To use it to predict what *will* happen assumes the following two points occur and if these conditions are met, historical data should work well:

1. The conditions and actions under which the process was originally performed are consistently repeated.
2. The actions to be performed will be performed exactly as those on which the historical data is based.

Since using historical data and guessing is clearly not an exact science, Frederick Taylor, one of the intellectual leaders of the efficiency movement, viewed work as something that could be engineered or controlled. Work did not have to be a haphazard repetition of what had gone on before. In fact, workers could be provided with instructions and training as to the best way to perform certain tasks. Tasks were then broken down into elements or short activities that could be arranged and managed to produce more productive and less fatiguing work. Each element was studied to determine which actions were productive and which were unproductive. Keeping only productive elements, a stopwatch was used to determine the time for each. The time recorded was the actual time taken by an individual to perform a certain task under specific conditions. To make such times transferable to other workers and other situations, time for the average skilled worker working under average conditions had to be determined. This was and is still accomplished by *performance rating* and *stopwatch time* studies.

The analyst determined the performance rating by observing the pace of the individual being studied compared to a worker working at a level of 100% (or average) skill and effort. If the worker observed was working with more skill and effort than the average worker, a rating of over 100% would be applied to the time from the stopwatch and the time would be increased to represent 100% or average performance. For example, if the stopwatch time is 1.00 minute and the overall performance rating is 115%, the stopwatch time of 1.00 minute is then increased by 15% for an allowed time of 1.15 minutes. Likewise, if the worker observed is not putting forth the effort or skill to be 100%, a rating of less than 100% would be applied to the time recorded from the stopwatch and the time would be decreased to reflect a 100% performance. For example, if the stopwatch time was 1.00 minute and the rating was 95%, then the stopwatch time would be decreased by 5% for an allowed time of 0.95 minutes. The scientific process of engineering a task using the time study method just described has two weak points:

1. The individual analyst must *subjectively* rate or compare the operator to an estimated 100% (average) performance standard.

2. No matter how sophisticated, expensive or precise the timepiece, a watch simply does not forecast, predict or accurately determine times for future situations; it can only determine the time for what has already occurred based on existing work conditions (methods, layouts, capacities, etc.).

It was discovered by Frank and Lillian Gilbreth, engineers and pioneers of time and motion study, that all manual operations are combinations of basic elements. The Gilbreths isolated and identified these elements primarily so that methods could be accurately documented, explained and improved. They reasoned that to reduce the motion content of a task was to reduce the effort and the time to perform the task. The result is higher production and an increased service level.

Followers of Frederick Taylor practiced *time study,* but followers of Frank and Lillian Gilbreth practiced *motion study.* Predetermined motion time systems were then developed and joined together the best of both techniques combining time and motion studies.

Predetermined motion time systems utilized time study and micromotion techniques to determine and assign times to specified basic motions. The motions and associated times were cataloged. Work measurement then became a matter of establishing the best basic motion pattern to perform a certain task and, from the catalog or data card, assigning the appropriate predetermined time for each basic motion in that pattern. Since the times for all motions were predetermined, one could now accurately predict future task times. The watch was needed only for timing equipment processes. But what about performance rating? The authors of the most common PMTS built their systems based upon the leveled times for 100% (average) performance. Therefore, with the catalogs of predetermined times already leveled to 100%, there was no longer a need to rate an associate. The analyst could now focus on the actual work being accomplished, not on the associate.

One of the first PMTS, Methods Time Measurement (MTM), was developed in 1948 by Harold B. Maynard, G. J. Stegemerten and J. L. Schwab. Because it is a very detailed system, MTM has been recognized as an accurate and widely accepted PMTS.

The MTM system features a detailed data card of basic motions (reach, move, grasp, position, release, body, leg and foot motions, etc.), each associated with particular variables. Basic motions are identified, the variables are considered and the appropriate times are chosen from the data card. Because of its detail, MTM can be a very exact system but also very slow to apply. Basic motion distances must be accurately measured in inches or centimeters and correctly classified. Because of the detail level needed with MTM, analyst errors can be a problem. The times that result from performing an MTM analysis reflect a 100% performance level and can be established for operations prior to their execution.

Modified versions of MTM-1 were developed to reduce analyst errors and lengthy analysis time. These two versions are MTM-2 and MTM-3. These systems group or blend together certain basic motions and/or variables to reduce the analyst effort required to apply the technique.

The analysis of work today, as practiced by industrial engineers using a PMTS, is performed by systematically breaking work down into very small and distinct units called *basic motions.* For highly-repetitive, short-cycle operations, this attention to detail is necessary and has been found to be effective in generating valuable methods improvements. For less repetitive operations or job shop production, however, this detailed approach is very tedious, requires a great deal of time and effort on the part of highly trained engineers and technicians and can be very costly. The benefits are often questionable when considering the amount of analysis effort required.

In addition to MTM-1, -2 and -3, several other simplified predetermined time systems, often based on MTM for specific application areas and companies, have been developed and are in use today.

BasicMOST, a revolutionary system that measures medium-cycle repetitive and non-repetitive work for general industrial applications, was introduced in 1972 in Sweden and in 1974 in the United States.

Since the 1970s, BasicMOST has been applied in retail, manufacturing, service and distribution industries. While BasicMOST is the most widely used system, MOST Systems was expanded in 1980 to include MiniMOST and MaxiMOST. MiniMOST was developed to give work measurement applicators a choice in measuring work for short-cycle, highly-repetitive operations. MTM had been the predominant system for these activities, but as stated earlier, it was too cumbersome and detailed to easily apply. MiniMOST, the most detailed system within the MOST Work Measurement Technique, is a simpler system than MTM but is still able to provide a high level of detail and accuracy.

On the other end of the spectrum, BasicMOST was considered too detailed for industries with long-cycle operations. To address this need, MaxiMOST was developed. MaxiMOST began in the shipbuilding industry and has expanded to cover maintenance and heavy assembly operations.

Since the inception of the MOST Work Measurement Technique, more than 40,000 individuals worldwide have been trained and certified in either BasicMOST, MiniMOST or MaxiMOST. Many analysts are certified in more than one MOST System based on the activities in their organizations. MOST is a straightforward, easy to understand and apply work measurement technique geared toward anyone wanting to improve workforce performance in their organization.

DEFINITION OF TERMS—MOST ANALYSIS

In order to facilitate the understanding of this text, the definition of several terms commonly used in connection with engineered standards and the MOST Work Measurement Technique is presented below. The first list of terms which are specific to the MOST Work Measurement Technique are listed below and shown in Figure 1.1:

- MOST analysis
- Activity
- Method step
- Sequence model
- Parameter
- Sub-activity
- Index value

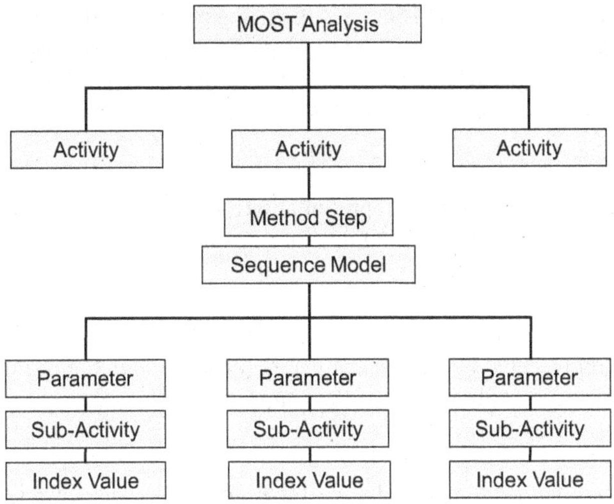

FIGURE 1.1 Terms associated with a MOST analysis. A MOST analysis includes one or more activities. Each activity is identified with a method step and one sequence model. Index values are applied for each sub-activity on each parameter in the sequence model.

MOST Analysis

A *MOST analysis* is a complete study of a task typically consisting of several method steps and corresponding sequence models. Appropriate parameter time values are assigned to the sequence model, resulting in a total *normal time* (defined below) for the task, which does not include allowances.

Activity

An *activity* is a series of logical events that take place when an object is moved, observed or treated by the hand or manipulated with a tool or handled with the aid of a transportation device. An *activity* begins when an operator reaches to gain control of an object *or* leaves the normal location (workplace) to perform these events and concludes when the operator has returned to the original location or releases the object. The word *activity* may also be used in a general sense designating a task or a series of events.

Method Step

A *method step* is a description of the activity being performed. A MOST analysis typically consists of 5–20 method steps describing the tasks performed. Method steps are determined by analyzing the movement of objects. There will only be one method step for each object movement and each method step is identified by a method description of the activity and a corresponding sequence model.

Sequence Model

Sequence models are used to represent the sequence of events that occur when an object is moved, a tool is used or a piece of equipment is used. One sequence model is applied to each method step and consists of a string of letters called parameters which represent different types of manual movements. Several predefined sequence models represent different types of activities including movement of an object through space, along a controlled path or the use of tools or equipment.

Parameter

A *parameter* is a one-character representation of a sub-activity. For example, the General Move Sequence Model contains four distinct parameters **A, B, G** and **P.**

Sub-activity

A *sub-activity* is a defined, discrete part of an activity or sequence model. Sub-activities are represented by parameters in a sequence model. For example, to collect a group of objects is a sub-activity of the Gain Control (**G**) parameter.

Index Value

Index values are numbers assigned to each parameter in a sequence model and represent the time it takes to perform the sub-activity. For example, to collect a group of objects is assigned an index value of '3' according to the rules of BasicMOST. Index values are used to generate the total time required to perform the task.

DEFINITION OF TERMS—STANDARD TIME

Engineered standards are developed systematically and scientifically and are based on method and time. The time is referred to as standard time. Figure 1.2 shows how a MOST analysis is used as the foundation for standard time. The terms associated with standard time are listed below:

- Normal time
- Sub-operation
- Operation
- Allowances
- Standard time

Normal Time

Normal time is the time required by an average skilled worker, working at a pace that is ordinarily used by workers when adequately trained and capably supervised to complete a task by following a prescribed method under specified work conditions and without interruptions. The result of a MOST analysis is normal time. Normal time includes manual time and process time. Manual time accounts for the motions performed by humans. Process time represents work controlled by machines.

Sub-operation

A *sub-operation* is a discrete, logical and measurable part of an operation. Sub-operations are often referred to as building blocks, or portions of work used to build an operation. The content of a sub-operation may vary depending on the type of work, accuracy requirements and application area. Typically, sub-operations represent a

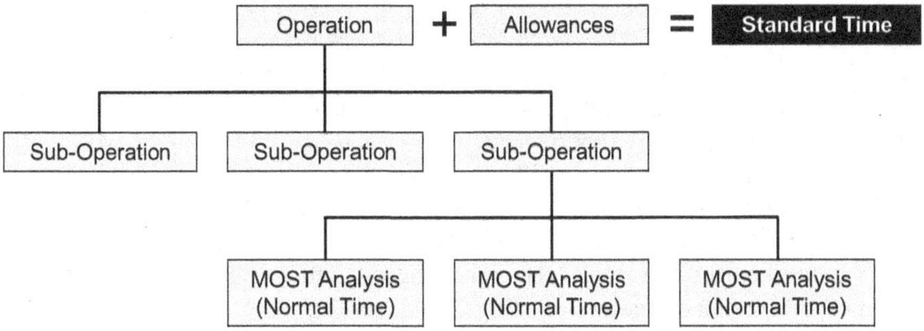

FIGURE 1.2 MOST analysis is used as the foundation for standard time. A MOST analysis provides the normal time (manual time + process time) for a sub-operation. One or more sub-operations make up an operation. Allowances are added to achieve the standard time.

building block or 'standard data' that can be used in all operations that contain this same piece of work. When using standard data, an example of a sub-operation may be: 'Process credit card transaction.'

Operation

An *operation*, when referenced with engineered standards, is a job or task consisting of one or more sub-operations. An example of an operation in a retail grocery environment may be: 'Complete customer check-out' which may consist of several sub-operations such as 'scan item,' 'key in item,' 'process credit card transaction' and 'bag item.'

Allowances

Allowances are the time added to the normal time to account for personal time, rest time and minor unavoidable delays. The allowances provide time that cannot be captured in the MOST analysis such as breaks and picking up dropped parts. The allowance '*factor*' is expressed as a percentage of normal time. Allowances are applied to the normal time to determine the standard time.

Standard Time

Standard time is the total allowed time that it should take an average skilled, trained operator working at a normal pace following a prescribed method under specified working conditions with capable supervision to perform an operation. The total time includes normal time (manual time + process time) and allowances based on established and documented work conditions and a specified work method.

THE CONCEPT OF THE MOST WORK MEASUREMENT TECHNIQUE

Because industrial engineers and those associated with productivity improvement are taught that with enough study, any method can be improved, many efforts have been made to simplify the task of measuring work. The result of this simplification was the creation of the concept MOST, known as the **M**aynard **O**peration **S**equence **T**echnique.

To many people, work means exerting energy to accomplish a task or to perform a useful activity. In the study of physics, it is learned that work is defined as the product of force multiplied by distance ($W = f \times d$), or more simply, work is the displacement of a mass or object. This definition applies quite well to the largest portion of the work accomplished every day at many organizations (e.g., pushing a button to open the cash register, lifting a box to load onto a pallet or placing a part into the dashboard of a vehicle). Thought process or thinking time is an exception to this concept, as no objects are being displaced. For the overwhelming majority of work, however, there is a common denominator from which work can be studied; the *displacement of objects*. All basic units of work should be organized for the purpose of accomplishing some useful result by simply moving objects. That is what work is. MOST is a system to measure work. Therefore, *MOST focuses on the movement of objects.*

Efficient, smooth, productive work is performed when the basic motion patterns are tactically arranged and smoothly choreographed with a method designed using the principles of methods engineering. The movement of objects follows a consistently repeating pattern such as reaching, grasping, moving and positioning of the object.

MOST identified and arranged these repeating patterns as a sequence of events (or sub-activities) followed in moving an object. The MOST Work Measurement Technique includes predefined sequence models that act as a standard guide in analyzing the movement of an object.

This concept provides the basis for the MOST sequence models. The primary work units are no longer individual basic motions as in MTM, but fundamental activities (collections of basic motions) dealing with moving objects from one location to another. To move an object, a standard sequence of events occurs. Consequently, the basic pattern of an object's movement is described by a standard sequence model instead of random, detailed basic motions. In analyzing these activities, it was identified that objects can be moved in only one of two ways:

1. Picked up and moved *freely* through space.
2. Moved while maintaining *contact with* another surface or along a *controlled path.*

For example, a box can be picked up and carried from one end of a workstation to the other end or it can be pushed across the top of the workstation. For each type of movement, a different sequence of events occurs; therefore, a different MOST sequence model applies.

The use of tools is analyzed through another sequence model that allows the analyst to follow the movement of a hand tool through a standard sequence of events, which is simply a *combination* of the two basic sequence models.

BasicMOST is the core system and will be used in this chapter to explain the concept of MOST. There are four sequence models used in BasicMOST for describing manual work.

- The *General Move Sequence Model* is used for the spatial movement of an object freely through the air.
- The *Controlled Move Sequence Model* is used for the movement of an object when it remains in contact with a surface or is attached to another object during the movement (e.g., the movement of the object is controlled).
- The *Tool Use Sequence Model* is used for the use of common hand tools as well as cognitive processes such as reading and inspecting.
- The *Equipment Use Sequence Model* is used to analyze common administrative activities such as using a keyboard or filing papers.

THE BASICMOST SEQUENCE MODELS

The BasicMOST System has four sequence models representing four types of activities. *General Move* is defined as the spatial movement of an object freely through the air. To account for the various ways in which a General Move can occur, the activity sequence is made up of four parameters:

A Action Distance (mainly horizontal motion)
B Body Motion (mainly vertical motion)
G Gain Control
P Placement

These parameters are arranged in a sequence model (Figure 1.3), consisting of a series of letters, organized in a logical sequence. The sequence model defines the events or actions that take place in a prescribed order when an object is being moved from one location to another.

BasicMOST Work Measurement System		
Activity	***Sequence Model***	***Parameters***
General Move	A B G A B P A	A - Action Distance B - Body Motion G - Gain Control P - Placement
Controlled Move	A B G M X I A	M - Move Controlled X - Process Time I - Alignment
Tool Use	A B G A B P * A B P A	F - Fasten L - Loosen C - Cut S - Surface Treat M - Measure R - Record T - Think
Equipment Use	A B G A B P * A B P A	W - Keyboard K - Keypad H - Letter / Paper Handling

FIGURE 1.3 Sequence models comprising the BasicMOST System.

The *General Move Sequence Model* is the most commonly used sequence model and is defined as:

A	**B**	**G**	**A**	**B**	**P**	**A**
Action Distance	Body Motion	Gain Control	Action Distance	Body Motion	Placement	Action Distance

The *parameters* (or letters) included in the sequence model above are then assigned time-related index values based on the motion content of the sub-activity. This approach provides complete analysis flexibility within the overall context of the sequence model. For each object moved, any combination of motions might occur, and using MOST, any combination may be analyzed. For the General Move Sequence Model, these index values are easily accessible from a data card (introduced in Chapter 3, Figure 3.1). A fully indexed *General Move Sequence Model* may look like:

$\mathbf{A_6 \quad B_6 \quad G_1 \quad A_1 \quad B_0 \quad P_3 \quad A_0}$

where:
$\mathbf{A_6}$ = Walk three to four steps to object location
$\mathbf{B_6}$ = Bend and arise to gain control of the object
$\mathbf{G_1}$ = Gain control of one light object
$\mathbf{A_1}$ = Move object a distance within reach
$\mathbf{B_0}$ = No body motion
$\mathbf{P_3}$ = Place object with adjustments
$\mathbf{A_0}$ = No return

This example could represent the following activities:

- Walk three steps to pick up a cap from a low shelf, arise and place the cap on an assembly.
- Walk four steps and pick up a lightweight package from the floor, arise and place the package with some adjustments on a scale to be weighed.

General Move is by far the most frequently used of the four sequence models. Roughly 50% of all manual work occurs as a General Move, with higher percentages for assembly, retail, warehouse distribution and material handling and lower percentages for machine shop operations.

The second type of move in BasicMOST is described by the *Controlled Move Sequence Model* (Figure 1.3). This sequence model is used when objects are restricted in some way and covers activities such as pressing a button, flipping a switch, pulling a lever, rotating a crank or simply sliding an object over a surface. In addition to the **A, B** and **G** parameters from the General Move Sequence Model, the sequence model for a Controlled Move contains the following parameters:

M	**X**	**I**
Move Controlled	Process Time	Alignment

As many as 30% of the activities occurring in machine shop operations may involve Controlled Moves. In assembly work, however, the fraction is usually much smaller. The use of the Controlled Move Sequence Model is also seen in many other environments where workers may be opening and closing tool boxes, sliding or pushing parts, pulling levers or pressing buttons to activate machines. A typical activity covered by the Controlled Move Sequence Model is to activate a machine to start a process such as pressing a button to open a door in a healthcare environment. The sequence model for this activity might be indexed as follows:

$\mathbf{A}_1 \quad \mathbf{B}_0 \quad \mathbf{G}_1 \quad \mathbf{M}_1 \quad \mathbf{X}_{10} \quad \mathbf{I}_0 \quad \mathbf{A}_0$

where:
$\mathbf{A}_1$ = Reach to the button a distance within reach
$\mathbf{B}_0$ = No body motion
$\mathbf{G}_1$ = Contact button
$\mathbf{M}_1$ = Push button
$\mathbf{X}_{10}$ = Process time of approximately 3.5 seconds
$\mathbf{I}_0$ = No alignment
$\mathbf{A}_0$ = No return

The third sequence model included in BasicMOST is the *Tool Use Sequence Model* (Figure 1.3). This sequence model covers the use of hand tools for activities such as fastening or loosening, cutting, surface treating, measuring and writing. Also, certain activities requiring mental processes can be classified as Tool Use such as reading and inspecting. The Tool Use Sequence Model is a combination of General and Controlled Move activities and was developed as a part of the BasicMOST System to simplify the analysis of activities related to the use of hand tools. Any hand tool activity is made up of General and Controlled Moves.

The use of a wrench might be described by the following sequence:

$\mathbf{A}_1 \quad \mathbf{B}_0 \quad \mathbf{G}_1 \quad \mathbf{A}_1 \quad \mathbf{B}_0 \quad \mathbf{P}_3 \quad \mathbf{F}_{10} \quad \mathbf{A}_1 \quad \mathbf{B}_0 \quad \mathbf{P}_1 \quad \mathbf{A}_0$

where:
$\mathbf{A}_1$ = Reach to wrench within reach
$\mathbf{B}_0$ = No body motion
$\mathbf{G}_1$ = Grasp wrench
$\mathbf{A}_1$ = Move wrench to a fastener within reach
$\mathbf{B}_0$ = No body motion
$\mathbf{P}_3$ = Place wrench on fastener
$\mathbf{F}_{10}$ = Tighten fastener with wrench (three wrist strokes)
$\mathbf{A}_1$ = Move wrench a distance within reach
$\mathbf{B}_0$ = No body motion
$\mathbf{P}_1$ = Lay wrench aside
$\mathbf{A}_0$ = No return

Another example of a Tool Use Sequence Model would be to pick up a part, inspect two points on the part and put it back on the conveyor. This analysis would be:

$\mathbf{A_1}$ $\mathbf{B_0}$ $\mathbf{G_1}$ $\mathbf{A_1}$ $\mathbf{B_0}$ $\mathbf{P_0}$ $\mathbf{T_3}$ $\mathbf{A_1}$ $\mathbf{B_0}$ $\mathbf{P_1}$ $\mathbf{A_0}$

where:
$\mathbf{A_1}$ = Reach to part within reach
$\mathbf{B_0}$ = No body motion
$\mathbf{G_1}$ = Gain control of part
$\mathbf{A_1}$ = Bring part within reach to eyes
$\mathbf{B_0}$ = No body motion
$\mathbf{P_0}$ = No placement
$\mathbf{T_3}$ = Inspect two points
$\mathbf{A_1}$ = Move part within reach
$\mathbf{B_0}$ = No body motion
$\mathbf{P_1}$ = Put part on conveyor
$\mathbf{A_0}$ = No return

The last sequence model included in BasicMOST is the *Equipment Use Sequence Model* (Figure 1.3). This sequence model includes values for certain administrative activities such as using a keyboard, keypad, stapling and filing. Like the Tool Use Sequence Model, Equipment Use is a combination of General and Controlled Moves. It was developed to assist the analyst in simplifying the process of analyzing administrative activities.

Pressing six keys on a security keypad to activate a safe may be analyzed as:

$\mathbf{A_0}$ $\mathbf{B_0}$ $\mathbf{G_0}$ $\mathbf{A_1}$ $\mathbf{B_0}$ $\mathbf{P_1}$ $\mathbf{K_3}$ $\mathbf{A_0}$ $\mathbf{B_0}$ $\mathbf{P_0}$ $\mathbf{A_0}$

where:
$\mathbf{A_0}$ = No reach to grasp an object
$\mathbf{B_0}$ = No body motion
$\mathbf{G_0}$ = No grasping motion
$\mathbf{A_1}$ = Move finger to security pad within reach
$\mathbf{B_0}$ = No body motion
$\mathbf{P_1}$ = Put finger to keypad
$\mathbf{K_3}$ = Press six keys on keypad
$\mathbf{A_0}$ = No move action
$\mathbf{B_0}$ = No body motion
$\mathbf{P_0}$ = No placing motion
$\mathbf{A_0}$ = No return

The General Move, Controlled Move, Tool Use and Equipment Use Sequence Models are the foundation of BasicMOST and are presented in Figure 1.3.

TIME UNITS

The time units used in MOST are identical to those used in the basic MTM system and are based on hours and portions of hours called *Time Measurement Units*

(TMU). One TMU is equivalent to 0.00001 hour. The following conversion table is provided for calculating standard times:

1 TMU = 0.00001 hour	1 hour = 100,000 TMU
1 TMU = 0.0006 minute	1 minute = 1667 TMU
1 TMU = 0.036 second	1 second = 27.8 TMU

The time value in TMU for each sequence model in BasicMOST is calculated by adding the index values and multiplying the sum by 10. Regardless of the sequence model used, the calculation is the same.

General Move Sequence Model: $\mathbf{A_6\ \ B_6\ \ G_1\ \ A_1\ \ B_0\ \ P_3\ \ A_0}$
Add index values: $6 + 6 + 1 + 1 + 0 + 3 + 0 = 17$
Multiply by 10: $17 \times 10 = 170$ TMU or approximately 6.1 seconds

Controlled Move Sequence Model: $\mathbf{A_1\ \ B_0\ \ G_1\ \ M_1\ \ X_{10}\ \ I_0\ \ A_0}$
Add index values: $1 + 0 + 1 + 1 + 10 + 0 + 0 = 13$
Multiply by 10: $13 \times 10 = 130$ TMU or approximately 0.08 minutes

All MOST index values and times reflect the effort of an *average skilled, trained operator* working at an *average performance level* or *normal pace following a prescribed method under normal working conditions with capable supervision.* This is often referred to as '100% performance level,' which in time study is achieved by using performance rating factors to level the times based on the skill and effort levels shown by the worker. When using MOST, it is *not* necessary to level times as the index values and resulting times are already leveled to 100%. Some organizations may choose to adjust the time to reflect a high task plan, but this is not recommended. This also means that a properly established time standard, using MOST, MTM or stopwatch time study will give nearly identical results in TMU.

The MOST analysis will consist of a series of sequence models describing the movement of objects to perform the activity. Total time for the complete MOST analysis is calculated by adding the times for each sequence model. The time for the activity may be left in TMU or converted to minutes or hours. Again, this time would reflect pure work content, referred to as normal time (no allowances) at a 100% performance level. To calculate the standard time, allowances would need to be applied to the normal time.

One of the features that differentiates MOST from other PMTS is the application of index values to parameters in predefined sequence models. This has proven to speed up the determination of a standard time for an operation. This is explained below along with other key components of the overall MOST Work Measurement Technique.

PARAMETER INDEXING

One objective of an effective work measurement system is to provide the documentation of a specified work method as a basis for the standard. This is accomplished in MOST by applying time-related index values to each sequence model parameter, based on the motion content of the sub-activity. Parameter indexing is the process of selecting the appropriate application rule from a data card (Figure 3.1) and applying the corresponding index value. With proper certification training and practice, over time the MOST analyst can memorize application rules and index values. Practically all analysis work can then be performed without any direct assistance from data cards. In addition, while MOST was developed as a manual system, software applications such as StandardsPro feature built-in data cards so in many cases, paper data cards and analysis forms are no longer used.

Time values for each application rule located on the data cards are based on detailed MTM-1, MTM-2 or MiniMOST backup analyses. These analyses are arranged or 'slotted' into fixed time ranges represented by an index value corresponding to the median time of each range. The time ranges or intervals were calculated using statistical accuracy principles (see Appendix A: Theory).

APPLICATION SPEED

MOST was designed to have a much faster application than conventional work measurement techniques such as time study. Several factors make MOST quicker to apply than time study. Properly administered time study generally requires that an operation is observed anywhere from 10 to over 100 times, depending upon duration and frequency of occurrence, in order to get a statistically accurate sample. Not only does it take time to make the observations, but often production schedules make it impractical to see a significant number of consecutive cycles of an operation at one point on the schedule. Analysts find themselves spending more time scheduling observations than actually making observations. Furthermore, time study requires additional time to subjectively break down the method into steps, to conduct performance rating and to relate method descriptions to times. Since MOST implicitly ties methods to times, it is much easier to review MOST analyses for validation and maintenance purposes. MOST uses clearly defined and easily understood rules and eliminates the subjective aspects of time study. The time required to review a MOST analysis with an operator, supervisor and/or union representative is much shorter than that required for time study. The differences described above focus on comparison of the actual analysis time. In fact, since time study is generally used as a direct measurement tool and MOST as a tool for developing standard data, the application time for creating complete engineered standards is significantly shorter using MOST.

There have not been any formal published studies to compare the application speed of MOST to time study. However, to ensure a statistically accurate time study, numerous observations of an activity are typically required, and a performance rating needs to be applied which can take a considerable amount of analysis time. Using MOST, the activity can be documented and the time for the analysis automatically calculated at a 100% performance level. Typically, this approach reduces the amount of analyst time required.

The simple structure of MOST also makes it quicker to apply than other PMTS. For example, in other PMTS, to arrive at a time for putting a part into a machine, each basic motion involved must be identified, recorded and assigned symbols and time values selected from tables. The time values are then added together to arrive at the time for performing the complete task.

Work Measurement System	Total TMU Produced Per Analyst Hour
MTM-1	300
MTM-2	1000
MTM-3	3000
MiniMOST	4000
BasicMOST	12,000
MaxiMOST	25,000

FIGURE 1.4 Comparison of application speeds.

MOST does not require that tasks be broken down into such detail. Instead, MOST groups together the basic motions that frequently occur into a predefined sequence. Arriving at a standard time with MTM for putting a part into a drill press might require the identification of as many as 15 separate basic motions followed by the assignment of symbols and time values to each motion from the MTM data card. Using MOST, the same analysis requires the identification directly from memory of only seven sub-activities in one sequence model. Using the predefined sequence models on the analysis form or in a software application leaves the analyst the task of filling in only the index values.

A comparison between the speed of MOST and other work measurement techniques is shown in Figure 1.4. In this study, 1 hour of analyst time yielded 300 TMU of measured work with MTM-1. MTM-2 and MTM-3 yielded 1000 and 3000 TMU, respectively. Using BasicMOST, the same amount of analyst time yielded 12,000 TMU. As a general rule, 1 hour of work can be measured using BasicMOST with an average of 10 hours of analyst time.

Note: The above analyses were performed under ideal conditions. Actual applications may yield a total TMU output other than the indicated numbers.

ACCURACY

The accuracy principles that apply to MOST are the same as those used in statistical tolerance control. That is, the accuracy to which a part is manufactured depends on its role in the final assembly. Likewise, with MOST, time values are based on calculations that guarantee the overall accuracy of the final time standard. Based on these principles, MOST provides the means for covering a high volume of manual work with accuracy comparable to existing PMTS. Since all index values were determined based on statistical accuracy calculations, they produce the same accuracy irrespective of the selected combination. Because of this, MOST is a very consistent system. A more detailed discussion of accuracy is presented in Appendix A: Theory.

METHOD SENSITIVITY

When using time study, a quantitative comparison of methods cannot be produced unless another time study is taken of the new method. MOST, like any PMTS, is concerned primarily with the motions that make up an activity. The times or index values for these motions have already been predetermined and are immediately available to the analyst from data cards, or after practice and experience, the analyst will have memory recall or even better, a computer's memory.

It is the analyst's responsibility to recognize the specific motion patterns and to assign the appropriate index values to each sequence model parameter. Since MOST index values are time-related, they provide a quick means for evaluating the relative length of time required for performing a specific method. The analyst's attention is automatically focused on motions requiring longer times, such as sub-activities with index values of six or greater. For example, if an analyst notices index values of six or higher, steps should be taken to improve the method and reduce index values since higher index values equal greater time. Such cases are easy to recognize because these higher index values will stand out among the lower values in a complete MOST analysis. Therefore, it is easy for the analyst to see the impact of improving a method, rearranging the workplace layout or introducing a new tool, fixture or procedure. By doing so, high index values may be reduced, sometimes even to zero. The analyst can change the appropriate index values and estimate the savings resulting from the improved method. This task can be done quickly and easily in most engineered standard software applications, saving the analyst even more time.

MOST is therefore considered a *method sensitive* technique. It is sensitive to the variations in time required by different methods. This feature is very effective in evaluating alternative methods of performing tasks with regard to time and cost. The MOST analysis will clearly indicate the more economical and less fatiguing method. By reviewing a completed MOST analysis, the analyst can quickly detect activities that may be considered non-value added, inefficient or unproductive (e.g., those having high index values or related to certain non-value added activities such as Action Distance and Body Motion parameters). A key feature of a standards software application should include effective reporting that can highlight method improvement analyses.

Because the MOST system is method sensitive, its value as a work measurement tool is greatly increased. Not only does it indicate the time needed to perform various activities, it also provides the analyst with an instant cue that a method should be reviewed. The results are clear, concise and easily understood time calculations that indicate opportunities for saving time, money and energy.

STRUCTURED METHOD DESCRIPTIONS

Writing a proper method description in any work measurement technique is critical to understanding the standard and what it includes. Using a structured approach to document the method description will:

1. Provide consistent and efficient method descriptions.
2. Allow other analysts to easily understand the method being used and index values assigned.
3. Avoid confusion when using the method description as an instructional tool to the person performing the job.

To further improve the consistency of method descriptions, descriptive words that can be used as a 'standard' when writing method descriptions are included throughout this book, on the data cards and in the StandardsPro software application. The sentence structure that will be used in this book follows a consistent pattern. The recommended sentence structure may be slightly different for each sequence model within BasicMOST, MiniMOST and MaxiMOST. The instructions for each will be included in their respective chapter.

This structure of writing method steps should be learned and practiced diligently by any user of MOST. By following this structure, the benefits of uniformity, consistency and clarity can be applied when MOST is used.

ANALYZING TECHNOLOGY WITH MOST

Organizations are constantly identifying new technologies in the workplace with the goal of improving productivity and making the workplace safer for associates. Although technology changes often, the constant is the methodology behind the MOST Work Measurement Technique which allows the system to be used to analyze any type of manual work. Using the fingers to press a touch screen is a manual action. Moving a wrist-based scanning device to an item is a spatial motion. Even the controls in using a drone can be analyzed with MOST. The saying "the only constant is change" is amazingly true and while methods, processes and technology change, the MOST Work Measurement Technique is a proven tool to measure these activities. By the time this edition has been published, hundreds of new technologies will have been developed and all manual actions associated with these new technologies can be analyzed with MOST.

INTEGRATION OF MOST STANDARDS

Organizations interested in developing a labor standards program to drive value in their organization need to consider the scope of the program and the tools and applications needed to make it a reality. Digital solutions for work measurement and labor planning provide companies with tools that reduce the time, and ultimately the cost, to develop and maintain engineered standards.

Integration of labor standards developed with MOST can be integrated into labor planning and warehouse management systems for maximum benefit for scheduling, labor management, long-term planning, budgeting and more. MOST was developed as a manual system, and can still be used that way, but has also transformed for use in software solutions like StandardsPro as well as digital solutions on multiple devices. These solutions provide a traceable, explainable back-up with times that are used to drive accurate activity-based standards to develop labor schedules, balance work flow and manage performance in any type of industry.

2 The MOST Systems Family

The MOST Systems family (see Figure 2.1) has grown significantly since the beginning of the General Move Sequence Model. It provides a comprehensive set of practical tools that are used to create activity-based engineered labor standards in virtually every industry. Because of its excellent reputation, MOST is highly accepted among industrial engineers, unions, management, cost analysts and productivity improvement analysts who have made these tools their top choice for tasks related to work measurement. The tools in the MOST family will be introduced here with brief descriptions and the types of applications for which they are best suited. More detailed descriptions are provided in the following chapters.

LEVELS OF WORK MEASUREMENT

MOST is used to cost-effectively measure work ranging from retail store and backroom operations, small electronic assembly and manufacturing to the building of ships and railroad cars. BasicMOST is the most widely applied work measurement technique and is used to analyze a vast range of manual activities in many industries.

MiniMOST provides detailed analysis of highly-repetitive activities such as small assembly and packing small items where accuracy over a short cycle time is needed. Even though the precision of MiniMOST is comparable to that of more detailed systems like MTM-1, the application speed is substantially higher.

MOST Systems	
BasicMOST	Use for **medium-cycle**, repetitive or non-repetitive work
MiniMOST	Use for **short-cycle**, repetitive work
MaxiMOST	Use for **long-cycle**, non-repetitive work

Applications	Industries
MOST can be applied in the following areas: - Retail operations - Material handling - Assembly - Machining and fabrication - Administrative tasks - Maintenance tasks - Any type of manual work	MOST is used in virtually every industry: - Retail - Distribution - Manufacturing - Supply chain - Healthcare - Aerospace - Utilities - Service

FIGURE 2.1 Overview of MOST Systems.

MaxiMOST is used for longer-cycle activities such as setups, maintenance, material handling, heavy assembly and job shop work. The section at the end of this chapter provides a more detailed explanation of the applicability of each version.

COMPATIBILITY OF MOST SYSTEMS

MOST Systems are designed to provide the optimal combination of speed, detail and accuracy of an analysis at all levels of application. Because of the consistent structure within and between the versions of MOST, as soon as an analyst is familiar with one activity sequence, the analyst is already acquainted with all other sequence models. Defined sub-activities, such as Action Distance analyzed with the **A** parameter, are used in similar sequence models in all versions. Every parameter is always indexed with a number from the set of MOST index values contained on data cards or in a software application such as StandardsPro: 0, 1, 3, 6, 10, 16, etc. The only difference among the MOST Systems is that the index value 3, for example, represents 3 TMU in MiniMOST, 30 TMU in BasicMOST and 300 TMU in MaxiMOST since the index value ranges differ by a factor of 10 from version to version.

Another feature of the compatibility of the MOST Systems is the consistency gained by the analyst when writing method descriptions following a prescribed sentence structure. This structure provides consistency and improves communication among analysts while allowing the MOST analysis to be used for operator or associate instructions. The sentence structure for each system will be described in its respective chapter.

APPLICATION OF MOST TO DEVELOP ENGINEERED STANDARDS

Work measurement as a tool has many applications including methods comparison, balancing work flow and developing times for weekly retail scheduling. The most common use of work measurement is for the development of engineered standards. Utilizing MOST is a quick and easy approach to measure work when developing an engineered standard. As discussed previously, an engineered standard is normal time with applied allowances. There are three approaches to developing engineered standards, all of which MOST supports:

1. Direct Measurement
2. Standard Data
3. Benchmark Standards

It is critical to identify the scope of the standards development effort. The scope should include the purpose of the standards, the extent of the effort and the content of the work being measured. In identifying the scope, the analyst needs to consider the use of the standards, the variation in activities, the number of standards, the detail needed in the standards and the potential for future changes in activities. Measuring the work with a system such as MOST, however, is just one step in developing a standard. It is important to understand all the components involved, including developing, storing, validating and maintaining the standards.

MOST SYSTEM SELECTION

The consistent, multi-level design of MOST has made it possible to establish simple guidelines for deciding which version is the most appropriate for measuring various work tasks. Appendix A provides a detailed explanation of the theory that supports these guidelines. The ability to decide which version of MOST to use for specific situations will increase as each version is studied in more detail in the following chapters. Just as a football field should not be measured with a micrometer or a warehouse area with a ruler, these guidelines will help the analyst choose the appropriate technique to match the proper situation.

BASICMOST

BasicMOST is the predominant version of MOST. As a general rule, activities that are likely to be performed more than 150 times but less than 1500 times per week should be analyzed with BasicMOST. An operation in this category may range from a few seconds to 10 minutes in length based on activities in the range of 200–2000 TMU. Distances in BasicMOST are typically analyzed as within reach to 10 steps. The majority of operations in most industries fall into this category. BasicMOST index ranges readily accommodate the cycle-to-cycle variations typical at this level. The method descriptions that result from BasicMOST analyses are practical and sufficiently detailed for use as operator instructions.

MINIMOST

To develop work measurement to be accurate for short-cycle operations, MiniMOST provides the best approach. In general, this level of detail and precision is required to analyze activities likely to be repeated *identically* more than 1500 times per week. Operations occurring this often usually have cycle times from a few seconds to 1.6 minutes based on activities in the range of 50–500 TMU. Such activities usually have little variation from cycle-to-cycle attributed to the operator's high level of practice and to management efforts to improve the design, layout and method. Opportunities for small, but significant improvements in these areas are often highlighted by a MiniMOST analysis. If method variations exist, BasicMOST may be used even though the cycle time is very short.

Regardless of the cycle length, MiniMOST should be used to analyze any operation in which nearly all reach and move distances for an operation are less than 10 inches (25 cm). Distances in MiniMOST are typically measured in inches or centimeters. Because high method levels (all manual motions with both hands are performed simultaneously) are predominant in MiniMOST analyses, the right and left hands are analyzed separately. Since its focus is on highly-repetitive work within reach of the operator, MiniMOST was not designed for analyzing activities in which the operator Action Distance exceeds two steps, Body Motions other than Bend and Arise occur or the weight or resistance per hand exceeds 10 pounds (5 kg). BasicMOST would normally be used to analyze these situations.

MAXIMOST

MaxiMOST is used to analyze activities that are likely to be performed fewer than 150 times per week. An operation in this category may be 2 minutes to more than several hours in length based on activities in the range of 2000–20,000 TMU. Distances in MaxiMOST are typically analyzed as walking two or more steps between work areas. MaxiMOST index ranges accommodate the wide cycle-to-cycle variations that are typical in work setup, heavy assembly, maintenance or utility activities. Even at this level, the method descriptions resulting from MaxiMOST analyses are very practical for instructional purposes.

Note: In the rare instances when the activity being analyzed does not clearly match the guidelines for choosing a specific version, select the version of MOST that would best fit the situation. For example, a daily preparation task for gathering tools and moving tools to the work area could be analyzed in MaxiMOST. The actual assembly tasks would be analyzed in BasicMOST. Activities analyzed with different versions of MOST can be combined to form an operation; however, each activity or sub-operation should be analyzed with only one version of MOST. It is not appropriate to combine a BasicMOST sequence model with a MaxiMOST sequence model within the same MOST analysis. For further information, see Appendix A.

DECISION DIAGRAM

The decision diagram in Figure 2.2 provides a simple procedure for selecting the appropriate MOST Work Measurement System. Note that the occurrence frequency numbers 150 and 1500 are based on an overall accuracy requirement of ± 5% with a 95% confidence level. If the required accuracy is only ± 10% with a 90% confidence level, these numbers should be increased to 770 and 7700, respectively. (More on this subject can be found in Appendix A.)

The two questions in the second column of the decision diagram (Figure 2.2) that reference the level of detail required in the analysis reflect the fact that MOST is method sensitive. Any version of MOST allows the analyst to focus attention on work methods, but a lower-level version requires a closer examination of the method than a higher-level system. So, if the emphasis is on improving methods, design or layouts, the analyst may choose BasicMOST instead of MaxiMOST (or MiniMOST instead of BasicMOST) to increase the opportunity for method improvements.

The decision diagram is used as a guideline to select a MOST System. When selecting a system and developing engineered standards, some of the variables to consider are the accuracy level required, amount of detail needed, the use of the standard and the cost involved in developing the standards.

SYSTEM SELECTION CHARTS

The system selection charts provide additional guidelines to determine which MOST System may be applicable for use. In reviewing the charts, the concepts of balancing time and balancing effect are introduced and will be explained in more detail in Appendix A but are presented here to provide the context for Figures 2.3 and 2.4.

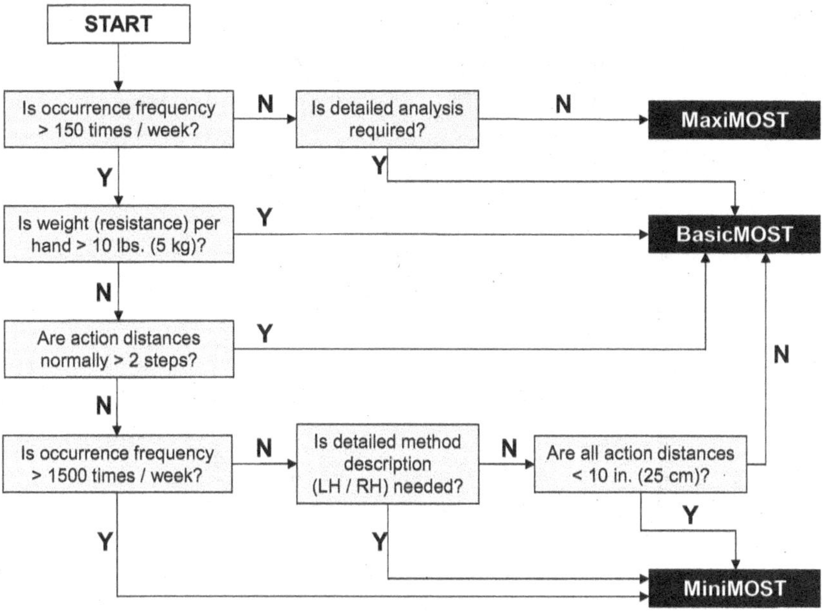

FIGURE 2.2 Process for selecting the appropriate MOST Work Measurement System.

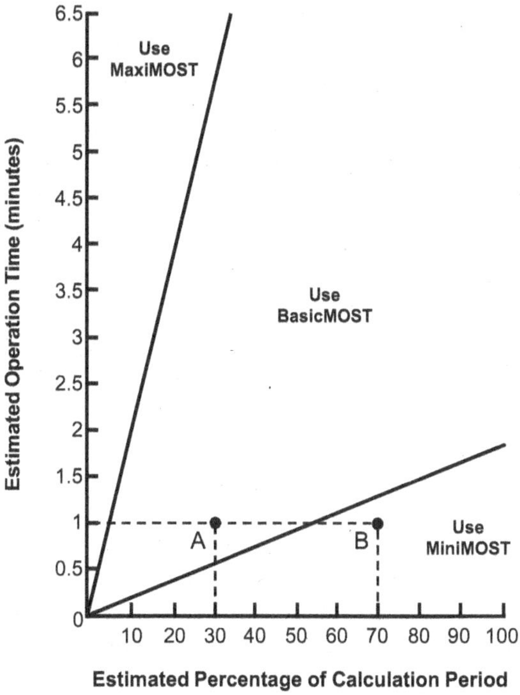

FIGURE 2.3 MOST Systems selection guidelines ± 5% accuracy at a 95% confidence level.

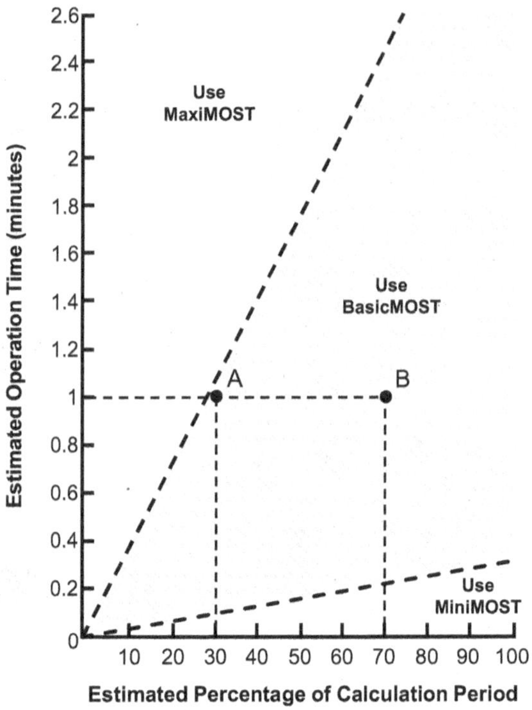

FIGURE 2.4 MOST Systems selection guidelines ± 10% accuracy at a 90% confidence level.

Balancing time is the time needed for the system's desired level of accuracy to be attained. Balancing effect is the leveling of individual deviations for a smaller overall deviation. Figures 2.3 and 2.4 provide another approach to applying these selection guidelines. These charts are based on two principles:

1. The longer the analyzed time, the more accurate the analysis because of the balancing effect.
2. The overall accuracy of a group of analyses improves as short-cycle analyses are properly combined.

These charts are designed to ensure that any set of standards that includes short-cycle analyses will have the expected level of accuracy.

Each chart covers one of the two levels of accuracy most often required in industry. With either chart, if the approximate length of the operation in minutes and the percentage of the standard calculation period occupied by repetitions of the operation can be *estimated,* it can quickly be determined which MOST version will be sufficiently accurate for the analysis. This provides a useful guideline for avoiding the extra work that would be required to analyze operations with a version of MOST that is more detailed than necessary.

For example, using Figure 2.3, if the operation is about 1 minute long and will be repeated enough times to occupy about 30% of the balancing period ('A' in Figure 2.3), a BasicMOST analysis will be sufficiently accurate. If repetitions of this same operation occupy 70% of the period ('B' in Figure 2.3), then MiniMOST would typically be used for the analysis. A similar determination is made for each analysis. When all analyses of the operations that fill the calculation period fall within the charted limits, overall accuracy within ± 5% is assured.

Using the same example, but now with an accuracy level desired of ± 10%, refer to Figure 2.4. For a 1-minute operation that occurs 30% ('A' in Figure 2.4) or 70% ('B' in Figure 2.4) of the period, BasicMOST can be used to analyze both operations and achieve the desired accuracy level of ± 10%.

To maintain overall accuracy when estimating the cycle time for the operation, do not include the time for any step or sequence of steps that is repeated *identically within* the operation cycle. For further details, see the Effect of Variations Within an Operation Cycle section in Appendix A.

APPLICATION OF A MOST WORK MEASUREMENT SYSTEM

Once a MOST System is selected, the process to properly apply MOST is to understand the rules for each system as explained in later chapters in this book. After a thorough understanding of the rules and MOST certification has been achieved, the key factors below should be considered when completing a MOST analysis, regardless of the system used.

IDENTIFYING THE BEST METHOD

Prior to developing the actual MOST analysis, the analyst should study the activity with the objective of establishing the most effective method of accomplishing the task. Although the 'best' method will not always be apparent, every job should be approached with the attitude that any method can be improved.

The starting point for a study is the information gathering or operation analysis phase. All important facts concerning the job such as the workplace layout, tools and equipment, materials and working conditions, should be collected and studied in detail. All data should be clearly documented and made easily accessible for future reference. This activity alone should point out many improvement possibilities.

In terms of parameter index values, MOST sequence models give a quantitative description of distances, types of placing activities, parameter frequencies and so on. During the course of completing sequence models, these index values can serve as indicators for evaluating potential improvements or comparing different methods. The MOST analyst should always strive to reduce the index values while not compromising safety or quality. Index values higher than six assigned to parameters should be investigated for possible method improvements. For sequence models using tools or equipment, index values should reflect the optimum time value based on the choice of tool or equipment used.

As the analyst asks the questions to identify the object and its movement, two additional questions should be asked for the analyst seeking method improvements:

1. Is this activity necessary to do the job (eliminate any unnecessary sub-activities from the analysis)?
2. What 'high' index values can be reduced by changing the workplace layout, method, tools, etc.?

Practical Analysis Procedures

Ideally, observation of several cycles will be sufficient to make a MOST analysis. If conditions permit, the operator should first perform the activity from start to finish, allowing the analyst to document the object and the object movement in the activity. This will identify the sequence model needed for the object movement. After watching additional cycles, the analyst can note the method descriptions with keywords that will identify the appropriate index values for each parameter. This procedure requires that the analyst be fully trained and certified, have experience with MOST application and be thoroughly familiar with the operation.

This approach is, of course, not always possible or even practical, and in many cases, cost analysts need to know the time before scheduling work and productivity improvement specialists want to know the time impacted when method changes are made. One of the benefits of MOST is that if the method is established and the analyst has complete knowledge of the operation and conditions, a MOST analysis can be prepared. This may require the use of workplace layouts that include the location and distances of tools, equipment and materials used. The completed analysis should be checked, if possible, by observing the actual operation along with the completed MOST analysis. This procedure is particularly useful for cost estimates of new components and products.

Another analysis procedure that works well is to record the operation. Since the MOST Work Measurement Technique is an easy-to-use system and a fast measurement method that does not require collection and specification of extremely detailed information, the MOST analysis can often be made directly from observing a recording of the operation.

Another efficient approach to documenting work methods used is an audio recording. Work area data and methods can quickly be recorded and played back for reference. Since it is quite possible to describe a process or method by talking faster than an operator can perform the work, one cycle may often be enough for the study. On the other hand, documenting a method by writing will typically take several cycles to complete.

Analyst Consistency

Since each parameter or variable pertaining to a MOST sequence model is predefined, the analyst will not easily omit or forget motions. Each parameter must be assigned an index value reflecting the selected sub-activity. This forces the analyst to decide and apply a value for all parameters. Even non-occurring sub-activities (index value 0) require a decision. For this reason, the analyst error of omitting motions is essentially eliminated. The result is a high level of consistency in the application of the MOST Technique.

GENERAL APPLICATION RULES FOR ALL MOST SYSTEMS

While MOST has transformed from a manual system to a fully integrated digital solution, the general application rules apply to any format used to document the analysis, and as such, the rules listed below must be followed when completing a MOST analysis:

- Each sequence model is fixed; *no letter may be added or omitted,* except as indicated in the Tool Use and Equipment Use Sequence Models of BasicMOST.
- Index values are fixed; no parameter may carry any index value other than 0, 1, 3, 6, 10, 16, 24, 32, 42, 54 and so on. For example, there is *no* index value 2.
- Each parameter variant must be supported by backup analysis. No index value for any parameter may be used unless this backup exists. All elements of the MOST Systems presented in this book are backed up by MTM-1 or MTM-2 analysis.
- Each method step in a MOST analysis will have one corresponding sequence model.

3 The BasicMOST System

The BasicMOST System can be used to measure many common work measurement situations in many industries. Every company has activities for which BasicMOST is the most logical and practical work measurement tool. Although MiniMOST or MaxiMOST could be applied exclusively in certain companies, they are considered supplements to BasicMOST.

The core sequence models of BasicMOST represent the two basic activities necessary to measure manual work: *General Move* and *Controlled Move*. The two remaining sequence models included in BasicMOST, *Tool Use* and *Equipment Use*, were added to simplify the measurement of hand tool use, activities with mental processes and administrative tasks. The fifth sequence model, *Manual Crane*, is used only where heavy objects are being moved within a workplace and is considered supplementary in BasicMOST.

Four primary sequence models arc used in BasicMOST to analyze manual motions:

1. *General Move Sequence Model*—Used to analyze the spatial movement of an object freely through the air.
2. *Controlled Move Sequence Model*—Used to analyze the movement of an object when it remains in contact with a surface or is attached to another object during the movement.
3. *Tool Use Sequence Model*—Used to analyze activities that use common hand tools.
4. *Equipment Use Sequence Model*—Used to analyze common administrative tasks.

The sequence models used in BasicMOST represent the complete activity of moving one or more objects from one location to another and the activity of using tools or performing administrative tasks. The analyst should always identify the object moved and the complete activity before selecting the appropriate sequence model and assigning the applicable index values.

The methodology behind the MOST Work Measurement Technique is that the system can be used for any type of manual work. The sequence models in BasicMOST provide a structure to analyze different types of work. Technology is advancing rapidly and can impact how work is performed. MOST, however, can still be used to analyze any manual motions affected by technology. Through proper certification training, it will become clearer how to measure new technologies with MOST.

The four main sequence models of BasicMOST are presented below.

Activity	Sequence Model	Parameter
General Move	A B G A B P A	**A** Action Distance
		B Body Motion
		G Gain Control
		P Placement
Controlled Move	A B G M X I A	**M** Move Controlled
		X Process Time
		I Alignment
Tool Use	A B G A B P * A B P A	**F** Fasten
		L Loosen
		C Cut
		S Surface Treat
		M Measure
		R Record
		T Think
Equipment Use	A B G A B P * A B P A	**W** Keyboard
		K Keypad
		H Letter/Paper Handling

Note: A blank space indicated with the asterisk (*) is used for the insertion of one of the Tool Action or Use Equipment parameters.

THE GENERAL MOVE SEQUENCE MODEL

General Move represents the spatial displacement of one or more objects. Under manual control, the object follows an unrestricted path through the air. If the object is in contact with, restricted by or attached to another object during the move, the General Move Sequence Model is not applicable. Such a move will be defined later in the chapter as a Controlled Move activity.

As defined in Chapter 1, MOST is based on the movement of objects. One or more objects can be moved with one or both hands. For simplification of the text, when one object is referenced it can mean one or more objects unless it specifically states only one object in the definition.

General Move follows a fixed sequence of sub-activities identified by the following steps:

1. *Reach* with one or two hands a distance to an object either directly or in conjunction with body motions or steps.
2. *Gain* manual *control* of the object.
3. *Move* the object a distance to the point of placement, either directly or in conjunction with body motions or steps.
4. *Put* the object in a temporary or final position.
5. *Return* to the original location.

These five sub-activities form the basis for the activity sequence describing the manual displacement of one or more objects freely through space. This sequence describes the manual events that can occur when moving an object freely through the air and is known as the *General Move Sequence Model.* The major function of the sequence model is to guide the analyst through a process of having a structured and standardized analysis format. The predefined sequence model provides increased analyst consistency and reduces sub-activity omission.

The Sequence Model

The sequence model consists of a fixed series of letters (called *parameters)* representing each of the various sub-activities of a General Move. The parameters of the General Move Sequence Model identify the sub-activities included in the five-step pattern already indicated:

A B G A B P A

where: **A** = Action Distance
B = Body Motion
G = Gain Control
P = Placement

Phases of the General Move Sequence Model

The displacement of an object through space occurs in three distinct phases as shown by the following General Move Sequence Model breakdown:

Get			Put			Return
A	**B**	**G**	**A**	**B**	**P**	**A**

The first phase, referred to as *Get,* describes the actions to reach the object with body motions (if necessary) and gain control of the object. The **A** parameter indicates the distance the hand or body must travel to reach the object, and **B** indicates the need for any body motions during this action. The degree of difficulty encountered in gaining control of the object is described by the **G** parameter.

The Put phase of the sequence model describes the action to move the object to another location. As before, the **A** and **B** parameters indicate the distance the hand or body travels with the object and the need for any body motions during the move before the object is placed. How the object is placed is described by the **P** parameter.

The third phase simply indicates the distance traveled by the operator to *Return* to the workplace following the placement of the object or to clear the hands for safety purposes to permit the next activity.

The MOST analyst should strictly adhere to the three-phase breakdown of the General Move Sequence Model. Following the sequence model and phases provides consistency in application and ease in communication.

Parameter Definitions

A Action Distance

This parameter is used to analyze all spatial movements or actions of the fingers, hands and/or feet, either loaded or unloaded (loaded means carrying an object, unloaded means the hands are free).

B Body Motion

This parameter is used to analyze either vertical motions of the body or the actions necessary to overcome an obstruction or impairment to body movement.

G Gain Control

This parameter is used to analyze all manual motions (mainly finger, hand and foot) used to obtain complete manual control of an object and release the object after placement. The **G** parameter may include one or more short move motions to gain full control of the object before it is to be moved to another location.

P Placement

This parameter is used to analyze actions at the final stage of an object's displacement to align, orient and/or engage the object with another object before control of the object is relinquished.

Parameter Indexing

Before assigning index values, the MOST analyst must ask two questions to determine the appropriate sequence model:

1. What is the object being moved?
2. How is the object moved (this question is key to determining the appropriate sequence model to use)?

Then, assuming a General Move:

- What does the operator do to *get* the item (determine index values for **A, B** and **G**—first phase)?
- What does the operator do to *put* the item (determine index values for **A, B** and **P**—second phase)?
- Does the operator *return* or 'clear' hands (determine index value for the final **A**—third phase)?

Asking these questions is vital to the effective application of BasicMOST. The answers will help the analyst:

- Avoid overlooking any operator activity or analyzing any unnecessary activity.
- Correctly divide a process into method steps and phases.

- Write accurate and clear method descriptions.
- Determine the index value for each parameter (sub-activity).
- Apply BasicMOST consistently.

Indexing each parameter of the General Move Sequence Model is accomplished by observing or visualizing the object movement during each phase of an activity and selecting the appropriate index value from the data card (Figure 3.1). The index value for each parameter is identified from the data card or a software application and is manually documented as $\mathbf{A_3}$. Software applications feature a built-in data card, and index values can be selected from a drop-down list or input from memory.

Consider the example of a worker grasping a box from a table, putting it on the floor and returning. Assume the worker is standing directly in front of the lightweight box when he gains control of the box. He then puts the box on the floor 10 steps away and returns to the original location. The sequence model for this activity is:

$$\mathbf{A_1 \ B_0 \ G_1 \ A_{16} \ B_6 \ P_1 \ A_{16}}$$

Since the worker is standing directly in front of the box, the first **A** parameter in the sequence model is indexed $\mathbf{A_1}$ because the box is located within reach. Refer to the Action Distance column of the data card (Figure 3.1) for Within Reach to note the corresponding index value. No Body Motion is needed to reach the box; therefore, a 0 is assigned to the Body Motion parameter ($\mathbf{B_0}$), and control of the object is gained with no difficulty ($\mathbf{G_1}$—Light Object in Gain Control column). The box is then moved 10 steps away ($\mathbf{A_{16}}$) and put on the floor ($\mathbf{B_6}$—Bend and Arise). No difficulty is encountered in putting the box on the floor; it is simply put aside ($\mathbf{P_1}$). The operator then walks back (returns) to the workplace, which is 10 steps away ($\mathbf{A_{16}}$).

BasicMOST System — General Move — A B G A B P A

Index x 10	A Action Distance	B Body Motion	G Gain Control	P Placement	Index x 10
0	≤ 2 in. (5 cm)	-	-	Pickup Toss	0
1	Within Reach	-	GRASP: Light Object Light Objects Simo	PUT: Lay Aside Loose Fit	1
3	1 – 2 Steps	Sit or Stand Bend and Arise 50% occ.	GET: Light Objects Non-Simo Heavy or Bulky Blind or Obstructed Disengage Interlocked Collect	PLACE: Loose Fit Blind or Obstructed Adjustments Light Pressure Double Placement	3
6	3 – 4 Steps	Bend and Arise		POSITION: Care or Precision Heavy Pressure Blind or Obstructed Intermediate Moves	6
10	5 – 7 Steps	Sit or Stand with Adjustments			10
16	8 – 10 Steps	Stand and Bend Bend and Sit Climb On or Off Through Door			16

FIGURE 3.1 General Move data card.

The time to perform this activity is calculated by adding all index values in the sequence model and multiplying by 10 to convert to TMU: $(1+0+1+16+6+1+16) \times 10=410$ TMU. Refer to Chapter 1 for a review of Time Measurement Units.

Action Distance (A)

Action Distance covers all spatial movement or actions of the fingers, hands and/or feet, either loaded or unloaded. Any control of these actions by the surroundings requires the use of other parameters.

A_0 ≤2 Inches (5 cm)

Any displacement of the fingers, hands and/or feet a distance less than or equal to 2 inches (5 cm) will result in a zero index value. Since the time for traveling these short distances is minimal, it is included within the Gain Control and Placement parameters.

Examples: Reach between numbers on a security keypad less than 1 inch (2.5 cm) apart.
Reach to place two batteries in a toy less than 1 inch (2.5 cm) apart.
Reach to place washers on bolts located less than 2 inches (5 cm) apart.

A_1 Within Reach

Actions are confined to an area within the arc of the outstretched arm pivoted at the shoulder. With body assistance such as a short bending or turning of the body from the waist, the 'within reach' area is extended somewhat such as reaching for a location on the far side of a desk. Taking a step for further extension of the area exceeds the limits of an $\mathbf{A_1}$ and must be analyzed with an $\mathbf{A_3}$ (1–2 steps).

In a well-defined workstation such as that shown in Figure 3.2, all parts and tools can be reached without displacing the body by taking a step.

The parameter value $\mathbf{A_1}$ also applies to the actions of the leg or foot reaching to an object, lever or pedal. If the trunk of the body is shifted, however, the action must be considered a step ($\mathbf{A_3}$).

Note: Reaching at the end of a walking distance is often simultaneous to the action of walking so a separate $\mathbf{A_1}$ may not be needed when a reach occurs during a step.

Examples: Reach to keyboard to enter customer information at hotel check-in.
Move item from conveyor to box.
Reach for scanner to scan price.
Move foot to pedal to check brake lights on a car.

Steps

Steps refer to the total number of times each foot hits the floor. The trunk of the body is shifted or displaced by walking, stepping to the side or turning the body around. While the index values for steps generally refer to the horizontal movement of the body, they also apply to walking up or down normally inclined stairs.

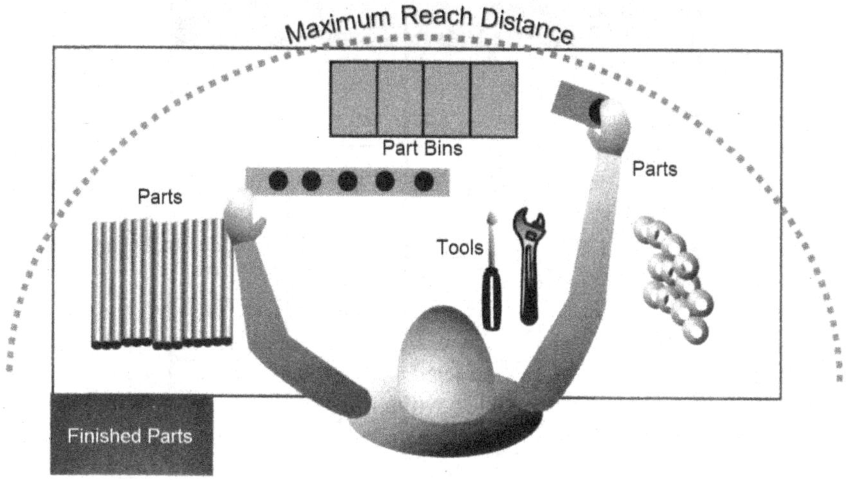

FIGURE 3.2 All parts and tools located within reach.

The preferred method to identify the correct index value is to count the number of steps taken (e.g., the number of times the foot hits the floor). Research has shown that the time required to take a step is relatively constant regardless of the size of the load carried. A worker uses the same amount of time to take five steps while carrying a heavy load as to take five steps with no load. However, the influence of the load may shorten the step length, thereby increasing the number of steps required to cover a specific distance. Therefore, the effect of any load is reflected in the Action Distance parameter. Whenever possible, Action Distance values should be based on the number of steps taken by the operator rather than the distance walked.

A_3–A_{16} 1 - 10 Steps

The index values for up to 10 steps are displayed on the data card as shown below.

$\mathbf{A_3}$ 1–2 Steps
$\mathbf{A_6}$ 3–4 Steps
$\mathbf{A_{10}}$ 5–7 Steps
$\mathbf{A_{16}}$ 8–10 Steps

Examples: A mechanic walks five steps to grasp tool: $\mathbf{A_{10}}$.
A retail associate walks four steps to put returned items in a bin: $\mathbf{A_6}$.

A_X 11 or More Steps

Index values for longer Action Distances involving walking are found in Figure 3.3. Index values are given in terms of steps, feet and meters. When using Figure 3.3, the preferred method is to count the number of steps taken.

Occasionally, it is not possible to observe the operator at work. If this is the case, Action Distance values can be determined from distances measured at the workplace

Action Distance			
Index Value	*Steps*	*Feet*	*Meters*
A_{24}	11 - 15	38	12
A_{32}	16 - 20	50	15
A_{42}	21 - 26	65	20
A_{54}	27 - 33	83	25
A_{67}	34 - 40	100	30
A_{81}	41 - 49	123	38
A_{96}	50 - 57	143	44
A_{113}	58 - 67	168	51
A_{131}	68 - 78	195	59
A_{152}	79 - 90	225	69
A_{173}	91 - 102	255	78
A_{196}	103 - 115	288	88
A_{220}	116 - 128	320	98
A_{245}	129 - 142	355	108
A_{270}	143 - 158	395	120
A_{300}	159 - 174	435	133
A_{330}	175 - 191	478	146

FIGURE 3.3 Extended Action Distance table. The values are read up to and including.

or obtained from drawings or layouts. The distances in Figure 3.3 are based on an average step length of 2 1/2 feet (0.75 m).

Note: The Action Distance values were generated to include walking in a normal working environment and, as a result, include an average step of 2 1/2 feet (0.75 m), obstructed and unobstructed walking, walking up or down normally inclined stairs and walking with or without weight. Should a particular task contain several long, unobstructed and unencumbered walking distances, the Action Distances provided may not be appropriate and the values should then be validated. Keep in mind that walking is a non-value added activity and should be kept to a minimum. Whenever possible, reduce steps through an optimized workplace layout.

Final A—Return or Retract

The final **A** parameter in the General Move Sequence Model is normally used to allocate time for an operator to the original workplace (starting position) *by walking*. This allows for a logical break point between sequence models. If all activities begin and end at the same location (regular workplace), gaps or overlaps can be avoided.

Time for retracting the hands without steps is normally not allowed in the last **A** parameter, since moving the hand to another object is part of the initial **A** parameter of the subsequent sequence model. An exception to this rule is a final **A** to retract one or both hands from inside a machine or moving one or both hands aside for *safety purposes* to allow the next activity to begin. This exception is primarily used when this is the final step of an analysis. Any movement of the hand to gain control of another object will be included in the Action Distance values of the next sequence model.

Body Motion (B)

Body Motion refers to either vertical motions of the body or the actions necessary to overcome an obstruction or impairment to body movement.

B_3 Sit or Stand

When the body is simply lowered onto a seat from an erect position without hand or foot motions required to manipulate the seat, or it is raised from a seated position without the aid of hand or foot motions, then Sit or Stand is appropriate. This value covers either Sit *or* Stand, not both.

Examples: Stand from a bench.
Sit on a stool in work area.

B_6 Bend and Arise

From an erect standing position, the trunk of the body is lowered by bending from the waist and/or knees to allow the hands to reach below the knees and subsequently return to an upright position. It is not necessary, however, for the hands to actually reach below the knees, only that the body be lowered sufficiently to allow the reach below the knees. $\mathbf{B_6}$ may be simply bending from the waist with the knees stiff, stooping down by bending at the knees or kneeling down on one knee. Figure 3.4 provides several different examples of Bend and Arise.

Examples: Kneel to check air in tire.
Bend to grasp box from pallet.

B_3 Bend and Arise, 50% Occurrence

When Bend and Arise is required only 50% of the time during a repetitive activity such as stacking or unstacking several objects, apply a $\mathbf{B_3}$. In stacking (Figure 3.5), the first few objects may require a full Bend and Arise to put the objects at floor level. As the stack becomes taller, the last objects for stacking require no body motions at all.

Note: When the bending activity occurs more or less than 50% of the time, the $\mathbf{B_6}$ (Bend and Arise) value would be applied with the appropriate percentage frequency.

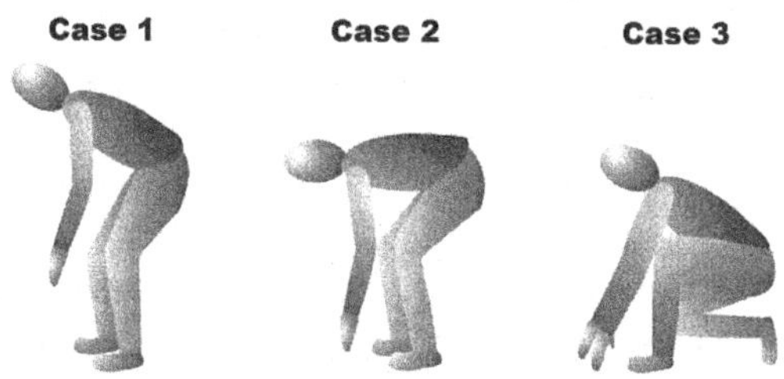

FIGURE 3.4 Examples of Bend and Arise. Notice that in each case the hands are able to reach below the knees.

FIGURE 3.5 Bend and Arise, 50% occurrence.

B_{10} Sit or Stand with Adjustments

When the act of sitting down or standing up requires a series of several hand, foot and body motions to move a chair or stool into a position that allows the body to either Sit or Stand, a $\mathbf{B_{10}}$ is applied. All the motions to manipulate the seat and body are included in the $\mathbf{B_{10}}$ Body Motion. If the chair or stool is stationary and several foot and body motions are necessary to either situate the body comfortably in the seat or to come down from the stool, a $\mathbf{B_{10}}$ would also apply. Note that $\mathbf{B_{10}}$ covers either Sit *or* Stand, not both.

Example: Associate pushes chair back from work area to stand.

B_{16} Stand and Bend

Occasionally a person sitting at a desk must stand up and walk to a location to gain control of an object below the knee level where a Bend and Arise is required. The index value for Stand and Bend most commonly appears within the **B** parameter in the Get phase of the sequence model. This combined Body Motion can be used as long as the actions are contained in a specific phase of the sequence model; in this case, the Get phase.

Note: $\mathbf{B_{16}}$ is simply a combination of $\mathbf{B_{10}}$, Stand with Adjustments and $\mathbf{B_6}$, Bend and Arise. Consequently, the time to arise from the bend is included in the $\mathbf{B_{16}}$ value.

Example: An associate stands from a chair, walks three steps and bends to open a file drawer and arises after obtaining the file.

B_{16} Bend and Sit

As with Stand and Bend, the combined body motion of Bend and Sit applies when a Bend and Arise is required followed by a Sit prior to or after placing the object. If the Sit occurs *after* the placement and walking is required, the return walking would be analyzed on the Final **A** of the sequence model. The index value for Bend and Sit most commonly appears on the **B** parameter in the Put phase of the sequence model. This combined Body Motion can be used as long as the actions are contained in a specific phase of the sequence model; in this case, the Put phase. While this activity may be found in some environments, this is not a common activity and should be examined to determine the ergonomic impact.

Example: A chemist bends to place a sample on the bottom shelf of a case, arises and then sits down at the desk five steps away.

B_{16} Climb On or Off

This parameter variant covers climbing on or off a work platform or any raised surface (approximately 3 feet or 1 m high) using a series of hand and body motions to lift or lower the body. Climbing onto a platform is accomplished by first placing one hand on the edge and then lifting the knee to the platform. By placing the other hand on the platform and bending forward, the weight of the body is shifted, allowing the other knee to be lifted onto the platform. The activity is completed by arising from both knees. Climbing off the platform consists of the same actions but performed in the reverse order. Note that $\mathbf{B}_{16}$ covers either Climb On *or* Climb Off, not both.

Example: Climb onto a truck frame on an assembly line to attach a bracket for the chassis.
A retail employee climbs onto a platform to change a window display.

B_{16} Through Door

Passing Through a Door normally consists of reaching for and turning the handle, opening the door, walking through the door and subsequently closing the door. This value will apply to virtually all hinged, double, sliding or swinging doors. Automatic doors do not require the same manual activities as other doors and would be assigned a $\mathbf{B}_0$ value.

The three or four steps required to pass through the doorway are included in the $\mathbf{B}_{16}$ value. These steps should not be added to or subtracted from the Action Distance. The proper application of a $\mathbf{B}_{16}$ in conjunction with an Action Distance is graphically shown in Figure 3.6.

Example: An operator walks five steps to a closed door, opens it, passes through the door and walks three steps to a desk where a light object is picked up and put on the floor beside the desk.

Note that the five steps to the door and the three steps beyond the door are all part of getting the object. The proper application of $\mathbf{B}_{16}$ requires adding the steps prior to and after the doorway to allow a single Action Distance value for eight steps ($\mathbf{A}_{16}$). The steps to actually pass through the doorway are included in the $\mathbf{B}_{16}$ value. The appropriate analysis for the Get phase for this example is:

Get			Put			Return
$\mathbf{A}_{16}$	$\mathbf{B}_{16}$	$\mathbf{G}_1$	$\mathbf{A}_x$	$\mathbf{B}_x$	$\mathbf{P}_x$	$\mathbf{A}_x$

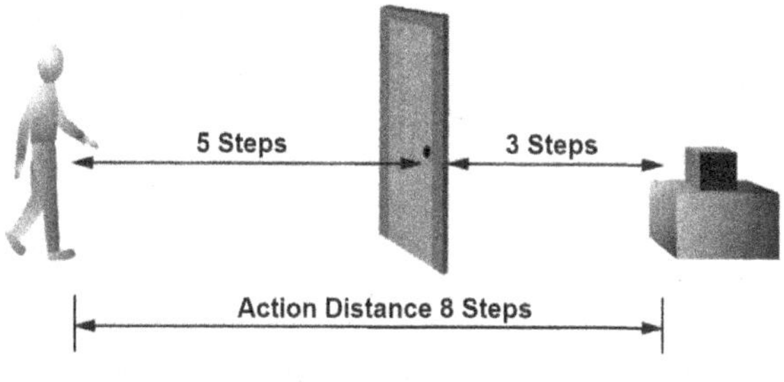

FIGURE 3.6 Application of $\mathbf{B}_{16}$ in conjunction with an Action Distance.

Gain Control (G)

Gain Control covers all manual motions (mainly finger, hand and foot) employed to obtain complete manual control of an object and release the object after placement. The **G** parameter can include one or several short motions (up to 2 inches or 5 cm in spatial movement) to gain full control of the object before it is moved to another location.

G_1 Light Object

Any type of grasp can be used as long as no difficulty is encountered as described by the $\mathbf{G_3}$ parameter variants. The object may be in a pile or grouped with other objects, lying close against a flat surface or simply lying alone. Control may be gained simply by contacting the object with the fingers, hand or foot (contact grasp), or a more difficult grasping action such as picking one object out of a pile of objects. One or two hands may be used as long as only one object is obtained and the object is accessible for both hands to simultaneously grasp it. If several objects are grouped together or arranged in a way that they may be picked up as one object, $\mathbf{G_1}$ will still apply (e.g., grasp two small boxes bound together with string).

Examples: Contact power button on computer.
Grasp lever before activating machine.
Using both hands, grasp box in distribution center.
Grasp documents from drawer grouped together with a clip.

G_1 Light Objects Simo

Simo refers to manual actions performed simultaneously by different body members. That is, one hand gains control of a light object ($\mathbf{G_1}$), while the other hand obtains another light object ($\mathbf{G_1}$). The total time is no more than that required to gain control of one light object.

Examples: Simultaneously grasp two completed assemblies from conveyor.
Grasp screwdriver with one hand and a screw with the other hand at the same time.

G_3 Light Objects Non-Simo

Because of the nature of the job or the conditions under which the job is performed, the operator is unable to gain control of two objects or of two accessible grasping points of one object simultaneously. With both hands, the operator reaches to the objects simultaneously and then, while one hand is grasping an object, the other hand will pause before it can grasp the other object. This can often happen when gaining control of items passing by on a conveyor. Gain control time must be allowed for both hands so the higher index value of $\mathbf{G_3}$ applies.

The ability of the operator to perform simultaneous motions is largely dependent on the amount of practice opportunity available. For example, an assembly operator who continuously gets parts from the same two locations will have no trouble performing the activity simultaneously. After repeating a number of cycles, the operator develops an automatic reaction to the exact location of each part.

On the other hand, simultaneous motions will sometimes be difficult for workers in a job shop. Because of the infrequent occurrence of many tasks, the operator will have little practice opportunity to gain the automatic skills necessary to perform simultaneous motions.

When determining a Simo versus Non-Simo index value, the analyst should observe the operator's method of gaining control of the object if possible. Normally, simultaneous actions can be easily recognized by their automatic appearance. (For further discussion, see the Application of the BasicMOST Work Measurement System section of this chapter.)

Example: At the same time, reach with both hands and then separately gain control of two full containers of cleaning solutions.

G_3 Heavy or Bulky

Control of heavy or bulky objects is achieved only after the muscles are tensed to a point at which the weight, shape or size of the object is overcome. This variant can be identified by the *hesitation* or *pause* needed to gain sufficient muscular force required to move the object.

This effect is influenced not only by the actual weight of the object but also by the location of the object with respect to the body, the existence of handles or grips for easy grasping or even the strength of the individual. Poorly located objects, even smaller or lighter ones, for example, may require some hesitation or movement of the body for balance or additional muscular control for leverage. With the existence of handles or other easy grasping devices located appropriately on the object, the effect of the weight can be significantly reduced.

When considering Heavy or Bulky for Gain Control, the major criterion is not the actual weight of the object, but the hesitation or pause needed for the muscles to tense or the body to stiffen prior to moving the object. See Figure 3.7.

Examples: Gain control of a chair to move to packing box.
Gain control of a large monitor before moving it to desk.
Brace arms around a large, empty dishwasher box.
Get glass window to replace broken window in car.

The weight or bulk of an object can also affect the method of gaining control. Before a heavy or bulky object can be completely controlled, it may be necessary to

FIGURE 3.7 Examples of G_3, Gain Control of heavy or bulky objects.

FIGURE 3.8 Gain Control of heavy object requiring intermediate moves.

move or reorient the object. This may require obtaining a temporary grip and sliding the object closer to the body before complete control of the object is obtained (see Figure 3.8). In extreme cases, calling for several 'intermediate moves' of the object, additional parameters or sequence models may be necessary. For example, use a Controlled Move Sequence Model to analyze sliding the object closer. If additional sequence models are necessary to analyze gaining control, the method should be reviewed and improved if possible.

G_3 Blind or Obstructed

The accessibility of the object is restricted because an obstacle either prevents the operator from seeing the object or creates an obstruction to the hand or fingers when attempting to gain control of the object. If the location is blind, the operator must *feel around* for the object before it can be grasped. When there is an obstruction, the fingers or hand must *work around* the obstacle before reaching the object. If the object is located on the person (shirt pocket or apron), it is probably not blind due to the operator's familiarity with its location. If the operator needs to work around other objects in the pocket or apron to gain control, however, it would be obstructed and a $\mathbf{G_3}$ would apply.

Examples: Work around other objects to gain control of a tool in a tool box (obstructed).
Reach behind the back of a computer to get a cord (blind).

G_3 Disengage

The application of muscular force is needed to free the object from its surroundings. *Disengage* is characterized by the application of pressure to overcome resistance followed by the sudden movement and recoil of the object. The recoil of the object must follow an unrestricted path through the air. If the object movement is along a controlled path, such as unseating a lever or a crank, the $\mathbf{G_3}$, Disengage value would not apply and a Controlled Move would be used.

Examples: Disengage a power cord from an outlet.
Remove a tight-fitting cap from the top of a part.

G_3 Interlocked

The object is intermingled or tangled with other objects and must be separated or worked free before complete control is achieved.

Examples: Remove a spring tangled with other springs from a box.
Gain control of a wiring harness intermingled with other parts.

G_3 Collect

Gaining control of more than one object may be performed with $\mathbf{G_3}$, Collect. The objects may be jumbled together in a pile or spread out over a surface. If jumbled, control of several objects is achieved by reaching down into the pile with the hand and bringing up a handful. When spread out, the objects may be swept together with the hand and fingers and picked up as one object.

Examples: Collect several coins from the cash register drawer.
Gather silverware spread out on table with one sweeping motion.
Collect fasteners from bin.

Placement (P)

Placement refers to actions occurring at the final stage of an object's displacement to align, orient and/or engage the object with another before control of the object is relinquished. The index value for the Placement parameter is chosen by the difficulty of the method encountered during the placement. An index value for **P** is never chosen by the weight of the object alone. Although weight may influence the difficulty in placement, it is the difficulty of the method that ultimately determines the value chosen for **P**, not the weight. For example, a heavy box may simply be put on the floor, in which case a $\mathbf{P_1}$ (Lay Aside) would be used, while a light package may have to be squeezed into a tight space between two other boxes on a shelf and a $\mathbf{P_6}$ (Heavy Pressure) is appropriate.

Placement also includes a limited amount of insertion (up to 2 inches, 5 cm) as part of the placing action. For insertions greater than this, both a General Move and Controlled Move must be used. This will be explained in more detail later in this section.

P_0 Pickup

For the Pickup rule to apply, the object is moved to an unspecified location and placement does not occur. The object is selected using a Gain Control followed by an Action Distance to physically move the object, and then held. Placement occurs in a later method step.

Examples: Pickup packing slip from table.
Pickup stapler from counter.

P_0 Toss

A specified placement does not occur with Toss. The object is released during the preceding move (Action Distance parameter) without placing motions or a pause to point the object toward the target. The time for the release motion to let go of the object is included in the **G** parameter.

Examples: Toss scrap material into trash bin.
Toss envelope into open mail bin.

P_1 Lay Aside

The object is simply put in an approximate location with no apparent aligning or adjusting motions. This placement requires low control by the mental, visual or muscular senses.

Examples: Lay aside wrench after using.
Put completed part on conveyor.
Put file on desk.

P_1 Loose Fit

The object is put in a more specific location than that described by the Lay Aside parameter, but tolerances are such that only a very modest amount of mental, visual or muscular control is necessary for the placement action. The clearance between the engaging parts is loose enough so that one adjustment, without the application of pressure, is required to put the object.

Examples: Put a washer on a bolt.
Put a shirt already on a hanger to a rack in the display area.

The use of stops at a workplace can make it possible for an operator to put an object to a precise location with little or no hesitation. For this reason, laying an object against stops can be considered a Loose Fit placement ($\mathbf{P_1}$).

Examples: Put scanner in holster.
Put box of cereal in open grocery bag.

P_3 Loose Fit Blind or Obstructed

Conditions are similar to those encountered by the Gain Control blind or obstructed parameter. What would normally be a $\mathbf{P_1}$, Loose Fit is now hidden or obstructed. In these situations, the operator must feel around or work around for the placement location before the placement can occur.

Examples: Place a company identification sticker on the back of a piece of equipment (blind).
Reach around soup cans on a shelf to place the freshest cans on the back of the shelf (obstructed).

P_3 Adjustments

Adjustments are defined as the corrective actions occurring at the point of placement caused by difficulty in handling the object, closeness of fit, lack of symmetry of the engaging parts or awkward working conditions. These adjustments are recognized as obvious efforts, hesitations or correcting motions at the point of placement to align, orient and/or engage the object.

Examples: Place a key in a lock.
Place a screw on a threaded junction and pick up the threads.[1]
Place a paper clip to a group of documents.

This parameter can also be applied to an object being lined up to two different marks following a General Move. For $\mathbf{P_3}$ to apply, however, these marks must be within 4 inches (10 cm) of each other. If there is more than 4 inches (10 cm) between each mark, special eye times are needed which require additional care in the placement ($\mathbf{P_6}$). (For more detailed information, see the definition for Alignment later in this chapter.)

Examples: Place a document to the corner of a photocopy machine.
Adjust a ruler to two points 3 inches (7.5 cm) apart after placing it on drafting paper.

P_3 Light Pressure

Because of close tolerances or the nature of the placement, the application of muscular force is needed to seat the object even if the initial positioning action could be classified as a $\mathbf{P_1}$, Loose Fit.

Examples: Snap socket on a ratchet.
Insert an electric plug into a socket.

P_3 Double Placement

Two distinct placements occur during the total placing activity. For example, place a bolt through a hole in two parts (Figure 3.9).

P_6 Care or Precision

Extreme care is needed to position an object within a closely defined relationship with another object. The occurrence of this variant is characterized by the obvious slow motion of the placement due to the high degree of concentration required for mental, visual and muscular coordination.

Examples: Position a thread through the eye of a needle.
Position a soldering iron to a circuit connection.
Position a full beaker of chemical solution on a lab table.

FIGURE 3.9 Example of $\mathbf{P_3}$, Double Placement.

[1] Threaded placements are nearly always a $\mathbf{P_3}$, unless they are either blind or obstructed ($\mathbf{P_6}$) or placed in a hole up to 2 inches (5 cm) deep where the threaded pickup action is not required. In the case of a deep, self-threaded fastener, the value will be a $\mathbf{P_1}$.

P_6 Heavy Pressure

As a result of very tight tolerances, not the weight of an object, a high degree of muscular force is needed to engage the object. Heavy Pressure can be easily recognized as the regrasping of an object, tensing of the muscles and the preparation of the body prior to the application of pressure. The tensing of the muscles and the use of both hands needed to position an object often differentiates a placement of $\mathbf{P_6}$, Heavy Pressure from $\mathbf{P_3}$, Light Pressure. The use of Heavy Pressure is uncommon, may exert a high level of stress on the worker and should be avoided if possible. In addition, once the object has been positioned with the $\mathbf{P_6}$, Heavy Pressure value, it may be followed by a Controlled Move to slide the object to its final destination. Controlled Move will be discussed later in this chapter.

Examples: Reposition a lid on oil drum with heavy pressure.
Position a tight-fitting cover on machine.

P_6 Blind or Obstructed

Conditions are similar to those encountered by the Gain Control, blind or obstructed parameter. Accessibility to the point of placement is restricted because an obstacle either prevents the operator from seeing the point of placement or creates an obstruction to the hand or fingers when attempting to position the object. If the location is blind, the operator must *feel around* for the placement location before the object can actually be positioned (normally with adjustments). When an obstruction presents itself, the fingers and/or hands must be *worked around* the obstacle before positioning the object with adjustments.

Examples: Position a plug from a monitor into a socket behind the workstation (blind).
Position a spark plug in an engine block after working the hands between the distributor wiring (obstructed).

P_6 Intermediate Moves

Several intermediate moves of the object are required before positioning it in a final location. These intermediate moves are necessary because the nature of the object or the conditions surrounding the object prevent direct placement. With heavy, bulky or difficult-to-handle objects, this parameter is recognized as a series of placing, shifting of grasps and moving actions occurring before final placement. This additional handling is needed to overcome the awkward nature of the object.

Examples: Position tire on frame with several shifting motions.
Position a large box down on its corner and 'walk it' into position.
Position a self-sticking label on an item, centering it into a recessed area.

A special case of this variant is encountered when placing one object from a handful of different objects from the palm of the hand. Before actually placing the object, several finger and hand movements are required to select and shift one of the objects from the palm to the fingertips. This 'unpalming' action is more than a simple regrasp. The hand must first be turned over allowing visual selection of the appropriate object. Several finger motions (intermediate moves) are then needed to shift the object up to the fingertips before placement can occur.

Note: This case of $\mathbf{P_6}$, Intermediate Moves applies only to a handful of *different* objects. If the objects held in the palm are all similar, visual selection is not necessary. A simple regrasp is then sufficient for unpalming any of the objects. As this regrasp normally occurs during the Action Distance to place the object, no additional regrasp time is needed. However, if the Action Distance in the Put phase is 2 inches (5 cm) or less ($\mathbf{A_0}$), then a regrasp ($\mathbf{G_1}$) should be allowed. The value for **P** is then chosen by the amount of difficulty required to place the object.

Examples: From a handful of change, use the thumb to push a coin to the fingertips and position it in a slot at a self-checkout terminal.
Using the thumb, select a 1/2 inch (12 mm) washer from a handful of assorted washers and nuts and position it on a bolt.

Placement with Insertion

In the introduction to Placement, it was stated that the Placement parameter value includes up to 2 inches (5 cm) of insertion.

For additional insertion, the Controlled Move Sequence Model must be used. While the application will be clearer once the section on Controlled Move has been reviewed, the following example illustrates the proper application of the data.

Example: A mechanic obtains an oil dipstick within reach and places it into the engine block with adjustments within reach. The dipstick is inserted 10 inches (25 cm). The analysis for this example is:

$$\mathbf{A_1\ B_0\ G_1\ A_1\ B_0\ P_3\ A_0\quad 60\ TMU}$$
$$\mathbf{A_0\ B_0\ G_0\ M_1\ X_0\ I_0\ A_0\quad \underline{10\ TMU}}$$
$$\mathbf{70\ TMU}$$

The $\mathbf{P_3}$ value covers the first 2 inches (5 cm) of insertion while the $\mathbf{M_1}$ value is used for the additional 8 inches (20 cm) of insertion. The $\mathbf{M_1}$ value in the Controlled Move Sequence Model covers an insertion of up to 12 inches (30 cm). Controlled Move will be discussed in the next section.

Parameter Frequencies

Frequency

Frequency is an entire sequence occurring more than once. If an activity occurs more than once, the frequency will be specified in a MOST analysis and the time for the activity multiplied by the frequency indicated. The time calculation for a frequency activity is:

1. Add all index values for the entire sequence model.
2. Multiply this total by the number of frequency occurrences.
3. Calculate the total TMU by multiplying by 10.

Example: A retail associate grasps a piece of clothing and puts it in a bag. She repeats this process for a total of five items. All items are within reach.

A_1 B_0 G_1 A_1 B_0 P_1 A_0 5

GET	A_1	Reach to item
	B_0	No body motion
	G_1	Grasp item
PUT	A_1	Move item to bag
	B_0	No body motion
	P_1	Put item in bag
RETURN	A_0	No return

The time calculation steps for the example above are:

1. $1 + 1 + 1 + 1 = 4$
2. $4 \times 5 = 20$
3. $20 \times 10 = 200$ TMU

Partial Frequency

Partial frequency is one or more parameters within a sequence model occurring more than once such as individually placing several objects from a handful. This activity is shown in the sequence model by placing parentheses around the parameters that are repeated and documenting the number of occurrences as a partial frequency in a MOST analysis, also within parentheses. The time calculation for partial frequency is shown below:

1. Add all index values for the parameters within parentheses.
2. Multiply this value by the number of occurrences (the number noted as a partial frequency at the end of the sequence model).
3. Add this total to the remaining parameter index values.
4. Convert the total to TMU by multiplying by 10.

Example: Collect three batteries within reach and place one each in three electronic toys that are 4 inches (10 cm) apart.

A_1 B_0 G_3 (A_1 B_0 P_3) A_0 (3)

GET	A_1	Reach to batteries
	B_0	No body motion
	G_3	Collect three batteries
PUT	A_1	Move battery to toy
	B_0	No body motion
	P_3	Place battery with light pressure
RETURN	A_0	No return

As indicated, only the parameters in the Put phase of this sequence model are repeated three times. The operator reaches ($\mathbf{A}_1$) with no body motions ($\mathbf{B}_0$) and places the battery in the toy ($\mathbf{P}_3$).

The time calculation is:

1. $(\mathbf{A}_1 \ \mathbf{B}_0 \ \mathbf{P}_3) = (1 + 0 + 3) = 4$
2. $4 \times 3 = 12$
3. $1 + 3 + 12 = 16$
4. $16 \times 10 = 160$ TMU

Note: More than one set of parentheses may be used in a sequence model provided the same frequency applies to all parameters within parentheses.

Frequency with a Partial Frequency

It is possible that a partial frequency activity and frequency can occur together. Using the example above where an associate collects three batteries and puts one in three toys, but now completes the activity (collecting and placing batteries) twice, the following analysis would be used:

$$\mathbf{A}_1 \ \mathbf{B}_0 \ \mathbf{G}_3 \ (\mathbf{A}_1 \ \mathbf{B}_0 \ \mathbf{P}_3) \ \mathbf{A}_0 \ \mathbf{(3)} \ \mathbf{2}$$

The calculation for this type of activity is:

1. Add all index values for the parameters within parentheses.
2. Multiply this value by the number of partial frequency occurrences (the number in parentheses noting partial frequency).
3. Add this total to the remaining parameter index values.
4. Multiply this total by the activity frequency (number at end of sequence model, not in parentheses).
5. Convert the total to TMU by multiplying by 10.

1. $(\mathbf{A}_1 \ \mathbf{B}_0 \ \mathbf{P}_3) = (1+0+3) = 4$
2. $4 \times 3 = 12$
3. $1+3+12 = 16$
4. $16 \times 2 = 32$
5. $32 \times 10 = 320$ TMU

Note: A frequency or partial frequency can be a whole number, decimal or fraction. For example, if a shipping box is sealed only when it is full and the box holds 16 items, then sealing the box only occurs once every 16 times.

Writing Method Descriptions

One of the advantages of MOST is using a standard sequence model to accurately determine time values. Another advantage is the method description that accompanies each sequence model can be written to consistently and clearly define the activity. It is recommended that the analyst follow a prescribed sentence structure and use consistent wording when writing method descriptions. This will provide other analysts and future readers of the analysis a clear understanding of the process.

Below is the recommended format for a clear and concise General Move method description. Additional words for Action Distance or Body Motion are optional and may be used to enhance the method description.

The recommended method description format for General Move is:

Gain Control Object ⟨From Location⟩ Placement To Location

⟨If the From Location is apparent, it is not necessary to indicate it in the method description.⟩

General Move Examples

Note: For the examples listed below, all distances are within reach unless specified otherwise.

1. A retail associate grasps two cans simo and puts each on the shelf one at a time.

   Grasp 2 cans and put on shelf

   $\mathbf{A_1\ B_0\ G_1\ (A_1\ B_0\ P_1)\ A_0\ (2)}$

   $[(1+1)\times(2)+1+1]\times 10=60$ TMU

2. An associate in a printing facility collects scrap paper from a paper cutter and tosses it into a garbage can.

   Collect scrap paper and toss into garbage can

   $\mathbf{A_1\ B_0\ G_3\ A_1\ B_0\ P_0\ A_0}$

   $(1+3+1)\times 10=50$ TMU

3. A postal carrier collects five envelopes from a bin with one hand and puts each in five different mail slots with the other hand.

   Collect letters and hold

   $\mathbf{A_1\ B_0\ G_3\ A_1\ B_0\ P_0\ A_0}$

   $(1+3+1)\times 10=50$ TMU

   Grasp letter and put in slot

   $\mathbf{A_1\ B_0\ G_1\ A_1\ B_0\ P_1\ A_0\ 5}$

   $(1+1+1+1)\times 5\times 10=200$ TMU

   50 TMU

   <u>200 TMU</u>

   250 TMU

4. A shipping associate walks three steps to grasp a tape gun, returns three steps and bends to put the tape gun to a box.

   Grasp tape gun and put to box
   $\mathbf{A_6\ B_0\ G_1\ A_6\ B_6\ P_1\ A_0}$
   $(6+1+6+6+1)\times 10=200$ TMU

5. An associate removes a heavy tray from a machine and places it with adjustments on a workstation one step away.

   Get heavy tray and place on workstation
   $\mathbf{A_1\ B_0\ G_3\ A_3\ B_0\ P_3\ A_0}$
   $(1+3+3+3)\times 10=100$ TMU

THE CONTROLLED MOVE SEQUENCE MODEL

Controlled Move describes the manual displacement of an object over a 'controlled' path. That is, movement of the object is restricted in at least one direction by contact with or attachment to another object or the nature of the work requires the object be deliberately moved along a specific or controlled path. A Controlled Move is performed under one of three conditions:

1. The object or device is restrained by its attachment to another object such as a button, lever, door or crank;
2. It is controlled during the move by the contact it makes with the surface of another object, such as pushing a box across a table; or
3. The object must be moved on a controlled path to accomplish the activity such as folding a cloth, coiling a rope, winding a spool, moving a balanced item or to avoid a hazard such as electricity, sharp edges or running machinery.

If the object can be moved freely through space and remains unaffected by any of the conditions listed above, it must be analyzed as a General Move.

Similar to the General Move Sequence Model, the Controlled Move Sequence Model follows a fixed sequence of sub-activities identified by the following steps:

1. *Reach* with one or two hands a distance to the object, either directly or in conjunction with body motions or steps.
2. *Gain* manual *control* of the object.
3. *Move* the object over a controlled path (within reach or with steps).
4. Allow time for a machine *process* to occur.
5. *Align* the object following the Move Controlled or at the conclusion of the Process Time.
6. *Return* to the workplace.

These six sub-activities form the basis for the activity sequence describing the manual displacement of an object over a controlled path.

The Sequence Model

The sequence model consists of a series of letters (parameters) representing each of the various sub-activities of Controlled Move. The **A, B** and **G** parameters shown below were introduced with the General Move Sequence Model. The three new parameters of **M, X** and **I** will be presented in this section.

A B G M X I A

where: **A** = Action Distance
B = Body Motion
G = Gain Control
M = Move Controlled
X = Process Time
I = Alignment

Phases of the Controlled Move Sequence Model

Similar to the General Move Sequence Model, object movement within the Controlled Move Sequence Model is broken down into three phases:

Get	Move or Actuate	Return
A B G	**M X I**	**A**

The Get and Return phases of Controlled Move use the same parameters found in the General Move Sequence Model and describe the same sub-activities. The fundamental difference is the activity immediately following the **G** parameter. The Move or Actuate phase describes actions either to simply *move* an object over a controlled path or to *actuate* a control device—often to initiate a machine process. Normally, 'Move' implies that the **M** and possibly the **I** parameter of the sequence model are involved. A move, for example, would occur when opening a car door or sliding a part to align it to another. 'Actuate' usually applies to situations involving the **M** and **X** parameters. Pushing a lever on a machine or pressing a button to start a machine process are examples of actuate. For either situation of Move or Actuate, any or all of the parameters in the sequence model could be used.

Parameter Definitions

The three new parameters introduced in Controlled Move are shown on the Controlled Move data card in Figure 3.10. The **A, B** and **G** parameters were discussed with the General Move Sequence Model and remain unchanged.

M Move Controlled

This parameter is used to analyze all manually guided movements or actions of an object over a controlled path.

BasicMOST System			Controlled Move			A B G M X I A	
Index x 10	M Move Controlled		X Process Time			I Alignment	Index x 10
	Push / Pull / Turn	Crank	Seconds	Minutes	Hours		
1	≤ 12 in. (30 cm) Button Switch Knob	-	.5 Sec.	.01 Min.	.0001 Hr.	1 Point	1
3	> 12 in. (30 cm) Resistance Seat or Unseat High Control 2 Stages ≤ 24 in. (60 cm) Total	1 Rev.	1.5 Sec.	.02 Min.	.0004 Hr.	2 Points ≤ 4 in. (10 cm)	3
6	2 Stages > 24 in. (60 cm) Total 1 – 2 Steps	2 – 3 Rev.	2.5 Sec.	.04 Min.	.0007 Hr.	2 Points > 4 in. (10 cm)	6
10	3 – 4 Stages 3 – 5 Steps	4 – 6 Rev.	4.5 Sec.	.07 Min.	.0012 Hr.	-	10
16	6 – 9 Steps	7 – 11 Rev.	7.0 Sec.	.11 Min.	.0019 Hr.	Precision	16

FIGURE 3.10 Controlled Move data card.

X Process Time

This parameter is used to account for the time for work controlled by electronic or mechanical devices or machines, not by manual actions.

I Alignment

This parameter is used to analyze manual actions following the Move Controlled or at the conclusion of Process Time to achieve the alignment of objects.

Parameter Indexing

Move Controlled (M)

Move Controlled covers all manually guided movements or actions of objects over a *controlled path*. That is, movement of the object is restricted in at least one direction by contact with or attachment to another object. Index values for the **M** parameter are listed under two separate categories on the Controlled Move data card. The most frequently occurring parameter variants of Move Controlled (**M**) are under the Push/Pull/Turn heading. The Crank category applies to a special type of Controlled Move dealing with cranks, handwheels or other devices requiring a circular cranking motion.

The following parameter variants apply to moves of an object or device that is hinged or pivoted at some point (e.g., a door, lever or knob), restricted because of its surroundings (e.g., by guides, slots or friction from surface) or restricted by other special circumstances requiring movement over a controlled path (e.g., using optical scanning devices).

Push/Pull/Turn

M_1 One Stage ≤ 12 Inches (30 cm)

The object is moved along a controlled path by movement of the fingers, hands or feet not exceeding 12 inches (30 cm).

Examples: Press foot pedal 6 inches (15 cm).
Pull lever 10 inches (25 cm).
Push box along roller conveyor 5 inches (12.5 cm).

M_1 Button/Switch/Knob

A device is actuated by a short pressing, moving or rotating action of the fingers, hands, wrist or feet.

Examples: Push power button on computer.
Flip switch on machine.
Turn knob on washing machine.

M_3 One Stage > 12 Inches (30 cm)

The object is moved along a controlled path by movement of the hands, arms or feet greater than 12 inches (30 cm). The maximum displacement covered by this parameter occurs with the extension of the arm plus body assistance.

Examples: Pull back plastic film on glass 30 inches (75 cm).
Open flap on box 14 inches (35 cm).
Pull vehicle hood 36 inches (90 cm) to close.

M_3 Resistance

Conditions surrounding the object or device require that resistance be overcome during the Controlled Move. This parameter variant can be identified by the muscular force needed to move the object with resistance.

Examples: Use carpet stretcher to install carpet.
Shove a heavy box along a roller conveyor to get it started.
Use muscular force to pull a lever which closes a valve.

M_3 Seat or Unseat

Conditions surrounding the object or device require that resistance be overcome prior to or following the Controlled Move. This parameter variant covers the application of muscular force with little or no movement to 'seat' or 'unseat' an object or, if necessary, the short manual actions employed to latch or unlatch the object. In the *unseating* of an object, the recoil must follow a restricted path (not to be confused with a $\mathbf{G_3}$ Disengage, which allows for the recoil of an object freely through the air).

Examples: Twist a radiator cap securely.
Push cap to secure on part.
Seat a latch on a tool box.

M_3 High Control

Care is needed to maintain or establish a specific orientation of the object during the Controlled Move. Characterized by a higher degree of visual concentration, this parameter variant is typically recognized by noticeably slower movements to keep within tolerance requirements or to prevent injury or damage. This Controlled Move demands that eye contact be made with the object and its surroundings during the move. This parameter may be followed by an Align value as in the case when turning a safe dial to a specific number and aligning it to a tick mark.

Examples: Gently push a fragile, glass tube into a holder.
With hand-held scanner, carefully scan a page of text.
Slide a carton full of glasses across the counter.

M_3 2 Stages ≤ 24 Inches (60 cm) Total

An object is displaced in two directions or increments a distance less than or equal to 24 inches (60 cm) for both stages *without relinquishing control.* If the movement is continuous and without an abrupt change of direction, it is *not* a two-stage move.

Examples: Open and close a cash register drawer 8 inches (20 cm) each way.
Push lever up 4 inches (10 cm) and then pull it down 4 inches (10 cm).

M_6 2 Stages > 24 Inches (60 cm) Total

An object is displaced in two directions or increments a distance exceeding a total of 24 inches (60 cm) for both stages *without relinquishing control.* If the movement is continuous and without an abrupt change of direction, it is *not* a two-stage move.

Examples: Open and then close a car door a total of 36 inches (90 cm) to test for a tight seal.
Shift a lever back 16 inches (40 cm) and then slide it to the side 10 inches (25 cm).

M_{10} 3 - 4 Stages

An object is displaced in three or four directions or increments *without relinquishing control.* If the movement is continuous and without an abrupt change of direction, it is *not* a multiple-stage move.

Example: Shift from first to reverse with a manual gearshift (Figure 3.11).

Push or Pull with Steps

One or more objects are manually moved along a controlled path (i.e., conveyor rollers or a cart) requiring steps to complete the move. The time to start the move of

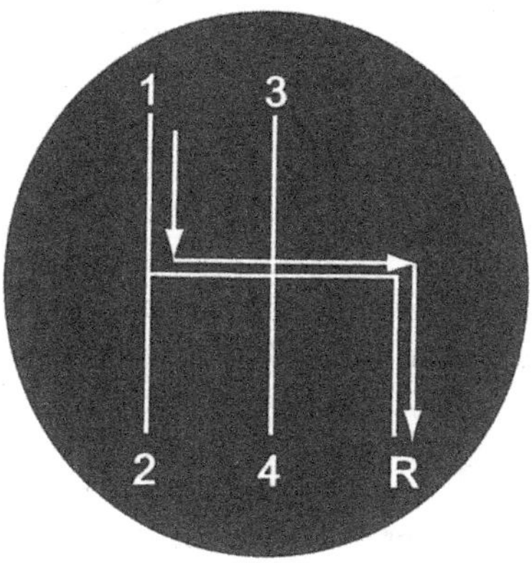

FIGURE 3.11 Moving a gear shift from first to reverse is an example of a three-stage move.

Push or Pull	
Index Value	*Steps*
M_{24}	10 - 13
M_{32}	14 - 17
M_{42}	18 - 22
M_{54}	23 - 28
M_{67}	29 - 34
M_{81}	35 - 41
M_{96}	42 - 49
M_{113}	50 - 57
M_{131}	58 - 67
M_{152}	68 - 77
M_{173}	78 - 87
M_{196}	88 - 98
M_{220}	99 - 110
M_{245}	111 - 122
M_{270}	123 - 135
M_{300}	136 - 149
M_{330}	150 - 163

FIGURE 3.12 Extended values for Push or Pull with Steps.

the object is included in the index value. If resistance occurs during the move, the number of steps taken will normally increase because shorter steps are often taken when resistance occurs. This will automatically allow the extra time to overcome resistance. In certain situations, pushing or pulling an object along a controlled path may require more than nine steps. A table with extended index values is shown in Figure 3.12 for these situations.

M_6–M_{16} 1 - 9 Steps

The index values to push or pull an object with steps are displayed on the data card as:

M_6 1–2 Steps
M_{10} 3–5 Steps
M_{16} 6–9 Steps

Examples: Push a box along a conveyor while taking two steps: **M_6**.
Push a cleaning cart down a hotel hallway while taking eight steps: **M_{16}**.

Summary of Foot Motions

Movement of the foot could appear in a Controlled Move Sequence Model under the Action Distance (**A**), the Gain Control (**G**) or the Move Controlled (**M**) parameter as shown:

Activity	Parameter and Index Value
Foot to pedal (without displacing the trunk of the body)	$\mathbf{A_1}$
Take one step	$\mathbf{A_3}$
Contact foot pedal	$\mathbf{G_1}$
Push pedal ≤ 12 inches (30 cm)	$\mathbf{M_1}$
Push pedal >12 inches (30 cm) or with resistance	$\mathbf{M_3}$
Operate pedal with high control (operate a variable speed pedal)	$\mathbf{M_3}$

Crank

This category of Move Controlled refers to the manual actions used to rotate objects such as cranks, handwheels and reels. This type of action is used when there are no obstructions in the circular path. These cranking actions are performed by moving the fingers, hand and/or forearm in a circular path more than half a revolution using one of the patterns pictured in Figure 3.13. Any motion less than half a revolution is not considered a crank and must be treated as a 'Push/Pull/Turn.' The overall distance the hand covers when making repetitive circular motions may be larger than any other motions described under the Move Controlled parameter. For this reason, a separate column is provided on the Controlled Move data card for Crank.

In addition to the actual 'cranking time,' index values for Crank also include a factor that covers the actions that sometimes occur before or after the cranking motion. These actions may involve the application of muscular force to seat or unseat the crank or the short manual actions employed to engage or disengage the device undergoing the cranking motion. Figure 3.14 lists the extended index values for cranking based on the number of revolutions completed, rounded to the nearest whole number.

Examples: Restaurant associate uses a cranking motion to close a patio awning with 10 cranks: $\mathbf{M_{16}}$.
Turn handle to open casement window with six rotations: $\mathbf{M_{10}}$.
Rotate the handle on a trailer jack with eight cranking motions to raise the trailer: $\mathbf{M_{16}}$.

Push–Pull Cranking

Occasionally, a method of cranking will result in back-and-forth movement of the elbow instead of pivoting at the wrist and/or elbow. This 'push–pull' cranking is analyzed by using the number of pushes plus pulls as a frequency for the $\mathbf{M_1}$ parameter. If there is substantial resistance during the cranking, an $\mathbf{M_3}$ would be assigned. Whenever possible, push–pull (reciprocal) cranking should be replaced by the more efficient pivotal cranking method.

Example: Grasp handle and move back and forth a total of six times to rotate crank.

$$\mathbf{A_x\ B_x\ G_x\ (M_1)\ X_x\ I_x\ A_x\ (6)}$$

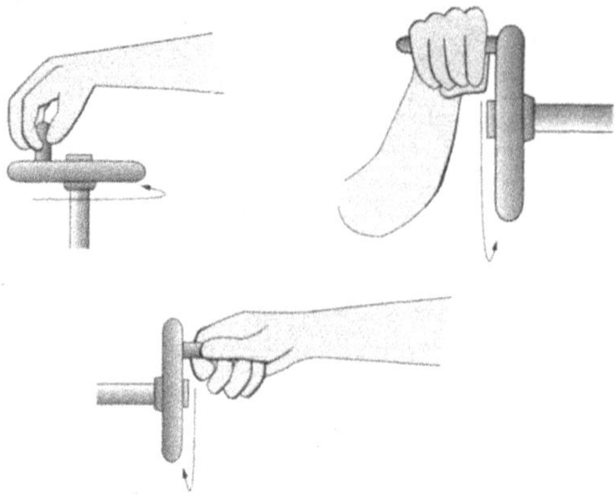

FIGURE 3.13 Examples of Crank.

Crank	
Index Value	*Number of Revolutions*
M_{24}	12 - 16
M_{32}	17 - 21
M_{42}	22 - 28
M_{54}	29 - 36

FIGURE 3.14 Extended index values for cranking based on the number of revolutions completed (rounded to the nearest whole number).

Process Time (X)

Process Time is defined as the portion of work that is controlled by electronic or mechanical devices or machines, not by manual actions. The **X** parameter of the Controlled Move Sequence Model is intended to cover process times of relatively short duration. These process times will normally have minor variations and are often difficult to time. Power fluctuations and minor differences in parts can affect the process time. The **X** parameter is applied by selecting the appropriate index value that corresponds to the observed or calculated 'actual time.'

Longer process times, such as machining times based on feeds and speeds, are normally calculated and entered separately as a process time in the analysis. The actual clock time is *never* placed on the **X** parameter of the sequence model. Only the index value that statistically represents the actual time should be placed in the sequence model. Figure 3.15 lists the extended index values for process times based on the actual clock time in seconds, minutes or hours.

Examples: A register receipt prints in 4 seconds: $\mathbf{X_{10}}$.

A double-folding machine runs for 6 seconds to make two folds in a piece of sheet metal: $\mathbf{X_{16}}$.

Process Time			
Index Value	**Seconds**	**Minutes**	**Hours**
X_{24}	9.5	.16	.0027
X_{32}	13.0	.21	.0036
X_{42}	17.0	.28	.0047
X_{54}	21.5	.36	.0060
X_{67}	26.0	.44	.0073
X_{81}	31.5	.52	.0088
X_{96}	37.0	.62	.0104
X_{113}	43.5	.72	.0121
X_{131}	50.5	.84	.0141
X_{152}	58.0	.97	.0162
X_{173}	66.0	1.10	.0184
X_{196}	74.5	1.24	.0207
X_{220}	83.5	1.39	.0232
X_{245}	92.5	1.54	.0257
X_{270}	102.0	1.70	.0284
X_{300}	113.0	1.88	.0314
X_{330}	124.0	2.06	.0344

FIGURE 3.15 Extended index values for Process Times (**X**). Values are read up to and including.

Alignment (I)

Alignment refers to manual actions following the Move Controlled (**M**) or at the conclusion of the Process Time (**X**) (i.e., adjust instrument setting) to achieve an alignment or specific orientation of objects.

Often, any adjusting motions required during a Controlled Move are covered in the $\mathbf{M_3}$ parameter variant for High Control. That index value, however, is not sufficient to cover the activity to line up an object to one or more points following the Move Controlled. This type of alignment is influenced by the ability (or inability) of the eyes to focus on one or more points in more than one area at a time.

The 'area of normal vision' is the average area covered by a single eye focus and is described by a circle 4 inches (10 cm) in diameter at a normal reading distance of about 16 inches (40 cm) from the eyes (Figure 3.16). Within this area of normal vision, the alignment of an object to two points can be performed without any additional 'eye times.' If one of the two points lies outside this area, two separate alignments are required because of the inability of the eyes to focus on both points simultaneously. In fact, an object would first be aligned to one point, the eyes would next shift to allow the alignment to the second point and then the object would finally be adjusted to correct for the minor shifting from the first point. The area of normal vision is therefore the basis for defining most of the Alignment parameter variants.

Whenever a Controlled Move involves the Alignment activity, the preceding **M** parameter is used to describe only the distance the object travels, either ≤ 12 inches (30 cm) ($\mathbf{M_1}$) or > 12 inches (30 cm) ($\mathbf{M_3}$).

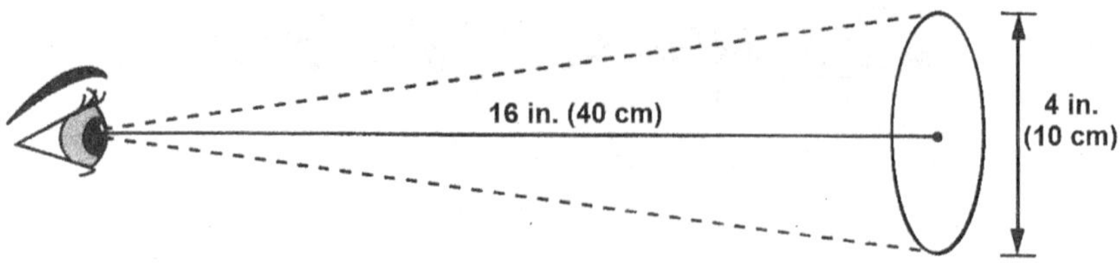

FIGURE 3.16 Area of Normal Vision.

I_1 Align to 1 Point

Following a Move Controlled, an object is aligned to one point. This is used when the demand for a precise alignment is modest and can be satisfied with a single correcting action. This variant is similar to the **P_1** variant except that **I_1** occurs following an **M** in Controlled Move; the **P_1** occurs following an **A** in General Move.

Examples: Align one corner to another corner on paper prior to folding it.
Align an index mark to a number on a dial.
Locate a mark on a wood block to a bandsaw blade prior to cutting.

I_3 Align to 2 Points ≤ 4 Inches (10 cm)

The object is aligned to two points less than or equal to 4 inches (10 cm) apart following a Move Controlled. For example, a straightedge is aligned to two marks located 3 inches (7.5 cm) apart, as shown in Figure 3.17. Both points are within the area of normal vision. An increasing demand for precision occurs in this situation. This also includes the time to make more than one correcting motion of the object within the area of normal vision.

Examples: A straightedge is aligned to two points on a blueprint 4 inches (10 cm) apart.
Align a pattern to two locating marks 4 inches (10 cm) apart in preparation for tracing.

I_6 Align to 2 Points > 4 Inches (10 cm)

The object is aligned to two points more than 4 inches (10 cm) apart following a Move Controlled action. For example, a straightedge is aligned to two marks located 8 inches (20 cm) apart as shown in Figure 3.17. One point is outside the area of normal vision; therefore, additional eye time must be allowed. Several correcting motions and eye focuses are included to allow the time for the hand-eye coordination to be accomplished.

Examples: A ruler is used to connect two points on a graph located 10 inches (25 cm) apart.
A 26×26 inch (65×65 cm) die set is aligned to two points at each corner on a press bed.

I_{16} Precision

The object is aligned to several points with extreme care or precision following a Move Controlled action.

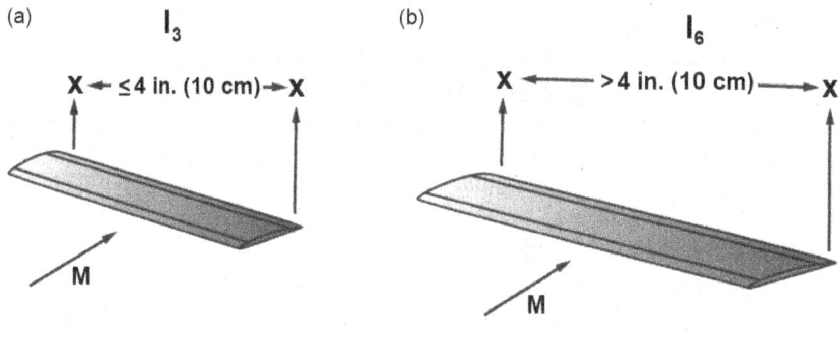

FIGURE 3.17 Align an object to two points ≤ 4 inches (10 cm) apart (a) and > 4 inches (10 cm) apart (b). The **M** parameter would be used only for the distance the ruler moved.

Examples: Align a French curve or a drawing template to several points.
Align a material template onto cloth before cutting.
Align a sheet metal template to several points over blank piece.

Machining Operations

A special group of Alignment parameter variants is frequently encountered in machine shop operations. Dealing with the alignment of 'machining tools,' these parameter variants cover the activity following the cranking action (**M**) to locate the cutting tool on a machine to the correct cutting position. Figure 3.18 lists the values for machining operations.

I_3 To Workpiece

The machining tool is aligned to the workpiece prior to making a cut. Following any cranking actions (**M**) to locate the tool near the cutting position, the crank or handwheel is manipulated so that the cutting edge of the tool just touches the workpiece.

I_6 To Scale Mark

The machining tool is aligned to a scale mark prior to making a cut. Following any cranking actions (**M**) to locate the tool near the cutting position, several taps with the fist of the hand (holding the handwheel) using the other hand may be observed to line up the cutting edge of the tool with a scale mark.

Alignment	
Index Value	***Align To***
I_3	Workpiece
I_6	Scale Mark
I_{10}	Indicator Dial

FIGURE 3.18 Index values for the Alignment of Machining Tools.

I_{10} To Indicator Dial

The machining tool is aligned to the correct indicator dial setting prior to making a cut. Following any cranking actions (**M**) to locate the tool near the cutting position, the machine operator must visually locate the indicator dial, read the indicator setting and carefully adjust the tool to the correct setting by tapping the hand that holds the handwheel several times with the other hand.

Alignment of Non-typical Objects

The final values for Alignment are for non-typical objects that are particularly flat, large, flimsy and sharp or require special handling and occur following the Move Controlled (**M**) parameter. Such activities are normally seen with press, shear or cutoff operations. Alignment will be observed as a series of short correcting motions (less than 2 inches or 5 cm of movement) following the Move Controlled. The alignment is normally made to stops, guides or marks. Figure 3.19 lists the index values for the Alignment of Non-typical Objects.

Alignment values are chosen based on the number of adjustments required to properly situate the object. The alignment value includes the short movement of the object along a controlled path and one or two visual checks for proper positioning. After each adjustment, the movement will stop.

Note: Stops or guides at the workplace may eliminate the need for adjustments. In this case, the Align value will be zero with the total time for the activity being covered by the Move Controlled (**M**) parameter.

If an object is heavy and the feet must be shifted prior to the next movement, the value for Alignment of Non-typical Objects will apply. In addition, separate Controlled Move Sequence Models should be used since the Alignment values do not include the time to move the body or gain control of the object. If the object can be realigned without shifting the feet, the original align values will apply.

Example: A press operator moves a 4 foot (1.2 m) by 8 foot (2.4 m) sheet of thin gauge steel, which is flimsy, a distance of 14 inches (35 cm). The steel sheet must be aligned to two stops on opposite ends of the sheet. It is not necessary for the operator to reposition the hands during the

Alignment		
Index Value	***Positioning Method***	***Non-typical Object Characteristics***
I_0	Against Stop(s)	• Flat • Large • Flimsy • Sharp • Difficult to handle
I_3	1 Adjustment to Stop	
I_6	2 Adjustments to Stop(s) 1 Adjustment to 2 Stops	
I_{10}	3 Adjustments to Stop(s) 2 - 3 Adjustments to Linemark	

FIGURE 3.19 Index values for the Alignment of Non-typical Objects.

activity. The operator must take one step back to gain control of the sheet. The correct analysis for this activity is:

$\mathbf{A_3\ B_0\ G_3\ M_3\ X_0\ I_6\ A_0}$ **150 TMU**

In the previous example, if the operator had to make separate grasps of the object and the object is moved and aligned to one stop with an adjustment and this process is repeated two times, the correct analysis would be:

$\mathbf{A_3\ B_0\ G_3\ M_3\ X_0\ I_3\ A_0}$ **2 240 TMU**

Note the frequency of two for the activity. This analysis assumes that a step had to be taken for the second movement of the object.

Selection of Placement (General Move) or Alignment (Controlled Move)

Placement and Alignment are similar in that both involve moving an object to a general or precise and predetermined location. Placement (General Move) provides for locating an object in space, bringing an object in contact with a surface and, when necessary, inserting one object into another. Alignment (Controlled Move) provides only for locating an object on the same surface. For example, placing a part on a workstation is a General Move. Sliding the part along the workstation is a Controlled Move.

The Alignment (**I**) parameter applies only when an alignment follows a Move Controlled. Should an object be moved freely without restrictions and then be 'aligned to two points,' the General Move Placement (**P**) parameter is the appropriate selection. In fact, there is a direct relationship between the Controlled Move and the General Move activities. That relationship is **M:I** as **A:P**. The Alignment (**I**) of an object occurs after the object is moved over a *controlled path* (**M**) and accounts for the time to align and/or orient the object just as the Placement (**P**) of an object occurs after the *spatial* displacement of an object (**A**) and accounts for the time to orient and/or position the object. Normally, the nature of the movement of the object, spatial or controlled, is the determining factor behind the selection of a sequence model.

Writing Method Descriptions

Below are the recommended method description formats for a clear and concise method description for Controlled Move. Additional words for Action Distance or Body Motion are optional and may be used to enhance the method description.

There are two recommended method description formats for Controlled Move: One for the movement of an object along a controlled path and the other for process time:

Gain Control	Object	⟨From Location⟩	Move	To Location
Gain Control	Object	Actuate	At Location	

⟨If the From Location is apparent, it is not necessary to indicate it in the method description.⟩

Controlled Move Examples

Note: For the examples listed below, all distances are within reach unless specified otherwise.

1. A worker contacts a ruler and pushes it 6 inches (15 cm) to two points that are 8 inches (20 cm) apart.

   Contact ruler and push to 2 points 8 inches (20 cm) apart

   $\mathbf{A_1\ B_0\ G_1\ M_1\ X_0\ I_6\ A_0}$

   $(1+1+1+6)\times 10=90$ TMU

2. A worker walks two steps to get a hand truck from a storage area, pushes it aside four steps and returns four steps to the original workplace.

   Get hand truck from storage area, push aside 4 steps and return

   $\mathbf{A_3\ B_0\ G_3\ M_{10}\ X_0\ I_0\ A_6}$

   $(3+3+10+6)\times 10=220$ TMU

3. A stockperson in a store grasps a freezer door handle and unseats it. The door is then opened 20 inches (50 cm).

   Grasp freezer door handle and unseat

   $\mathbf{A_1\ B_0\ G_1\ M_3\ X_0\ I_0\ A_0}$

   $(1+1+3)\times 10=50$ TMU

   Pull door 20 inches (50 cm)

   $\mathbf{A_0\ B_0\ G_0\ M_3\ X_0\ I_0\ A_0}$

   $3\times 10=30$ TMU
   50 TMU
   30 TMU
   80 TMU

4. Using the foot pedal to activate the machine, a sewing machine operator makes a stitch requiring 3.5 seconds process time. (The operator must reach to the pedal with the foot.)

   Push pedal to activate 3.5 second process time at sewing machine

   $\mathbf{A_1\ B_0\ G_1\ M_1\ X_{10}\ I_0\ A_0}$

   $(1+1+1+10)\times 10=130$ TMU

5. An operator grasps a handwheel and cranks it with eight revolutions to align a tool to a scale mark.

   Grasp handwheel and crank 8 revolutions to align tool to scale mark

   $\mathbf{A_1\ B_0\ G_1\ M_{16}\ X_0\ I_6\ A_0}$

   $(1+1+16+6)\times 10=240$ TMU

THE TOOL USE SEQUENCE MODEL

Manual work is not always performed with the hands alone. The use of tools extends the strength and capabilities of the hands through leverage. Even though automation and mechanization has occurred in almost every industry, a large and very critical portion of work remains literally 'in the hands of the worker.' Because of the desirability of having the MOST Work Measurement Technique apply to all manual work and since the analysis of the frequent use of certain tools through a series of General and Controlled Moves could take additional time and result in inconsistent applications, a third manual sequence model was developed—the *Tool Use Sequence Model.*

General and Controlled Moves can be used to measure the multiple actions needed to loosen threaded fasteners using the hand or hand tools such as screwdrivers, wrenches or ratchets. To simplify the analysis process, however, special Fasten or Loosen parameter variants and a special Tool Use Sequence Model were created to measure these multiple actions in terms of the body member performing the action (i.e., finger, wrist or arm). For example, running a nut down with the fingers is considered a finger action, but tightening a wood screw with a screwdriver requires a wrist action. These actions are, by literal definition, a series of Controlled Moves.

Any activity involving a hand tool can be analyzed as a series of General and/or Controlled Moves. For example, get and place screwdriver (General Move), fasten screw (a series of Controlled Moves) and lay screwdriver aside (General Move). However, special Tool Action parameters have been developed not only for fastening and loosening using common hand tools, but also for activities related to cutting, surface treating, measuring, recording and even thinking! Because of the ease of use, the consistency provided and the analysis time saved, such sets of multiple moves should be analyzed with the Tool Use Sequence Model when possible.

The development of the Tool Use Sequence Model not only increased consistency and application speed, but also provided analyses that were more accurate than those using a series of sequence models to analyze the use of tools. By repeating individual analyses, deviations between the allowed time (assigned index value) and the 'actual time' could occur. By developing specific elements using the statistically determined index ranges and assigning one index value representing Tool Use, the compounding of these deviations was eliminated. Accuracy was therefore maintained through the system design independent of the nature or complexity of the manual actions being performed. (This is substantiated by the system theory explained in Appendix A.) For these reasons, the Tool Use Sequence Model should be used in MOST analyses whenever appropriate. When the existing Tool Use index values will not cover a special tool or a tool with an identical or similar motion pattern, the procedure in the Development of Elements for Special Tools or Situations section can be followed to develop new elements for such tools.

The Tool Use Sequence Model is comprised of phases and sub-activities from the General Move Sequence Model along with specially designed parameters describing the actions performed with hand tools or, in some cases, mental processes required when using the senses as a tool. As noted above, Tool Action parameters have been developed for many manual tool applications, and in most cases, the use of the following tools can be analyzed with the Tool Use Sequence Model:

Fasten or Loosen	Measure
• Fingers or Hand (when used as a tool)	• Fixed Scale
• Screwdriver	• Steel Tape
• Ratchet	• Caliper
• Wrench	• Micrometer
• T-Wrench	• Gauge
• Hammer	
• Power Tool	
Cut	**Record**
• Pliers	• Pen/Pencil
• Scissors	• Marker
• Knife	• Stylus
	• Scribe
Surface Treat	**Think**
• Air Nozzle	• Eyes
• Brush	
• Wiping Cloth	

Other hand tools that utilize the same methods as above can also be analyzed by comparison to the tools in the tables. For example, using a rubber mallet to tap ceramic tile into place uses a similar method to tapping with a hammer. The index values for the hammer could then be used to analyze the activity.

The Sequence Model

The sequence model consists of a series of letters or parameters representing each of the various sub-activities of the *Tool Use Sequence Model.* This sequence model uses the parameters from the General Move Sequence Model and a blank space indicated by an asterisk (*) symbol is used for one of seven Tool Use parameters.

A B G | A B P | * | A B P | A

where: **A** = Action Distance
B = Body Motion
G = Gain Control
P = Placement

The blank space in the sequence model is used for the insertion of one of the following Tool Action parameters. These parameters, which refer to the tool action being used, are listed as follows:

where: **F** = Fasten
L = Loosen
C = Cut
S = Surface Treat
M = Measure
R = Record
T = Think

Phases of the Tool Use Sequence Model

Tool Use follows a fixed sequence of sub-activities, which occur in five phases:

Get tool or object	Put tool or object in place	Tool action	Put tool or object aside	Return operator
A B G	**A B P**	*	**A B P**	**A**

1. Get Tool or Object:
   - Reach with hand a distance to tool or object, either directly or in conjunction with body motions or steps.
   - Gain manual control of the tool or object.
2. Put Tool or Object in Place:
   - Move the tool or object a distance to where it will be used, either directly or in conjunction with body motions or steps.
   - Place the tool or object in position for use.
3. Tool Action:
   - Apply number or extent of tool actions.
4. Put Tool or Object Aside:
   - Retain the tool or object for further use (hands and fingers are of course *always* retained), toss or lay the tool aside, return the tool to its original location or move it to a new location for disposition, either directly or in conjunction with body motions or steps.
5. Return:
   - Return to the workplace.

Parameter Definitions

Other than the Tool Action parameters, the Tool Use Sequence Model contains only parameters from the General Move Sequence Model. The **A**, **B**, **G** and **P** parameters were discussed with the General Move Sequence Model and remain unchanged.

F Fasten

This parameter is used to establish the time for manually or mechanically assembling one object to another using the fingers, hand or a hand tool.

L Loosen

This parameter is used to establish the time for manually or mechanically disassembling one object from another using the fingers, hand or a hand tool.

C Cut

This parameter covers the manual actions used to separate, divide or remove part of an object using a sharp-edged hand tool such as pliers, scissors or a knife.

S Surface Treat

This parameter covers the activities used to remove unwanted material or particles from, or apply a substance, coating or finish to, the surface of an object.

M Measure

This parameter includes the actions used in determining a certain physical characteristic of an object using a standard measuring device.

R Record

This parameter covers the manual actions performed with a pencil, pen, marker, stylus, scribe or other marking tool for the purpose of recording information.

T Think

This parameter refers to the eye actions and mental activity used to obtain information (read) or to inspect an object, including reaching to touch, when necessary, to feel the object.

Parameter Indexing

With the exception of the Tool Action parameters, the Tool Use Sequence Model contains only parameters from the General Move Sequence Model. Index values for these parameters are found on the General Move data card (Figure 3.1). Additional data cards are provided for the Tool Action parameters. Figure 3.20 contains index values for tools covered by the Fasten or Loosen parameters. Figure 3.21 covers activities such as cutting, surface treating and measuring and Figure 3.22 includes values for recording and thinking. These tables for indexing the Tool Action parameters are used following the same procedure outlined in the General and Controlled Moves sections.

Consider, for example, an assembly operation in which a bolt is used to fasten one object to another. The operator picks up a bolt from a bin located within reach, places it in the required location also within reach and runs it down with three finger spins. The sequence model would be indexed:

Grasp bolt, place and fasten with 3 finger spins

$\mathbf{A_1\ B_0\ G_1\ A_1\ B_0\ P_3\ F_6\ A_0\ B_0\ P_0\ A_0}$

$(1+1+1+3+6)\times 10=120$ TMU

BasicMOST System			Tool Use					A B G A B P * A B P A				
	F Fasten or **L Loosen**											
Index x 10	Finger Action	Wrist Action				Arm Action					Power Tool	Index x 10
	Spins	Turns	Strokes	Cranks	Taps	Turns		Strokes	Cranks	Strikes	Screw Diam.	
	Fingers, Screwdriver	Hand, Screwdriver, Ratchet, T-Wrench	Wrench	Wrench, Ratchet	Hand, Hammer	Ratchet	T-Wrench, 2-Hands	Wrench	Wrench, Ratchet	Hammer	Power Wrench	
1	1	-	-	-	1	-	-	-	-	-	-	**1**
3	2	1	1	1	3	1	-	1	-	1	1/4 in. (6 mm)	**3**
6	3	3	2	3	6	2	1	-	1	3	1 in. (25 mm)	**6**
10	8	5	3	5	10	4	-	2	2	5		**10**
16	16	9	5	8	16	6	3	3	3	8		**16**
24	25	13	8	11	23	9	6	4	5	12		**24**
32	35	17	10	15	30	12	8	6	6	16		**32**
42	47	23	13	20	39	15	11	8	8	21		**42**
54	61	29	17	25	50	20	15	10	11	27		**54**

FIGURE 3.20 Tool Use data card for Fasten or Loosen. Values are read up to and including.

BasicMOST System		Tool Use				A B G A B P * A B P A			
	C Cut				**S Surface Treat**			**M Measure**	
Index x 10	Cutoff	Secure	Cut	Slice	Air-Clean	Brush-Clean	Wipe	Measure	Index x 10
	Pliers		Scissors	Knife	Nozzle	Brush	Cloth	Measuring Tool	
	Wire		Cuts	Slices	sq. ft. (0.1 m²)	sq. ft. (0.1 m²)	sq. ft. (0.1 m²)		
1	-	Grip	1	-	-	-	-	-	**1**
3	Soft	-	2	1	-	-	1/2	-	**3**
6	Medium	Twist Form Loop	4	-	1 Spot or Cavity	1	-	-	**6**
10	Hard	-	7	3	-	-	1	Profile Gauge	**10**
16		Secure Cotter Pin	11	4	3	2	2	Fixed Scale Caliper $\le$ 12 in. (30 cm)	**16**
24			15	6	4	3	-	Feeler Gauge	**24**
32			20	9	7	5	5	Steel Tape $\le$ 6 ft. (2 m) Depth Micrometer	**32**
42			27	11	10	7	7	OD-Micrometer $\le$ 4 in. (10 cm)	**42**
54			33					ID-Micrometer $\le$ 4 in. (10 cm)	**54**

FIGURE 3.21 Tool Use data card for Cut, Surface Treat and Measure. Values are read up to and including.

BasicMOST System			Tool Use			A B G A B P * A B P A			
	R Record				T Think				
Index x 10	Write			Mark	Inspect	Read			Index x 10
	Pencil / Pen / Stylus			Marker	Eyes / Fingers	Eyes			
	Digits	Words	Copy	Digits	Points	Digits, Single Words	Text of Words	Compare	
1	1	-	-	Check Mark	1	1	3	1	**1**
3	2	-	1	1 Scribe Line	3	3 (Gauge)	8 (Gauge)	2	**3**
6	4	1	3	2	5 Feel for Heat	6 (Scale Value Date or Time)	15 (Scale Value Date or Time)	4	**6**
10	6	-	5	3	9 Feel for Defect	12 (Vernier Scale)	24 (Vernier Scale)	8	**10**
16	9 (Signature or Date)	2 (Signature or Date)	8	5	14	(Table Value)	38 (Table Value)	13	**16**
24	13	3	10	7	19		54		**24**
32	18	4	14	10	26		72		**32**
42	23	5	18	13	34		94		**42**
54	29	7	22	16	42		119		**54**

FIGURE 3.22 Tool Use data card for Record and Think. Values are read up to and including.

In this example, the 'Get' and 'Put' phases of the sequence model are used for getting and placing the bolt. Placement of a threaded fastener will nearly always be a $\mathbf{P}_3$ (with adjustments) unless it takes place in a blind or obstructed location ($\mathbf{P}_6$). Since this is a fastening activity, the **F** parameter is chosen and inserted in the sequence model. The appropriate index value is determined by considering the body member performing the fastening activity (in this case, the fingers) and the number of actions performed. In Figure 3.20, it can be determined that three finger actions require an index value of 6. The remaining parameters in the sequence (**A**, **B**, **P** and **A**) carry zero index values since no activity was performed to set aside a tool or object.

After the fastening activity, if the operator picks up a small box end wrench lying on the table within reach and tightens the bolt with three wrist strokes and asides the tool within reach, this sequence model would be analyzed:

Grasp wrench and fasten bolt with 3 wrist strokes and aside

$\mathbf{A}_1\ \mathbf{B}_0\ \mathbf{G}_1\ \mathbf{A}_1\ \mathbf{B}_0\ \mathbf{P}_3\ \mathbf{F}_{10}\ \mathbf{A}_1\ \mathbf{B}_0\ \mathbf{P}_1\ \mathbf{A}_0$

$(1+1+1+3+10+1+1)\times 10 = 180$ TMU

Again, using the Fasten or Loosen data card, the index value is taken from the 'Strokes' column below Wrist Actions. Index values in this column reflect the way in which a wrench is normally used. That is, after each wrist action, the wrench must be repositioned on the fastener before any subsequent actions are made. In the example, three wrist actions are performed with the wrench. The corresponding index value applied is $\mathbf{F}_{10}$.

In addition to the Tool Action phase of the sequence model, the remaining parameters in this sequence apply to handling the tool. The $\mathbf{P}_3$ prior to the Tool Action in the previous example covers the initial placement of the wrench on the bolt. The parameters following the Tool Action Phase—$\mathbf{A_1}$ $\mathbf{B_0}$ $\mathbf{P_1}$ $\mathbf{A_0}$—indicate that the wrench is put aside following the fastening activity and there is no activity in the Return phase.

Use of the Tool Action data card (Figure 3.21) for Cut can be demonstrated with the following example. Suppose that during a sewing operation a seamstress picks up a pair of scissors within reach and makes three cuts to remove the excess material from around a stitch. This activity would be described as follows:

Grasp scissors, cut material with 3 cuts and put scissors aside

$\mathbf{A_1\ B_0\ G_1\ A_1\ B_0\ P_1\ C_6\ A_1\ B_0\ P_1\ A_0}$

$(1+1+1+1+6+1+1)\times 10=120$ TMU

The appropriate Tool Action parameter for this example would be Cut, which is represented by the letter **C**. Looking down the column titled 'Cut' in Figure 3.21, three cuts with scissors carries the index value $\mathbf{C_6}$. The initial placement of the scissors prior to the cutting action is assumed to be $\mathbf{P_1}$ in this case. Applying index values for the placement of tools will be discussed later in this section.

The remainder of this section examines in detail each of the Tool Action parameters and application.

Fasten or Loosen (F or L)

Fasten or Loosen includes manually or mechanically assembling or disassembling one object to or from another using the fingers, hand or a hand tool. Index values for the **F** and **L** parameters are primarily grouped according to the body member (e.g., finger, wrist or arm) performing the Tool Action. An additional category is provided for power-operated hand tools.

With the exception of power tools, all of the data in Figure 3.20 refers to the number of actions performed by the respective body member during either a Fasten (**F**) or Loosen (**L**) activity. An action is defined as the back-and-forth or up-and-down movement of the fingers, wrist or arm to perform one Turn, Stroke or Tap with the tool. In the case of the Crank data, action refers to one revolution of the tool.

Finger Actions (Spins)

Finger Spins include the movements of the fingers and thumb to run a threaded fastener down or out. These short finger movements are characterized by rolling or spinning an object between the thumb and index finger. Examples include running a nut down with the fingers or turning a machine screw with a small screwdriver. Because of the limited strength in the fingers, the muscular force (pressure) exerted on the fastener while performing spins is minimal. The Finger Spin data, however, includes a light application of pressure for seating and unseating the fastener. This light pressure includes up to three wrist turns (see below), which often occur at the end of a finger spin activity when the resistance increases as in replacing a cap on a bottle. If more than three wrist turns occur, the appropriate index value for Wrist Turns should be applied separately. This situation describes the use of Multiple Tool Actions which will be discussed later in this section.

In some situations, the finger spin action converts into a finger crank action typified by turning a wing nut on a bolt with the forefinger held straight and pivoted at the base joint. Each 360 degree turn would be counted as one spin.

Example: A small screwdriver is used with six finger spins to fasten eye glass frame: $\mathbf{F}_{10}$.

Wrist Actions

A wrist action refers to the twisting motion of the wrist about the axis of the forearm or the pivoting of the hand from the wrist with either a circular or back-and-forth motion. As Figure 3.20 indicates, the data is classified according to how the wrist actions are being performed.

Wrist Turn

Tool actions covered under Wrist Turns include using the hand, screwdriver (Figure 3.23), ratchet (Figure 3.24) or small T-wrench (Figure 3.25). These tools are not removed from the fastener during use and are not repositioned on the fastener after an action. The values for Wrist Turns include the time for repositioning the hand on a tool handle after each action. As a result of the added strength when using the larger muscles of the hand and forearm, a final tighten or initial loosen action can be accomplished with a Wrist Turn when using a tool. The wrist itself does not have enough muscular force to completely tighten a nut or bolt to the needed torque. A Wrist Turn using the hand can be used for tightening a fastener for the purpose of securing it. Final tightening with a tool is used to tighten the fastener to the defined specifications. The index values assigned from the Wrist Turn column include the time for final tightening or initial loosening of a fastener. Figure 3.31 illustrates which Tool Use actions allow time to final tighten or initial loosen.

Note: In the case where the hand is the tool and another tool is used to final tighten, a second Tool Use Sequence Model is then used to show the final tightening activity.

Example: A ratchet is used to secure a nut on a completed assembly with six wrist turns: $\mathbf{F}_{16}$.

FIGURE 3.23 Example of a Screwdriver.

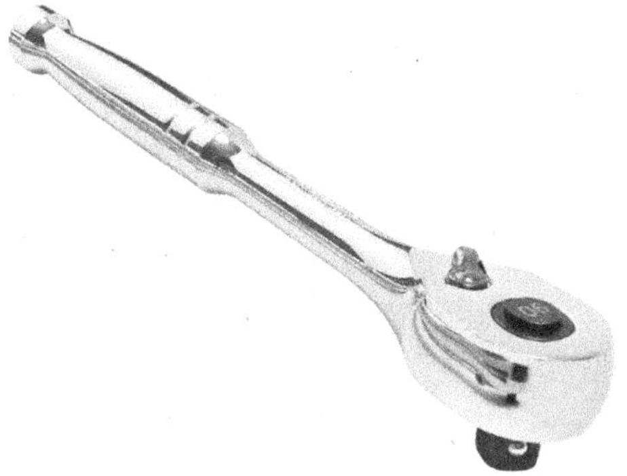

FIGURE 3.24 Example of a Ratchet.

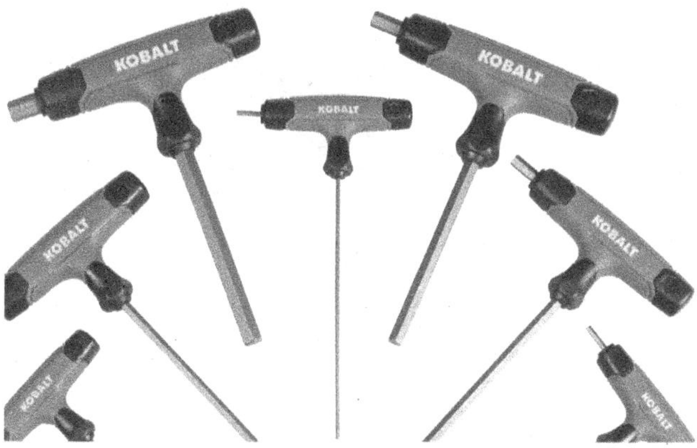

FIGURE 3.25 Example of a T-wrench.

Wrist Stroke

The Wrist Stroke column covers the method normally employed when using a wrench. That is, after each stroke with the tool and before making each subsequent stroke, the wrench must be removed from and repositioned on the fastener. Index values in this column apply to the number of strokes (actions) performed with the wrench. The time for the wrench to be removed from and repositioned on the fastener between strokes is included in the index values. The repositioning of the wrench includes up to 2 inches (5 cm) of Action Distance to reposition the tool. The data for Wrist Stroke allows for the final tightening or initial loosening activity. Tools covered by this parameter include the following types of wrenches: Combination wrench (Figure 3.26) which includes an open end and a closed end, hexagon (Figure 3.27) and adjustable (Figure 3.28). These tools are normally *repositioned* on a fastener during use.

Example: Six wrist strokes are used with a wrench to tighten a nut on the leg of a chair during the assembly process: $\mathbf{F_{24}}$.

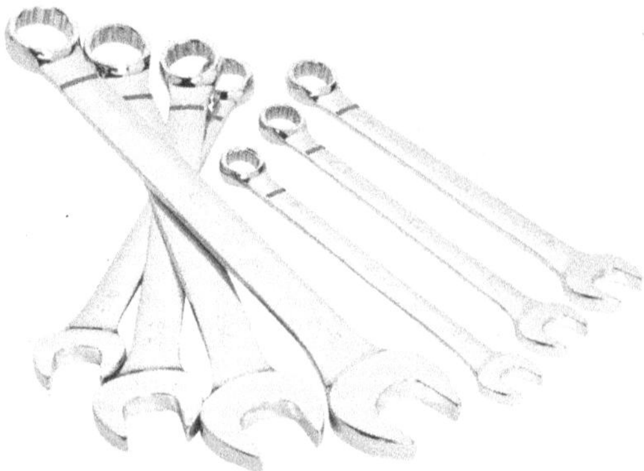

FIGURE 3.26 Combination Wrench.

FIGURE 3.27 Hexagon Wrench.

Wrist Crank

Data from the Wrist Crank column applies to tools that are spun or rotated around a fastener while remaining affixed to it. They are guided with a circular movement of the hand as it is pivoted from the wrist (Figure 3.29). This type of wrist action is sometimes used with either wrenches or ratchets when there are no obstructions in the circular path of the tool. After the initial placement of the tool, the fingers and hand are used to push or crank the tool completely around the fastener. However, these wrist actions are employed by operators only when little or no resistance is encountered; therefore, data in the Wrist Crank column does *not include* the time for final tightening or initial loosening of a fastener. If, after a number of wrist cranks, a fastener is final tightened, the normal type of tool action (Wrist Turn or Wrist Stroke) will be used to analyze the final tightening activity. Usually, one or several of these

FIGURE 3.28 Adjustable Wrench.

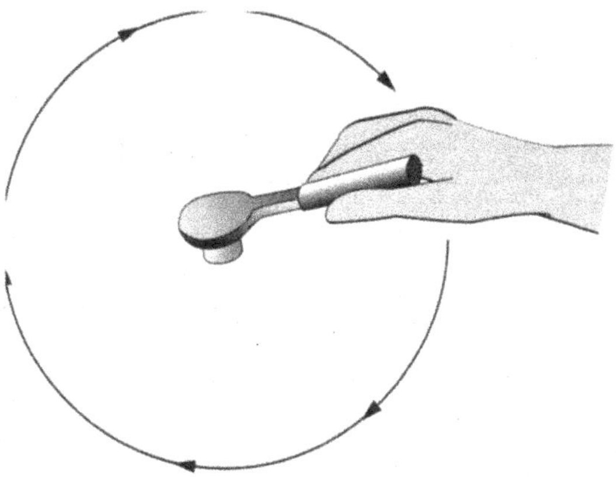

FIGURE 3.29 Example of a wrist crank.

actions will be needed and will be analyzed in a separate Tool Use Sequence Model. Index values for Wrist Crank cover the number of revolutions performed with the tool. If a partial revolution is observed, round to the nearest whole number.

Fasten or Loosen with continuous cranking motions is the most economical way of running down a screw. One cranking motion results in running down one thread on the screw while other methods produce only one-third to one-sixth of a thread per action.

Example: An Allen key (hexagon wrench, hex key) is used to make eight circular motions on a screw to align a belt on a machine: $\mathbf{F}_{16}$.

Tap

The use of the hand, a small hammer (Figure 3.30) or other similar tools is covered by the data under the heading Taps. Index values from the Tap column refer to the short up-and-down tapping motions performed with the hand as it is pivoted at the

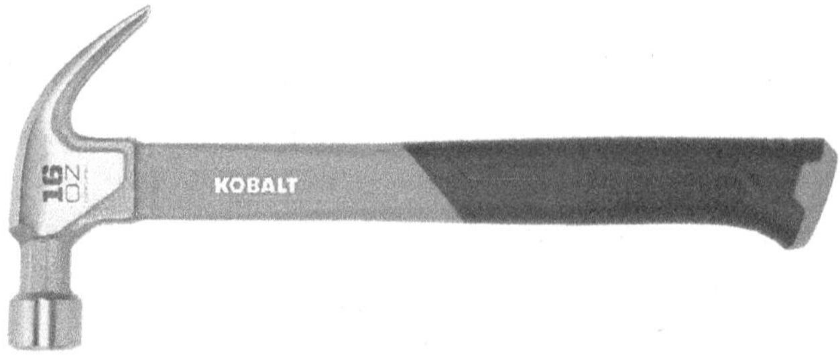

FIGURE 3.30 Example of a Hammer for taps or strikes.

wrist. The number of actions on the data card is based on down motions or taps. The time to retract the hand, or the up motion, is included in the index values.

Example: A hammer is used with six wrist taps to drive a nail into a wall: $\mathbf{F_6}$.

Arm Actions

Arm actions include the motions of the hand requiring elbow and shoulder movements. With the wrist relatively rigid, the forearm is pivoted from the elbow with an up-and-down, circular or back-and-forth motion. These forearm motions may be assisted by the pivoting of the upper arm from the shoulder.

Arm Turn

In the first column, the tools covered under the heading Arm Turns include only the use of a ratchet. Arm actions of this type are employed when the ratchet is held near the end of the handle resulting in a pulling action on the tool. Index values from the Arm Turn column include time for the final tightening or initial loosening that may occur in the complete fastening or loosening activity. The data in the second column under Arm Turns is provided to analyze the use of a large T-wrench with two hands. Each arm action involves a 180 degree turn of the T-wrench. All subsequent two-handed arm actions include the reach of each hand to the opposite handle before making the next turn. The data for T-wrench, two hands also allows for the final tightening or initial loosening involved in the complete fastening or loosening activity. This would also be appropriate for turning a large valve or other such item with both hands.

Example: Use a T-wrench with two hands to loosen a water main valve with four arm turns: $\mathbf{L_{24}}$.

Arm Stroke

Similar to the Wrist Stroke data, the Arm Stroke column applies to the normal method of using a wrench. That is, following each stroke or pull with the tool, the wrench must be removed and repositioned on the fastener before making a subsequent pull. Index values in this column apply to the number of arm actions (pulls) performed with the wrench. Index values for Arm Stroke allow for the final tightening or initial loosening activity that may occur in the complete fastening or loosening. Tools covered by this parameter include a wrench (box end, open end, hexagon and adjustable).

Example: Using a box end wrench, fasten a nut on an assembly with six arm strokes: $\mathbf{F}_{32}$.

Arm Crank

The data from the Arm Crank column applies to tools used with a circular movement of the forearm as it is pivoted at the elbow or the shoulder. Arm actions of this type are occasionally used with either wrenches or ratchets when there are no obstructions in the circular path of the tool. The hand is used to push or crank the tool around the fastener. Like the wrist actions under the same heading, this type of action is used only when resistance is minimal; therefore, the values in the Arm Crank column *do not* include the time for final tightening or initial loosening of a fastener. The data in this column refers to the number of revolutions performed with the tool. If a partial revolution is observed, round to the nearest whole number.

Example: Using a ratchet, loosen a bolt on a steel beam with four arm cranks: $\mathbf{L}_{24}$.

Strike

The use of a hammer with an arm action is accounted for under the heading Strike. The data in this column refers to the up-and-down motions performed with the hand as it is pivoted from the elbow. The number of actions on the data card is based on down motions or strikes. The time to retract the arm, or the up motion, is included in the index values (Figure 3.20).

Example: Using a hammer, strike a nail into a piece of plywood with eight arm strikes: $\mathbf{F}_{16}$.

Power Tools

Power Tools include the use of power-operated hand tools. The data provided in Figure 3.20 covers electric (Figure 3.33) and pneumatic power wrenches (Figure 3.34). Index values are based on the time required to run a standard threaded

Final Tighten / Initial Loosen	
Tool Action	***Do the Tool Action Values Include Time to Final Tighten or Initial Loosen?***
Finger Actions	No
Wrist Turns with Hand	No
Wrist Turns with Tool	Yes
Wrist Strokes	Yes
Wrist Cranks	No
Arm Turns	Yes
Arm Strokes	Yes
Arm Cranks	No

FIGURE 3.31 Summary of actions for Final Tighten and Initial Loosen.

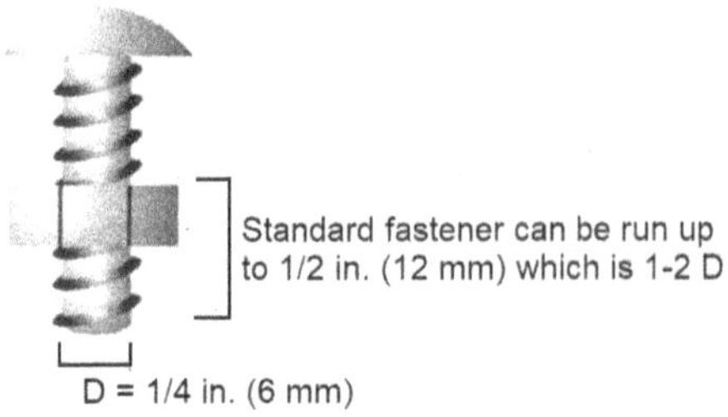

FIGURE 3.32 The index values for Power Tools are based on the time required to run a standard threaded fastener down or out a length equal to one or two times the diameter of the fastener.

fastener down or out a length equal to one or two times the bolt diameter of the fastener. Two values are found in Figure 3.20: $\mathbf{F_3}$ or $\mathbf{L_3}$ for a screw diameter of 1/4 inch (6 mm) or smaller and $\mathbf{F_6}$ or $\mathbf{L_6}$ for larger screws up to and including 1 inch (25 mm) in diameter. To apply **F** or **L** to a power tool, choose the Fasten or Loosen value based on the diameter of the fastener.

Note: These index values apply to standard fasteners where the length of holding threads is one to two times the diameter only. When running down or out longer fasteners where more threads are needed to hold the item or threads are fine, a frequency can be applied to the **F** or **L** value chosen. Refer to the example in Figure 3.32 which shows that a bolt with 1/4 inch (6 mm) diameter can be run in up to 1/2 inch (12 mm). That meets the definition of a standard threaded fastener and the index value applied would be an $\mathbf{F_3}$. If the bolt is run in 1 inch (25 mm), then the analyst has two options:

1. Frequency the $\mathbf{F_3}$ value by two; or
2. Analyze the activity with General and Controlled Moves (the process time will need to be developed using a stopwatch, process specifications or engineering calculations related to machine speed and feed rates).

FIGURE 3.33 Electric Power Tool.

Note: The basic values for Fasten or Loosen with a power tool must be compared and validated to the time required by the brands of power tools used. Guidelines for validating the process times for Power Tools can be found in the Application of the BasicMOST Work Measurement System section. If there is a difference in the BasicMOST values for Fasten or Loosen with a power tool and those studied, new elements for the tools must be created using the formula outlined in the Development of Elements for Special Tools or Situations section for developing new elements.

Torque Wrench

Supplementary values for a torque wrench, which are not found on the data card, have been developed and are presented below.

F_6 Torque Wrench

Tighten a bolt or nut with a torque wrench (Figure 3.35) with a handle length of up to 10 inches (25 cm). The value is for one arm action and includes the time to either align the dial or to await the click.

F_{10} Torque Wrench

Tighten a bolt or nut with a torque wrench with a handle length of 10–15 inches (25–37.5 cm). The value is for one arm action and includes the time either to align the dial or to await the click.

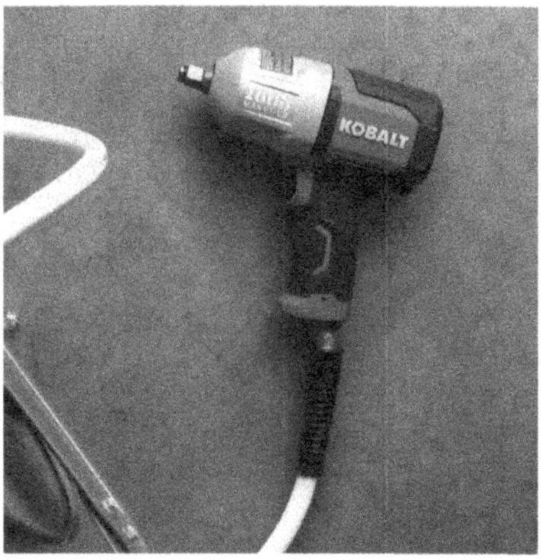

FIGURE 3.34 Pneumatic Power Tool.

FIGURE 3.35 Example of a Torque wrench.

F_{16} Torque Wrench

Tighten a bolt or nut with a torque wrench with a handle length of 15–40 inches (37.5 cm–1 m). The value is for one arm action and includes the time either to align the dial or to await the click.

Tool Placement

The **P** parameter preceding the Tool Action parameter is used to indicate the index value for the placement of a tool or object in a working position prior to the tool action. The index value for the placement of the tool should be selected using the guidelines in the General Move section. However, as a general rule, the **P** parameter for the Fasten or Loosen tools will carry the index values indicated in Figure 3.36. This Tool Placement chart has been developed to speed up application when using the Tool Use Sequence Model.

Notice that the placement of the fingers or hands when used as a tool is typically considered a $\mathbf{P_1}$. This would normally be considered as a $\mathbf{G_1}$ Gain Control activity. However, since the fingers or hands are used in the same way as a fastening or loosening tool, the activity is considered the placement of a tool instead of a grasp. For example, if an operator were to contact a nut on a bolt within reach and loosen it with three finger spins, the sequence model would be analyzed:

Get tool	Put tool in place	Tool action	Aside tool	Return	
A_0 B_0 G_0	A_1 B_0 P_1	L_6	A_0 B_0 P_0	A_0	**80 TMU**

Tool Placement	
Tool	***Index Value***
Hammer	P_0 (1)
Fingers or Hand	P_1 (3) (6)
Pliers	P_1 (3)
Scissors	P_1 (3)
Knife	P_1 (3)
Surface Treating Tool	P_1
Measuring Tool	P_1
Writing Tool	P_1
Screwdriver	P_3
Ratchet	P_3
T-Wrench	P_3
Wrench	P_3
Power Tool	P_3
Adjustable Wrench	P_6 (3)

FIGURE 3.36 Index values for tool placement.

If the fingers or hands are placing a fastener, such as a nut or bolt, immediately preceding the action to fasten it, the **P** parameter refers to the placement of the fastener. The placement of a threaded fastener nearly always requires a $\mathbf{P_3}$ placement unless the placement occurs in a blind or obstructed location; under those conditions, $\mathbf{P_6}$ would be appropriate. For example, if an operator were to grasp and place a nut on a bolt within reach and fasten it with three finger spins, the sequence model would be:

Get fastener	Place fastener	Tool action	Aside tool	Return	
A_1 B_0 G_1	A_1 B_0 P_3	F_6	A_0 B_0 P_0	A_0	**120 TMU**

There may or may not be an initial placement of a hammer prior to any tapping or striking actions. Normally, if a hammer is being used to drive small nails or tacks, the hammerhead will be positioned over the nail ($\mathbf{P_1}$) to line the hammer up with the nail prior to performing any actions. In many cases, however, no initial placement of the hand or hammer is necessary ($\mathbf{P_0}$). Simply tapping or striking a larger object or surface area without lining the hammer up is an example of $\mathbf{P_0}$ placement for a hammer.

In Figure 3.36, the standard placement value for pliers, scissors and knife is a $\mathbf{P_1}$. This placement allows for one adjustment of the tool and will cover the majority of operations performed by the average operator. If a more exact placement is needed (cutting material to be exactly 1 yard in length, for example), a $\mathbf{P_3}$ would normally apply. This larger value is shown on the data card to cover the additional adjustments in placement of these tools, if necessary.

Notice from Figure 3.36 that the placement of an adjustable wrench is a $\mathbf{P_6}$. This larger index value is required to cover the additional actions necessary to adjust the jaws of the wrench (with intermediate moves) to the size of the fastener. A value of $\mathbf{P_3}$ (also noted in Figure 3.36) is used for subsequent placements once the wrench has been adjusted to the proper fastener size.

Writing Method Descriptions

Below is the recommended format for a clear and concise method description for the Tool Use Sequence Model. Additional words may be used to enhance the method description. These could be Action Distances, Body Motions or adjectives. Examples of this structure can be found in the Tool Use examples listed below and throughout the Tool Use section.

The recommended method description format for Tool Use is:

Gain Control	Tool	Tool Action	Number of Fasteners (items)	Activity	At Location	Aside

Tool Use Examples for Fasten or Loosen

Note: For the examples listed below, all distances are within reach unless specified otherwise.

1. Obtain a nut from a parts bin, place it on a bolt and run it down with seven finger actions.

   Grasp nut and fasten on bolt with 7 spins

   $\mathbf{A_1\ B_0\ G_1\ A_1\ B_0\ P_3\ F_{10}\ A_0\ B_0\ P_0\ A_0}$

   $(1+1+1+3+10)\times 10=160$ TMU

2. Grasp hammer and bend to tap shelving unit with four wrist taps. Aside hammer when finished.

   Grasp hammer, fasten shelving unit with 4 wrist taps and put aside

   $\mathbf{A_1\ B_0\ G_1\ A_1\ B_6\ P_0\ F_6\ A_1\ B_0\ P_1\ A_0}$

   $(1+1+1+6+6+1+1)\times 10=170$ TMU

3. Pick up a screwdriver from a workbench and loosen a screw on a remote control with three wrist turns.

   Grasp screwdriver and loosen screw on remote control with 3 wrist turns

   $\mathbf{A_1\ B_0\ G_1\ A_1\ B_0\ P_3\ L_6\ A_0\ B_0\ P_0\ A_0}$

   $(1+1+1+3+6)\times 10=120$ TMU

4. Obtain a ratchet from an open tool box four steps away, walk eight steps to loosen one bolt with four arm cranks on an air-conditioning unit and put the ratchet back in the tool box.

   Grasp ratchet, loosen bolt with 4 arm cranks on air conditioning unit and aside

   $\mathbf{A_6\ B_0\ G_1\ A_{16}\ B_0\ P_3\ L_{24}\ A_{16}\ B_0\ P_1\ A_0}$

   $(6+1+16+3+24+16+1)\times 10=670$ TMU

5. Operator grasps a wrench from the tool bin, walks two steps back and makes one wrist stroke on a bolt located on a piece of farm equipment. The wrench is held when the activity is complete.

   Grasp wrench, fasten bolt on farm equipment with 1 wrist stroke and hold

   $\mathbf{A_1\ B_0\ G_1\ A_3\ B_0\ P_3\ F_3\ A_0\ B_0\ P_0\ A_0}$

   $(1+1+3+3+3)\times 10=110$ TMU

Tool Use Frequencies

Occasionally an activity may involve the fastening or loosening of several fasteners in succession using the same tool. By using a special convention where an extra

A parameter is inserted between the **P** and **F** or **L** (or any Tool Action parameter) to allow for the Action Distance between fasteners, the entire activity can then be analyzed using only one Tool Use Sequence Model. For example, an operator picks up a screwdriver within reach and tightens four screws with eight wrist turns each and then sets aside the screwdriver. The first step in making an analysis of this activity is to look at the situation as if only one screw were fastened and then repeat the appropriate parameters to tighten the second screw. The analysis for fastening one screw would be:

For one screw:

$$\mathbf{A_1\ B_0\ G_1\ A_1\ B_0\ P_3\ F_{16}\ A_1\ B_0\ P_1\ A_0}$$

What must be repeated to fasten the second screw? First, there is a reach to the second screw, then the tool must be placed and the screw fastened; therefore, the Action Distance to the fasteners, the Placement and the Fastening must be repeated.

Covering the Action Distance of the tool to each fastener requires that an additional **A** parameter be added to the sequence model between the **P** and **F** parameters. For example:

Add an 'A' to cover the reach between the fasteners:

$$\mathbf{A_1\ B_0\ G_1\ A_1\ B_0\ P_3\ A\ F_{16}\ A_1\ B_0\ P_1\ A_0}$$

Parentheses are then placed around those parameters that are repeated (e.g., **P**, **A** and **F**) as shown below:

$$\mathbf{A_1\ B_0\ G_1\ A_1\ B_0\ (P_3\ A\ F_{16})\ A_1\ B_0\ P_1\ A_0}$$

If the distance between the screws is ≤ 2 inches (5 cm), an $\mathbf{A_0}$ is assigned between the **P** and **F** parameters. For example, if using a screwdriver to tighten four screws with eight wrist turns each. The distance between the screws is ≤ 2 inches (5 cm). The multiplier for the parameters (the number of fasteners included in the fastening activity) is noted as a partial frequency in the MOST analysis, also within parentheses.

$$\mathbf{A_1\ B_0\ G_1\ A_1\ B_0\ (P_3\ A_0\ F_{16})\ A_1\ B_0\ P_1\ A_0\ (4)}$$

Note: '**A**' must be added to the Tool Action section of the sequence model to account for the distance between the screws.

If the distance between the screws is > 2 inches (5 cm), an $\mathbf{A_1}$ must be placed in the parentheses. Since the Action Distance to each fastener is covered by the **A** parameter within the parentheses, *the A following the Gain Control will now carry a zero index value.* This is to avoid counting an 'extra' Action Distance value. For example, using a screwdriver to tighten four screws with eight wrist turns each. The distance between the screws is 5 inches (12.5 cm). The correct sequence model is:

$$\mathbf{A_1\ B_0\ G_1\ A_0\ B_0\ (P_3\ A_1\ F_{16})\ A_1\ B_0\ P_1\ A_0\ (4)}$$

Note: When the distance between fasteners is > 2 inches (5 cm), the $\mathbf{A_1}$ placement value must be dropped since it is included in the frequency value. As illustrated in the previous example, there are two Action Distances, one to the first screw and one to the second. The number in parentheses at the end of the sequence model multiplied by the **A** in the parentheses will account for all of the needed reaches.

The *incorrect* sequence model would be:

$$\mathbf{A_1\ B_0\ G_1\ A_1\ B_0\ (P_3\ A_1\ F_{16})\ A_1\ B_0\ P_1\ A_0\ (4)}$$

Notice the $\mathbf{A_1}$ after the Get phase. By keeping the $\mathbf{A_1}$ in the sequence model, the analyst will have an added Action Distance that is not needed.

The time calculation for the fastening or loosening activity is performed by adding all index values contained within the parentheses and then multiplying this sum by the number of fasteners involved (the partial frequency value). The sequence model total is calculated by adding the index values from the remaining parameters to the partial frequency total. The total TMU is determined by multiplying this sum by 10. For example,

$$\mathbf{A_1\ B_0\ G_1\ A_0\ B_0\ (P_3\ A_1\ F_{16})\ A_1\ B_0\ P_1\ A_0\ (4)}$$

$$[(3+1+16)=20\times4=80+1+1+1+1=84\times10]=840 \text{ TMU}$$

The Tool Action partial frequencies are most commonly used with the Fasten or Loosen parameters but can be applied to any Tool Action or Equipment Use parameter.

Tool Use Frequency Examples

Note: For the examples listed below, all distances are within reach unless specified otherwise.

1. An operator grasps a wrench from a table and fastens six bolts that are 5 inches (12.5 cm) apart with three arm strokes each. The operator then asides the wrench to the table.

   Grasp wrench, fasten 6 bolts 5 inches (12.5 cm) apart with 3 arm strokes each and aside wrench on table

   $$\mathbf{A_1\ B_0\ G_1\ A_0\ B_0\ (P_3\ A_1\ F_{16})\ A_1\ B_0\ P_1\ A_0\ (6)}$$

   $$[(3+1+16)=20\times6=120+1+1+1+1=124\times10]=1240 \text{ TMU}$$

2. A worker grasps a screwdriver and fastens three screws that are 1 inch (2.5 cm) apart with nine wrist turns each. The screwdriver is then placed with adjustments in a box under a bench.

   Grasp screwdriver, fasten 3 screws 1 inch (2.5 cm) apart with 9 wrist turns each and place in box

   $$\mathbf{A_1\ B_0\ G_1\ A_1\ B_0\ (P_3\ A_0\ F_{16})\ A_1\ B_6\ P_3\ A_0\ (3)}$$

   $$[(3+16)=19\times3=57+1+1+1+1+6+3=70\times10]=700 \text{ TMU}$$

3. Get a heavy power tool from three steps away, return to loosen five 1/2 inch (12 mm) diameter nuts. The nuts are 6 inches (15 cm) apart. Put the tool aside.

   Get power tool, loosen 5 nuts [1/2 inch (12 mm)] 6 inches (15 cm) apart and put aside

   $\mathbf{A_6\ B_0\ G_3\ A_6\ B_0\ (P_3\ A_1\ L_6)\ A_1\ B_0\ P_1\ A_0\ (5)}$

   $[(3+1+6)=10\times5=50+6+3+6+1+1=67\times10]=670$ TMU

Multiple Tool Actions

The data found in the Tool Use Fasten or Loosen data card (Figure 3.20) is organized according to the body member predominantly performing the tool action, not necessarily by the tool itself since the tool can be used with more than one type of tool action. In fact, an operator may use a combination of different finger, wrist or arm actions during a fastening or loosening activity with a *single* tool. This can happen quite often when Finger Spins, Wrist or Arm Cranks are involved because the values in those columns on the Fasten or Loosen data card *do not include* the time for final tightening or initial loosening of a fastener. Therefore, as previously explained, when a fastener is finally tightened or initially loosened in conjunction with any of the above activities, another activity (e.g., wrist or arm action) is performed and should be analyzed with a separate sequence model.

For example, when using a screwdriver, the initial tool actions to run down a screw may be performed with finger spins if no resistance is encountered. The final tightening (more than the finger pressure to seat the screw) may require the use of wrist actions. Or, a ratchet may be used first with cranking actions followed by Wrist Turns to final tighten the fastener.

These and other similar fastening or loosening activities are described in two sequence models. Consider the following examples:

1. The operator grasps a screwdriver within reach and one screw is fastened with 18 finger spins using the screwdriver. Four additional wrist turns are necessary to tighten the screw and then the operator asides the screwdriver to a table. The tool is never removed from the fastener. The completed sequence models would be:

   Grasp screwdriver and fasten screw with 18 finger spins

   $\mathbf{A_1\ B_0\ G_1\ A_1\ B_0\ P_3\ F_{24}\ A_0\ B_0\ P_0\ A_0}$

   $(1+1+1+3+24)\times10=300$ TMU

   Tighten screw with 4 wrist turns and put aside

   $\mathbf{A_0\ B_0\ G_0\ A_0\ B_0\ P_0\ F_{10}\ A_1\ B_0\ P_1\ A_0}$

   $(10+1+1)\times10=120$ TMU

2. Grasp a ratchet and run down a nut three revolutions with a wrist crank followed by six wrist turns. Aside ratchet when finished. All distances are within reach. The completed sequence models would be:

   Grasp ratchet and fasten with 3 wrist cranks

   $\mathbf{A_1\ B_0\ G_1\ A_1\ B_0\ P_3\ F_6\ A_0\ B_0\ P_0\ A_0}$

   $(1+1+1+3+6)\times 10=120$ TMU

   Tighten with 6 wrist turns and aside

   $\mathbf{A_0\ B_0\ G_0\ A_0\ B_0\ P_0\ F_{16}\ A_1\ B_0\ P_1\ A_0}$

   $(16+1+1)\times 10=180$ TMU

Cut, Surface Treat, Measure, Record and Think

The index values for common activities within the parameters of Cut, Surface Treat, Measure, Record and Think are found in Figures 3.21 and 3.22. The list of values is not meant to be comprehensive. In fact, should special or supplementary activities or tools be required to analyze a particular situation, the analyst is encouraged to develop those values under the guidelines included in the Development of Elements for Special Tools or Situations section. Following this procedure, the analyst tailors the data card to their particular situation or industry.

Cut (C)

Cut describes the manual actions employed to separate, divide or remove part of an object using a sharp-edged hand tool. As Figure 3.21 indicates, index values for the **C** parameter cover the use of pliers, scissors or a knife for general cutting activities. In addition, pliers are used for gripping and bending activities. These cutting tools and uses are described below.

Pliers—Cutoff

The use of pliers is broken down into two categories: Cutoff and Secure. The Cutoff values are used to cut through wire.

Three different methods may be used to cut through a wire using pliers (Figure 3.37). The particular method used largely depends on the firmness of the

FIGURE 3.37 Example of Pliers.

wire material and the diameter or gauge of the wire. The data (Figure 3.21) for cutting with pliers includes three index values for cutting wire.

C_3 Soft

This parameter applies to cutting a soft steel, copper or other small-gauge wire and is recognized by using the pliers with one hand and making one cut.

Example: An electrician uses one hand to make one cut on a thin wire cutting it in half.

C_6 Medium

This parameter applies to cutting a steel wire or cable and can be recognized by using the pliers with one hand and making two cuts. Following the initial cut, the pliers are rotated around the wire and repositioned over the cut before completely cutting through the wire.

Example: A technician uses one hand and makes two cuts on a medium-gauge cable.

C_{10} Hard

This parameter applies to cutting a heavier wire (approximately 10 gauge) and can be recognized where both hands are needed to apply sufficient force to cut through the wire to make two cuts. Similar to medium wire, following the initial cut, the pliers are rotated around the wire and repositioned over the cut before completely cutting through the wire with two hands.

Example: A worker uses two hands to make two cuts on a heavy gauge piece of wire.

Pliers—Secure

The Secure values are used for the general use of pliers for activities such as gripping, twisting and bending.

C_1 Grip

Following the initial placement of the pliers, the operator squeezes the pliers to simply hold an item and subsequently releases the pressure on the item.

Example: Using pliers, hold a wire in place for soldering.

C_6 Twist

Following the placement of the pliers on two wires, the jaws are closed and two twisting motions of the pliers join the wires together. Should more than two twisting actions be needed, divide the number of actions observed into groups of two and apply this as a partial frequency to the C_6 value.

Example: Using pliers, twist the ends of two wires together.

C_6 Form Loop

Following the initial placement of the pliers, the operator closes the jaws and, using two actions, forms a loop or eye in the end of a wire.

Example: Using pliers, form an eye in the end of a wire to fit over a terminal in a junction box.

C_{16} Secure Cotter Pin

Following the initial placement, an operator bends both legs on a cotter pin to secure it in position.

Example: Using pliers, bend legs on a cotter pin to secure it through a small shaft.

Scissors

The data for scissors (Figure 3.38) applies to cutting paper, fabric, light cardboard or other similar lightweight material using scissors. Index values are selected according to the number of cuts or scissor actions used during the cutting activity. To cut off a piece of plastic binding on a box, for example, only one cutting action is required. The appropriate index value from the Tool Use data card (Figure 3.21) is $\mathbf{C_1}$ (one cut with scissors). If an associate in a retail fabric store cuts through a piece of fabric with four cutting actions, the value of $\mathbf{C_6}$ (four cuts with scissors) would apply. The placement of scissors is normally a $\mathbf{P_1}$ ($\mathbf{P_3}$ if accurate placement is required, such as along a dotted line).

Note: If the scissors are held open following an initial cut to make one long cut (e.g., cutting through a piece of plastic), a Controlled Move Sequence Model should be used to analyze the long cut.

Example: A worker makes an initial cut on a piece of fabric and then continues to slide the open scissors through the length of the fabric to cut off a section.

Grasp scissors and make 1 cut on fabric

$\mathbf{A_1\ B_0\ G_1\ A_1\ B_0\ P_1\ C_1\ A_0\ B_0\ P_0\ A_0}$

Slide scissors along fabric

$\mathbf{A_0\ B_0\ G_0\ M_3\ X_0\ I_0\ A_0}$

Knife

A sharp knife (Figure 3.39) can be used for cutting string, material and light cord or to cut through corrugated material or cardboard. The length of one cut can be up to 32 inches (80 cm) if the motion is continuous without changing directions. For the multiple action index values to apply for a knife, the knife must remain in contact

FIGURE 3.38 Example of Scissors.

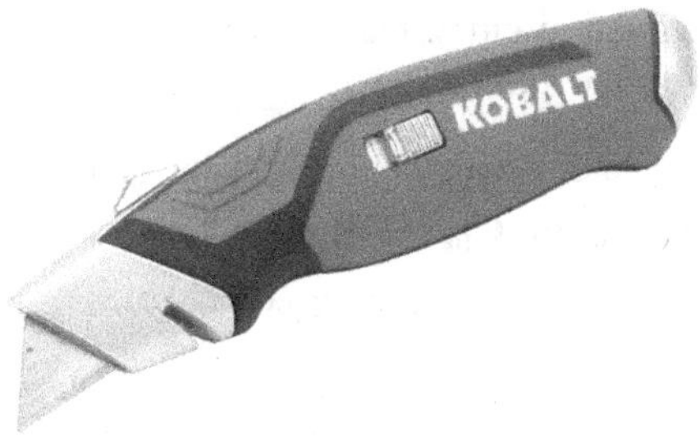

FIGURE 3.39 An example of a Utility Knife for cutting.

with the object and change directions. If four sides of a box are cut with four slices *without* lifting the knife, the value would be $\mathbf{C_{16}}$ for four slices.

If a knife is lifted to cut through tape on two sides on the top of a box, the value of $\mathbf{C_3}$ for one cut would be applied two times using the tool action partial frequency convention as shown below:

$$\mathbf{A_1\ B_0\ G_1\ A_0\ B_0\ (P_1\ A_1\ C_3)\ A_1\ B_0\ P_1\ A_0\ (2)\ 140\ TMU}$$

The criterion for selecting the index value to account for the initial placement of a knife is the same as was discussed in the General Move section for Placement. As a general rule, a $\mathbf{P_1}$ is typically used for a knife. If the slice must be accurate, such as placing a knife on a specific mark to begin slicing, $\mathbf{P_3}$ is appropriate.

Example: A distribution center employee uses a knife to slice open a carton with three continuous slices: $\mathbf{C_{10}}$.

Tool Use Examples for Cut

Note: For the examples listed below, all distances are within reach unless specified otherwise.

1. An electrician grasps a pair of pliers from her tool belt and cuts off a piece of medium-gauge wire. The pliers are put back in the tool belt.

   Grasp pliers, cut medium wire and put in tool belt

   $\mathbf{A_1\ B_0\ G_1\ A_1\ B_0\ P_1\ C_6\ A_1\ B_0\ P_1\ A_0}$

   $(1+1+1+1+6+1+1)\times 10=120$ TMU

2. With scissors in hand, a worker cuts a thread from the sewing machine. The scissors are held in the palm upon completion.

   Use scissors to make 1 cut on a thread and hold

   $\mathbf{A_0\ B_0\ G_0\ A_1\ B_0\ P_1\ C_1\ A_0\ B_0\ P_0\ A_0}$

   $(1+1+1)\times 10=30$ TMU

3. An associate picks up a knife, makes two slices across the top of a cardboard box and sets the knife aside.

   Grasp knife, slice box with 2 slices and put aside

   $\mathbf{A_1\ B_0\ G_1\ A_1\ B_0\ P_1\ C_{10}\ A_1\ B_0\ P_1\ A_0}$

   $(1+1+1+1+10+1+1)\times 10=160$ TMU

Surface Treat (S)

Surface Treat covers the activities used to clean material or particles from or applying a substance, coating or finish to the surface of an object. Many types of activities may be included in the Surface Treat category such as lubricating, painting, cleaning, polishing, gluing, coating and sanding. However, the data in the Surface Treat section of the Tool Use data card (Figure 3.21) covers only general cleaning activities performed with a rag or cloth (Wipe), an air hose (Air-Clean) or a brush (Brush-Clean). Other kinds of surface treating activities, if encountered, may be treated as special tools (see Development of Elements for Special Tools or Situations section) and supplementary elements may be developed for those particular activities. Index values for the cleaning tools described below are based primarily on the amount of surface area being cleaned.

Air-Clean

Using an air hose or nozzle to blow small particles or chips out of a hole, cavity or from a surface. The index value is chosen by the size of the surface area cleaned with the exception of a hole or cavity in a part, jig or fixture where the value $\mathbf{S_6}$ (Spot or Cavity) is appropriate.

Example: A woodworker uses an air hose to remove wood chips from a 10 square foot (1 m^2) work surface: $\mathbf{S_{42}}$.

If more than one cavity is cleaned in this manner, the $\mathbf{S_6}$ value along with the **P** parameter and an Action Distance (**A**) to account for the distance between cavities will be multiplied by the number of cavities. For example, air-clean five holes with an air hose where the holes are > 2 inches (5 cm) apart. The sequence model would be:

$\mathbf{A_1\ B_0\ G_1\ A_0\ B_0\ (P_1\ A_1\ S_6)\ A_1\ B_0\ P_1\ A_0\ (5)}$ **440 TMU**

Brush-Clean

Using a brush to remove particles, chips or other debris from an object or surface. The index value is chosen by the size of the surface area cleaned. To brush-clean a small object, an $\mathbf{S_6}$ is appropriate because the object is most likely less than 1 square foot (0.1 m^2) in size. A small object refers to brushing a jig, fixture or cavity.

Example: A technician uses a brush to remove metal shavings from a 4 square foot (0.4 m^2) piece of aluminum sheeting: $\mathbf{S_{32}}$.

Wipe

Using a rag or cloth for wiping light oil or a similar substance from a surface. The index value is chosen by the size of the surface area cleaned.

Example: A deli clerk uses a towel to wipe up oil from a 2 square foot (0.2 m^2) cutting board: $\mathbf{S_{16}}$.

Tool Use Examples for Surface Treat

Note: For the examples listed below, all distances are within reach unless specified otherwise.

1. An associate contacts a cloth already on a glass case and wipes the 3 square feet (0.3 m^2) area.

   Contact cloth on glass case and wipe 3 square feet (0.3 m^2)

   $\mathbf{A_1\ B_0\ G_1\ A_0\ B_0\ P_0\ S_{32}\ A_0\ B_0\ P_0\ A_0}$

   $(1+1+32)\times 10=340$ TMU

2. An operator grasps a brush to clean a 6 square foot (0.6 m^2) area and then tosses the brush into a can.

   Grasp brush, clean a 6 square foot (0.6 m^2) area and toss into can

   $\mathbf{A_1\ B_0\ G_1\ A_1\ B_0\ P_1\ S_{42}\ A_1\ B_0\ P_0\ A_0}$

   $(1+1+1+1+42+1)\times 10=470$ TMU

3. A worker gets an obstructed air hose and spot cleans one cavity behind the machine, asiding the hose when finished.

   Get air hose, spot clean cavity behind machine and aside

   $\mathbf{A_1\ B_0\ G_3\ A_1\ B_0\ P_3\ S_6\ A_1\ B_0\ P_1\ A_0}$

   $(1+3+1+3+6+1+1)\times 10=160$ TMU

Measure (M)

Measure includes the actions employed to determine a certain physical characteristic of an object using a standard measuring tool.

Index values for the Measure (**M**) elements cover all actions necessary to align, adjust and examine both the measuring tool and the object during the measuring activity. Therefore, the initial placement of the tool will normally be analyzed with a $\mathbf{P_1}$. The data from the Measure section of the Tool Use data card (Figure 3.21) covers the following measuring tools.

Note: Measuring tools continue to evolve and may include easier to read scale values or digital readouts. In those cases, validation of the applied index value is required.

M_{10} Profile Gauge

This value covers the use of an angle, radius, level or screw-pitch gauge to compare the profile of the object to the gauge. The $\mathbf{M_{10}}$ value includes adjusting the gauge to the object plus the visual actions to compare the configuration of the object with the gauge. A level and a square are shown as examples of a profile gauge in Figures 3.40 and 3.41.

Example: A builder uses a carpenter's square to check the angle of a window frame.

M_{16} Fixed Scale

This parameter value covers the use of a linear [12 inch (30 cm) ruler, yardstick, meter stick, etc.] or an angular (protractor) measuring device as shown in Figures 3.42 and 3.43. The $\mathbf{M_{16}}$ value includes adjusting and readjusting the tool to two points and the time to read the actual dimension from the graduated scale.

Example: A quality assurance associate uses a yardstick to check the distance between two drilled holes on an assembly.

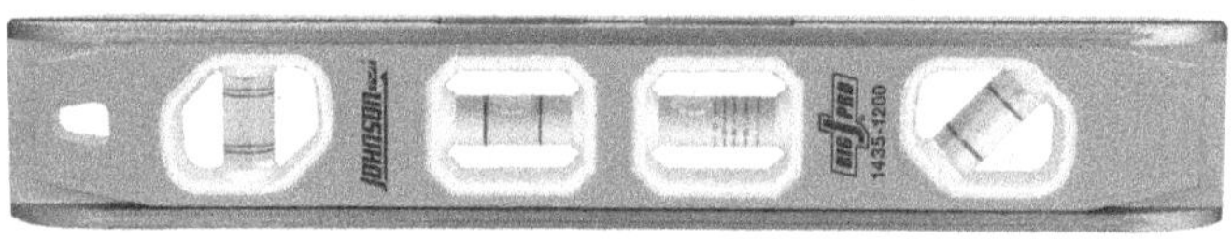

FIGURE 3.40 A level is an example of $\mathbf{M_{10}}$, Profile Gauge.

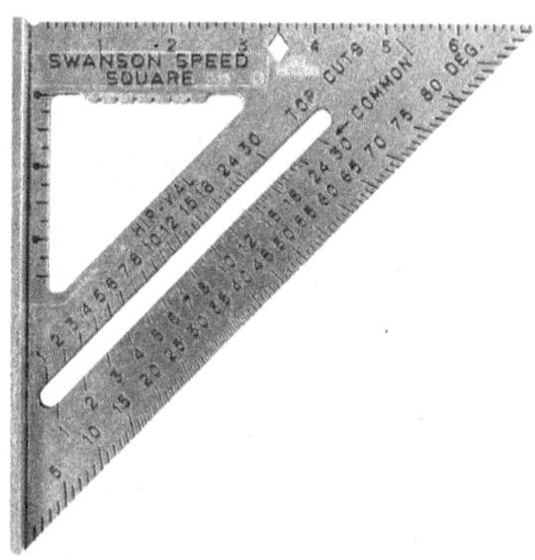

FIGURE 3.41 A square can be used as an $\mathbf{M_{10}}$, Profile Gauge.

FIGURE 3.42 Example of $\mathbf{M_{16}}$, Fixed Scale.

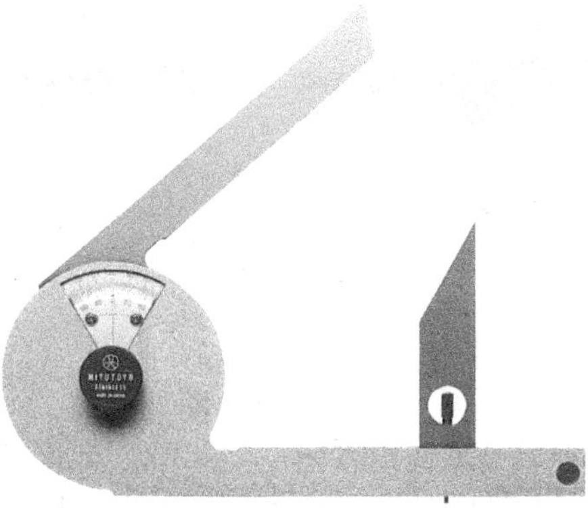

FIGURE 3.43 A protractor is an example of an $\mathbf{M_{16}}$, Fixed Scale.

M_{16} Caliper ≤ 12 Inches (30 cm)

This parameter covers the use of a caliper with a Vernier scale (Figure 3.44) with a maximum measurement capacity of up to 12 inches (30 cm). The $\mathbf{M_{16}}$ value includes setting the caliper legs to the object dimension, locking the legs in place and reading the Vernier scale to determine the measurement.

Note: If the caliper includes a dial (Figure 3.45) or digital readout (Figure 3.46) of the measurement, the index value *must* be validated.

Example: Worker confirms the thickness of a brake pad is 0.125 inches (3.17 mm) using a caliper.

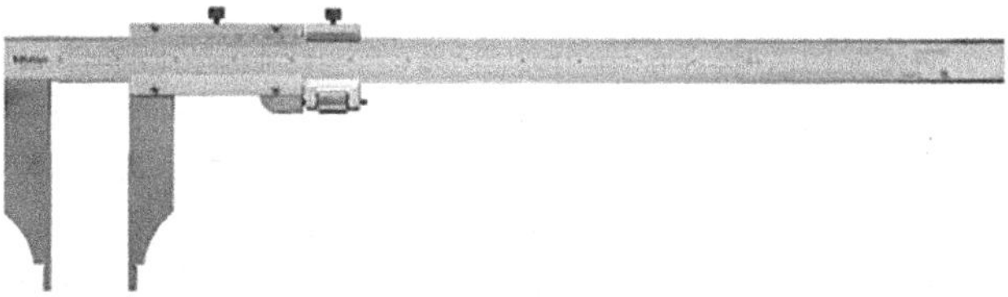

FIGURE 3.44 Example of $\mathbf{M_{16}}$, Caliper with a Vernier scale.

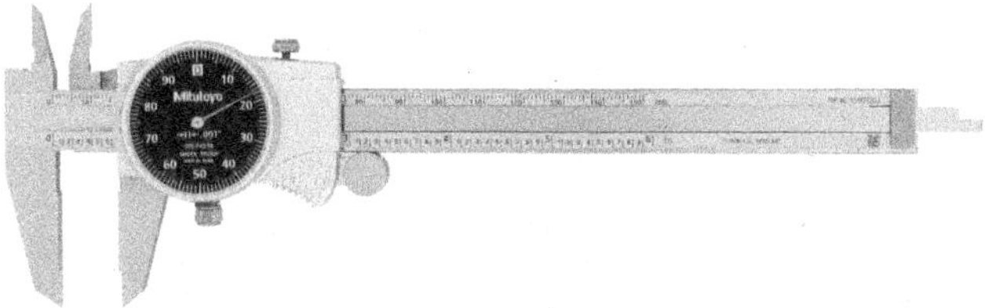

FIGURE 3.45 Caliper with dial indicator.

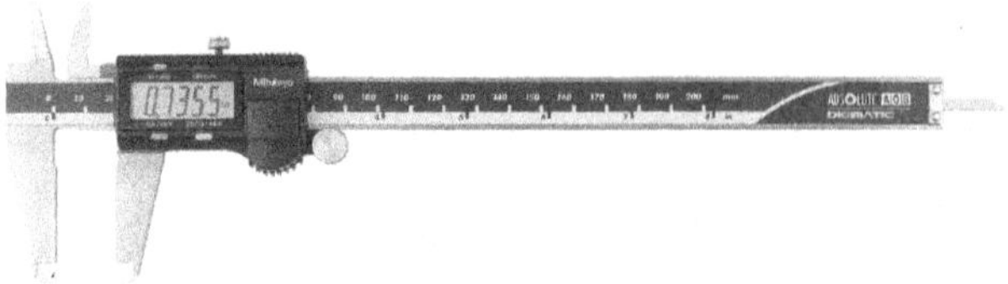

FIGURE 3.46 Caliper with digital readout.

M_{24} Feeler Gauge

This parameter covers the use of a feeler gauge (Figure 3.47) to measure the gap between two points. The $\mathbf{M_{24}}$ value includes fanning out the blades, reading and selecting the appropriate blade size and positioning the blade to the gap to check for fit.

Example: A mechanic uses a feeler gauge to measure the end gap on a piston ring after it is inserted into an engine cylinder.

M_{32} Steel Tape/Tape Measure ≤ 6 Feet (2 m)

This parameter covers the use of a steel tape, often referred to as a tape measure (Figure 3.48), to measure the distance between two points. The $\mathbf{M_{32}}$ value includes pulling the tape from the reel, positioning the end of the tape, adjusting and readjusting the tape between the two points, the time to read the dimension from the scale and finally pushing the tape back into the reel. This value is confined to the use of a steel tape from a fixed position and includes *no walking* between the two points to adjust the tape.

Example: A retail associate uses a steel tape to determine the length of a display shelf.

$M_{32}/M_{42}/M_{54}$ Micrometers ≤ 4 Inches (10 cm)

These three index values cover the use of three different micrometers: $\mathbf{M_{32}}$ for measuring depth (Figure 3.49), $\mathbf{M_{42}}$ for measuring outside diameter (OD) (Figure 3.50) and $\mathbf{M_{54}}$ for measuring inside diameter (ID) (Figure 3.51). These values are based on micrometers designed for maximum dimensions of 4 inches

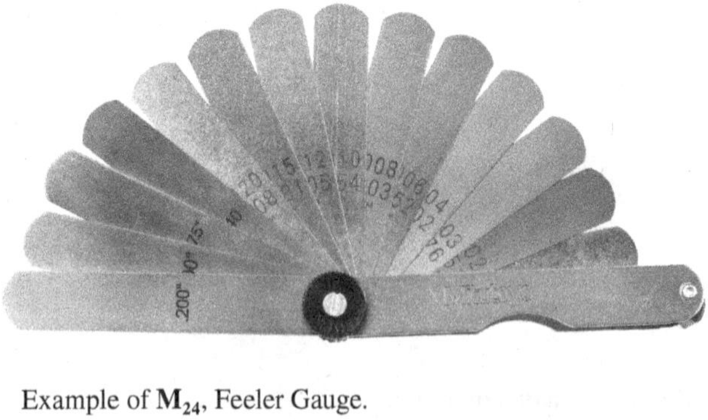

FIGURE 3.47 Example of $\mathbf{M_{24}}$, Feeler Gauge.

FIGURE 3.48 Example of $\mathbf{M_{32}}$, Steel Tape/Tape Measure.

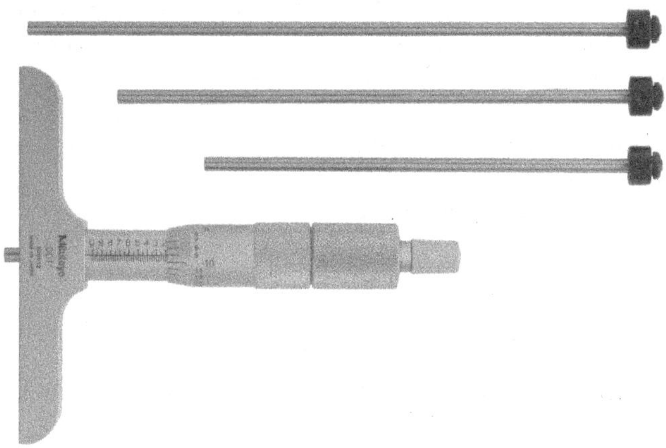

FIGURE 3.49 Example of $\mathbf{M_{32}}$, Depth Micrometer.

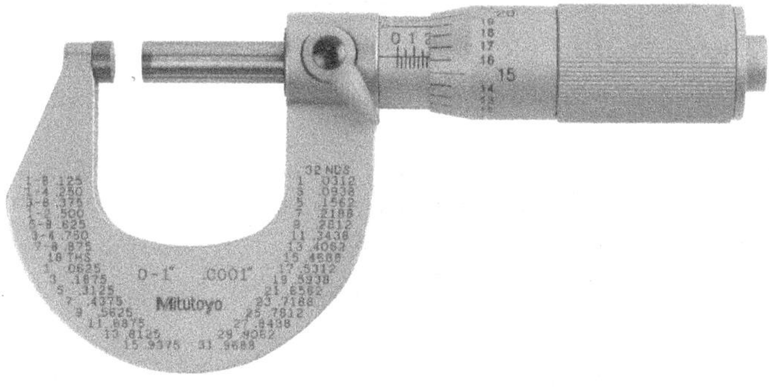

FIGURE 3.50 Example of $\mathbf{M_{42}}$, Outside Micrometer.

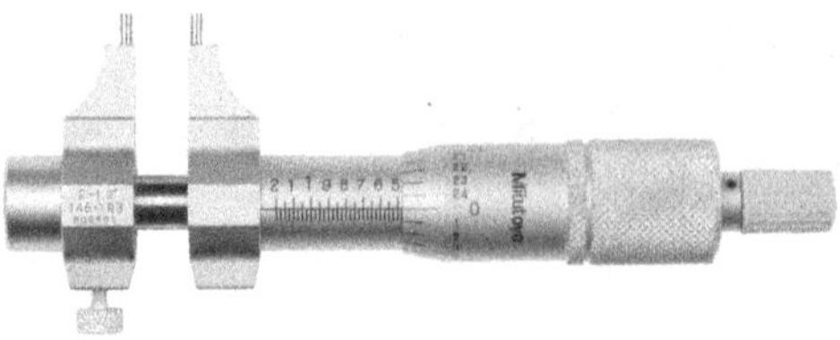

FIGURE 3.51 Example of $\mathbf{M_{54}}$, Inside Micrometer.

(10 cm) in diameter. The values include setting the micrometer to the part, adjusting the thimble for fit, locking the device and finally reading the Vernier scale to determine the dimension.

Notice that for these index values, all the placing and adjusting motions are included in the Measure parameter. The result is that the adjusting motions following the initial placement of the measuring device are covered by each index value for **M**. For this reason, the placement parameter prior to the tool action will normally carry an index value of $\mathbf{P_1}$ whenever the Measure parameter is involved.

Note: If the micrometer includes a dial or digital readout of the measurement, the index value *must* be validated.

Example: Worker uses an inside micrometer to measure the bore diameter of a cylinder: $\mathbf{M_{54}}$.

Tool Use Examples for Measure

Note: For the examples listed below, all distances are within reach unless specified otherwise.

1. A welder walks three steps to grasp a square from a workbench, walks three steps to a set of plates, checks the angle between the plates and asides the square back at the workbench.

   Grasp square from workbench, measure angle and aside at workbench

   $\mathbf{A_6\ B_0\ G_1\ A_6\ B_0\ P_1\ M_{10}\ A_6\ B_0\ P_1\ A_0}$

   $(6+1+6+1+10+6+1)\times 10=310$ TMU

2. A machinist walks two steps to grasp an OD micrometer from a table and checks the diameter of a small shaft on a lathe. He then walks two steps back to the table to aside the micrometer.

   Grasp OD micrometer from table, measure shaft at lathe and aside on table

   $\mathbf{A_3\ B_0\ G_1\ A_1\ B_0\ P_1\ M_{42}\ A_3\ B_0\ P_1\ A_0}$

   $(3+1+1+1+42+3+1)\times 10=520$ TMU

3. An operator obtains a steel tape from a tool pouch and measures a line 4 feet (1.2 m) long. The tape is put back to the tool pouch.

   Grasp steel tape from tool pouch, measure 4 feet (1.2 m) and put tape in tool pouch

   $\mathbf{A_1\ B_0\ G_1\ A_1\ B_0\ P_1\ M_{32}\ A_1\ B_0\ P_1\ A_0}$

   $(1+1+1+1+32+1+1)\times 10=380$ TMU

Special Measuring Tools

Several supplementary values for various measuring devices have been developed that do not appear on the data card and are as follows.

M_6 Snap Gauge

Measure with a snap gauge an outer diameter up to 2 inches (5 cm).

M_{10} Snap Gauge

Measure with a snap gauge (Figure 3.52) an outer diameter up to 4 inches (10 cm).

M_{16} Plug Gauge

Measure with a plug gauge, GO+NOGO ends, up to 1 inch (2.5 cm).

M_{24} Thread Gauge

Measure with a thread (plug or ring) gauge, GO+NOGO ends, internal or external threads up to 1 inch (2.5 cm).

M_{32} Plug Gauge

Measure with a plug gauge, GO+NOGO ends, up to 2 inches (5 cm).

M_{24} Depth Gauge

Measure with a depth gauge with a Vernier scale (Figure 3.53) up to 6 inches (15 cm).

Note: If the depth gauge includes a dial or digital readout (Figure 3.54) of the measurement, the index value selected *must* be validated.

FIGURE 3.52 Example of $\mathbf{M_{10}}$, Snap Gauge.

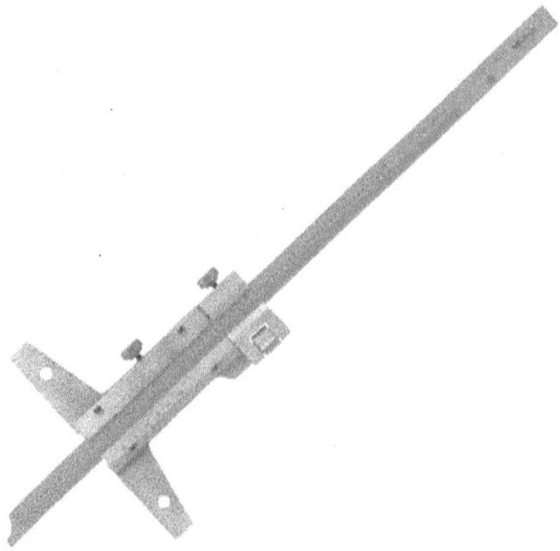

FIGURE 3.53 Example of a Depth Gauge with a Vernier scale.

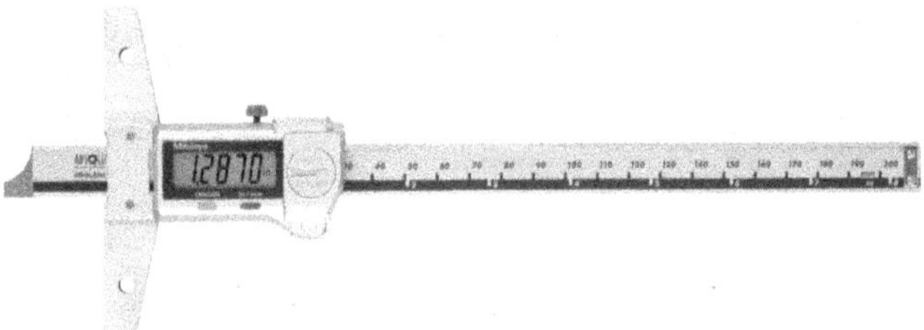

FIGURE 3.54 Example of a Depth Gauge with a digital readout.

M_{42} Thread Gauge

Measure with a thread (plug or ring) gauge, GO+NOGO ends, internal or external threads 1–2 inches (2.5–5 cm).

RECORD (R)

Record covers the manual actions performed with a writing or marking tool for the purpose of recording information. Two categories of data are found on the Record section of the Tool Use data card (Figure 3.22) for the **R** parameter. The index values for Write apply to the normal-size [up to 1 inch (2.5 cm) in height] handwriting operations (script or print) performed with a pen, pencil or other writing instrument such as a stylus (Figure 3.55). The Mark values cover the use of marking tools such as a scribe or marker for the purpose of identifying or making a larger mark (1–3 inches, 2.5–7.5 cm) on an object. The initial placement of a recording instrument before writing or marking usually occurs as a $\mathbf{P_1}$. A possible exception may be the placement of a marking device prior to scribing a line. If the beginning point of the line is critical, a $\mathbf{P_3}$ would be used to cover the necessary adjustments to place the tool accurately.

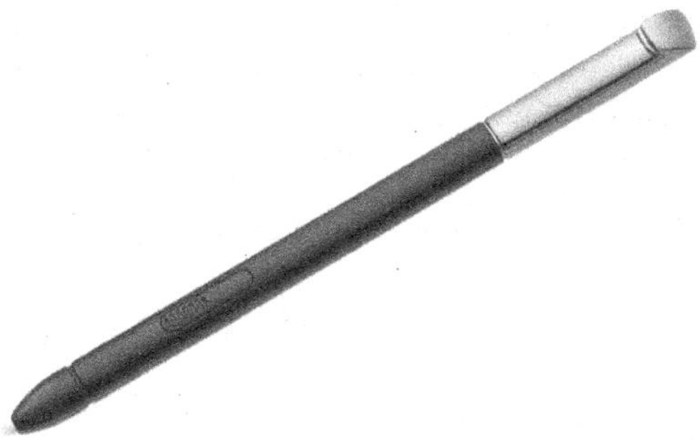

FIGURE 3.55 Example of a Stylus.

Write

The Write data is provided to cover the routine administrative activities encountered in virtually every industry. These activities may include filling out forms, writing a part number or even writing a signature. Index values for the **R** parameter are selected primarily on the basis of the number of digits (letters or numerals) or the number of words written. Consider the values for writing a date (either in the form 03-14-20, March 14, 2020 or 14 March 2020) or writing one's signature as writing two words and assign an $\mathbf{R}_{16}$ for either item.

Record also includes values to copy numbers. The index values to Copy numbers include the use of the eyes, mental processes and writing instruments to transfer data from one source to another. The index values are based on the number of digits that are copied at each observation.

Examples: A customer service representative uses a pen to write a five-digit order number on a notepad: $\mathbf{R}_{10}$.
An associate copies a six-digit part number from a screen to a return authorization form: $\mathbf{R}_{16}$.

Mark

The Mark data applies to marking or identifying an object or container using a marking tool such as a scribe or marker. Each mark is counted as a 'digit.' The index values for marking digits apply to printed characters (letters and numerals) of 1–3 inches (2.5–7.5cm) in height. Other common marking values include making a check mark ($\mathbf{R}_1$-✓) and scribing a line ($\mathbf{R}_3$-__________).

Example: An assembler uses a marker to make a checkmark on the bottom of a finished part: $\mathbf{R}_1$.

Tool Use Examples for Record

Note: For the examples listed below, all distances are within reach unless specified otherwise.

1. An operator grasps a tablet and stylus (simo) from the counter, fills out the completion date on the screen and signs his name beside it which is 1 inch (2.5 cm) apart. He then simultaneously puts the tablet and stylus back on the counter.

   Grasp tablet and stylus (simo), write date and signature on screen and aside both items (simo)

   $\mathbf{A_1\ B_0\ G_1\ A_1\ B_0\ (P_1\ A_0\ R_{16})\ A_1\ B_0\ P_1\ A_0\ (2)}$

   $[(1+16)\times(2)+1+1+1+1+1]\times 10=390$ TMU

2. A clerk grasps a marker and makes a check mark on a dry erase board.

   Grasp marker, make a check mark on dry erase board

   $\mathbf{A_1\ B_0\ G_1\ A_1\ B_0\ P_1\ R_1\ A_0\ B_0\ P_0\ A_0}$

   $(1+1+1+1+1)\times 10=50$ TMU

3. A delivery associate grasps a stylus, writes an eight-digit number on a touch screen and puts the stylus in her pocket.

   Grasp stylus, and write 8-digit part number and put to pocket

   $\mathbf{A_1\ B_0\ G_1\ A_1\ B_0\ P_1\ R_{16}\ A_1\ B_0\ P_1\ A_0}$

   $(1+1+1+1+16+1+1)\times 10=220$ TMU

Think (T)

Think refers to the use of sensory mental processes, particularly those involving visual perception, and may also include 'reaching to feel an object.' The Think data on the Tool Use data card (Figure 3.22) is designed to cover only those types of reading and inspection activities that occur as a necessary part of a worker's job. Although these operations usually occur internally to the manual work and therefore have no effect on the duration of the work cycle, on some occasions these activities must be considered in the overall work content of the job. The analyst should exercise care in determining the extent to which these activities affect the total analysis time.

Inspect

The data in this column applies to inspection work designed for making simple decisions regarding certain characteristics of the object under inspection. The activity involves first locating the inspection points and then making a quick yes-or-no decision concerning the existence of a defect. These mental processes presume that the inspector possesses a clear understanding of the characteristic being judged and that the presence of any defect, such as a scratch, stain, scar or color variance, is readily apparent to the inspector.

The index values for Inspect refer to the number of inspection points examined on the object. For each point, a yes-or-no decision is made concerning the presence or absence of readily distinguishable characteristics.

Except for reaching to feel an object, these parameter values *do not cover* the manual handling of the object that may occur during the inspection. Caution should be exercised in using these or any inspection values. In practical work situations, inspection time is rarely external but usually occurs during the manual handling of objects. Whenever possible, work should be designed to make inspections internal to other activities.

Along with inspecting a number of points, values are provided for activities to Feel for Heat ($\mathbf{T_6}$), where the hand is moved to the object, moved over the surface of the object and removed, and Feel for Defect ($\mathbf{T_{10}}$), where the hand is moved to the object, moved over up to three surfaces of the object and removed.

Example: A baker checks to see if the oven indicator light is on or off: $\mathbf{T_1}$.

Read

To read is to locate and interpret characters or groups of characters. The data for Read is divided into three sections: Digits or Single Words, Text of Words and Compare.

Digits or Single Words

The column Digits or Single Words is to be used for reading data such as item numbers, codes, quantities or dimensions from a document or screen. A digit is considered a letter, a number or a special character. To index the **T** parameter, simply count the number of digits or single words read and choose the appropriate index value from the Read section of the Tool Use data card (Figure 3.22).

Example: Mechanic reads four-character code to identify type of oil needed: $\mathbf{T_6}$.

Text of Words

The column Text of Words is used when analyzing situations in which the operator is required to read words arranged in sentences or paragraphs. The data is based on an average reading rate of 330 words per minute or 5.05 TMU per word which most average workers accomplish. These index values may be applied to reading a set of instructions in a manual or job aid or gathering general information from reading tabular data.

Example: Worker reads safety instructions (45 words) before operating slicer: $\mathbf{T_{24}}$.

Compare

The values for Read also include time to Compare numbers. This refers to the ability to momentarily retain a character and to examine its similarities or differences to other numbers or to a series of numbers. The data allows for scanning several characters, reviewing them and either accepting or rejecting the comparison. The index value is chosen by the number of letters or digits compared.

Example: Associate compares eight-digit reward number on screen to membership card: $\mathbf{T_{10}}$.

Additional values that apply to more specific reading activities, such as reading gauges, scales and tables are also provided on the data card.

T_3 Gauge

Use when a device is read to see if the pointer is within a clearly marked tolerance range.

Example: A pointer is showing the fuel tank is close to full as shown in Figure 3.56.

T_6 Scale Value

A specific quantity is read from a graduated scale such as a measuring stick, temperature gauge or pressure gauge. This does not apply to digital scales.

Example: Read a pressure gauge to determine the psi as shown in Figure 3.57.

T_6 Date or Time

The month, day and year are read from a document, screen or calendar. The $\mathbf{T_6}$ value for date includes the time to turn the wrist to read the date from a wristwatch or look at a calendar. The format for the date can be in the form of 03-14-20, March 14, 2020 or 14 March 2020. The $\mathbf{T_6}$ value can also be used to read the time of a day from a clock or wristwatch.

Example: Retail associate reads the date on return receipt.

FIGURE 3.56 Example of $\mathbf{T_3}$, Gauge value.

FIGURE 3.57 Example of $\mathbf{T_6}$, Scale Value.

T_{10} Vernier Scale

Visually locate and read an exact value from a micrometer, caliper or similar device using a Vernier scale. This value does not contain time for placing and setting the device to an object—only the time to read the scale. A Vernier scale is not a specific tool, it is however a type of scale used on many measurement tools. An example is shown in Figure 3.58 in which the scale is noted on an outside micrometer.

T_{16} Table Value

A specific value is located and read from a table after scanning the table horizontally and vertically.

Examples: The correct machine setting is read from a feed-speed table.
An index value is read from the BasicMOST data card.

Tool Use Examples for Think

Note: For the examples listed below, all distances are within reach unless specified otherwise.

1. An airline employee views a monitor to check the flight number (four digits) for a passenger.

   Read 4-digit flight number on monitor

   $\mathbf{A_0\ B_0\ G_0\ A_0\ B_0\ P_0\ T_6\ A_0\ B_0\ P_0\ A_0}$

   $6 \times 10 = 60$ TMU

2. An operator picks up a work order and reads a paragraph that describes the method to be followed which contains an average of 30 words. When finished, he puts the order aside on a workbench.

   Grasp work order, read 30 words and put on workbench

   $\mathbf{A_1\ B_0\ G_1\ A_1\ B_0\ P_0\ T_{16}\ A_1\ B_0\ P_1\ A_0}$

   $(1+1+1+16+1+1) \times 10 = 210$ TMU

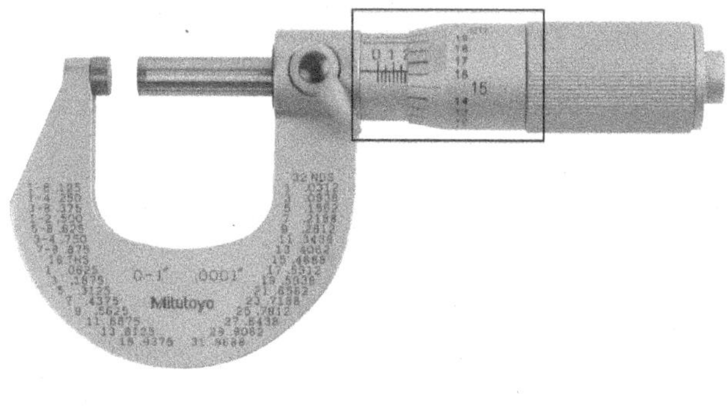

FIGURE 3.58 Example of a Vernier Scale.

3. A pharmacist grasps a medicine bottle, inspects two points on the bottle and bends to put the bottle on the bottom shelf.

   Grasp bottle, inspect 2 points and put on shelf

   $\mathbf{A_1\ B_0\ G_1\ A_1\ B_0\ P_0\ T_3\ A_1\ B_6\ P_1\ A_0}$

   $(1+1+1+3+1+6+1)\times 10=140$ TMU

THE EQUIPMENT USE SEQUENCE MODEL

The Equipment Use data card contains values for administrative activities such as using a keyboard and keypad, stapling, stamping and filing. Similar to Tool Use, Equipment Use consists of a series of General and Controlled Moves. The application of the Equipment Use Sequence Model is designed to assist a MOST-certified applicator to analyze certain administrative activities with only one sequence model. The index value chosen from the Equipment Use data card is based on the activity such as number of keys pressed on a security pad, the number of flipping actions when filing or clicks of a mouse. While the automation of some tasks will continue to evolve, administrative tasks will always be some part of many organizations. The use of the Equipment Use Sequence Model simplifies the analysis of these tasks.

The Sequence Model

The sequence model consists of a series of letters or parameters representing each of the various sub-activities of the *Equipment Use Sequence Model.* This sequence model uses the parameters from the General Move Sequence Model and a blank space (indicated by the asterisk '*' symbol) is used for one of three Equipment Use parameters.

A B G | A B P | * | A B P | A

where: **A** = Action Distance
B = Body Motion
G = Gain Control
P = Placement

The blank space in the sequence model ('Use Equipment' phase) is provided for the insertion of one of the following Use Equipment parameters. These parameters are as follows:
where: **W** = Keyboard
K = Keypad
H = Letter/Paper Handling

Phases of the Equipment Use Sequence Model

Get equipment or object	Put equipment or object in place	Use equipment	Put equipment or object aside	Return operator
A B G	**A B P**	*	**A B P**	**A**

Equipment Use follows the same fixed sequence of sub-activities as Tool Use which occurs in five phases:

1. Get Equipment or Object:
   - Reach with hand a distance to the equipment or an object, either directly or in conjunction with body motions or steps.
   - Gain manual control of the equipment or object.
2. Put Equipment or Object in Place:
   - Move the equipment or object a distance to where it will be used, either directly or in conjunction with body motions or steps.
   - Place the equipment or object in position for use.
3. Use Equipment:
   - Use the equipment.
4. Put Equipment or Object Aside:
   - Retain the equipment or object for further use (hands and fingers are of course *always* retained), toss or lay the equipment or object aside, return the equipment to its original location or move it to a new location for disposition, either directly or in conjunction with body motions or steps.
5. Return:
   - Return to the workplace.

Parameter Definitions

Other than the specific Equipment Use parameters, the Equipment Use Sequence Model contains only parameters from the General Move Sequence Model. The **A**, **B**, **G** and **P** parameters were discussed with the General Move Sequence Model and remain unchanged.

W Keyboard

Refers to the use of the fingers and the hands to click a mouse, type words, sentences, the date or an address.

K Keypad

Covers the use of the eyes, fingers and hands to read a figure, keying it into an adding or calculating-type machine and depressing a function key or to depress a key into an electronic keypad.

H Letter/Paper Handling

Refers to the use of the fingers, hands or office equipment to perform the actions necessary to change or prepare documents, envelopes, etc., for office distribution and handling.

Parameter Indexing

With the exception of the special Use Equipment parameters, the Equipment Use Sequence Model contains only parameters from the General Move Sequence Model. Index values for these parameters are found on the General Move data card

(Figure 3.1). The data card for Equipment Use is found in Figure 3.59. The tables for indexing the Use Equipment parameters are used following the same procedure outlined in the General Move, Controlled Move and Tool Use sections.

Consider, for example, an activity in which a document needs to be folded into thirds for a mailing. The assistant grasps the letter from a stack, puts it on the desk and folds it in thirds. The assistant then puts the folded letter aside.

Grasp letter, put on desk and fold and crease

$\mathbf{A_1\ B_0\ G_1\ A_1\ B_0\ P_1\ H_{16}\ A_1\ B_0\ P_1\ A_0}$

$(1+1+1+1+16+1+1)\times 10=220$ TMU

In this example, the 'Get' and 'Put' phases of the sequence model are used for getting and putting the letter on a desk in preparation for folding. The distance to reach the letter is within reach so an $\mathbf{A_1}$ is assigned. A $\mathbf{G_1}$ is assigned because it is a light object. An $\mathbf{A_1}$ is used to move the letter closer to the worker. A $\mathbf{P_1}$ is used to simply put the letter on the desk and an $\mathbf{H_{16}}$ is used to fold and crease the letter. The letter is moved within reach ($\mathbf{A_1}$) and set aside ($\mathbf{P_1}$).

Use of the Equipment Use data card (Figure 3.59) can be demonstrated with another example. Suppose that an accountant was reconciling the daily bank deposit and keyed in 48 digits. This activity would be analyzed as:

Key in 48 digits using keypad data

$\mathbf{A_0\ B_0\ G_0\ A_1\ B_0\ P_1\ K_{32}\ A_0\ B_0\ P_0\ A_0}$

$(1+1+32)\times 10=340$ TMU

The appropriate Equipment Use parameter for this example would be Keypad–Data, which is represented by the letter **K**. In the column titled Data under **K** in Figure 3.59, 48 digits would be assigned an index value of $\mathbf{K_{32}}$. The initial placement of the hands prior to the data entry action is $\mathbf{P_1}$ to put the hand on the adding machine.

BasicMOST System			Equipment Use					A B G A B P * A B P A					
Index x 10	W Keyboard	K Keypad		H Letter / Paper Handling									Index x 10
	Words	Digits	Data	Operations	Jog or Tap	Staple	Stamp	Leaf Through Paper	Filing: Select	Filing: Open / Close Select	Filing: File	Filing: Open / Close File	
1	Click Mouse	2	2	-	1	Electric	-	1	-	-	-	-	**1**
3	1	6	6	Open Envelope	3	Hole Punch Hand Remove	-	4	-	-	-	-	**3**
6	2 Date	11	12	Interleaf	6		1	7	1	-	-	-	**6**
10	4	18	20	Seal Envelope	10		2	12	3	-	1	-	**10**
16	6	28	32	Fold and Crease	16		3	20	6	2	4	1	**16**
24	8	39	46				5	28	9	6	7	5	**24**
32	11	52	60					37	12	9	10	8	**32**
42	15 Address	68	79					47	17	12	15	11	**42**
54	19	85	100					61					**54**

FIGURE 3.59 Equipment Use data card for Keyboard, Keypad and Letter/Paper Handling.

Equipment Use Placement	
Equipment	***Index Value***
Keyboard	P_1
Keypad	P_1
Letter / Paper Handling	P_1

FIGURE 3.60 Placement index values for Equipment Use.

The Placement values for Equipment Use are typically assigned a $\mathbf{P_1}$ as shown in Figure 3.60.

The remainder of this section examines the Equipment Use parameters and application in more detail.

Keyboard (W)

The data in the Keyboard category refers to the use of the fingers and hands performing multiple moves to click a mouse, type words, sentences, letters, headings, etc. The data also includes time to type the date or an address. The data for the **W** parameter should always be considered within reach and the hand placement is $\mathbf{P_1}$.

Words

The Words section of the Equipment Use Sequence Model defines words as an average of 5.5 characters per word plus time for spacing and any punctuation or capitalization needed. To use these values, it is assumed that the operator is familiar with the keyboard and able to type an average of 52 corrected words per minute. This allows time to correct any typing mistakes. This is achievable with the vast majority of individuals performing data entry on a regular basis. The extended values for Words are shown in Figure 3.61. In addition to typing Words, there are three other elements that can be used for common activities.

Example: A pharmacy technician puts his hands to a keyboard and enters in a customer's allergy specifications (12 words): $\mathbf{W_{42}}$.

W_1 Click Mouse

The value to Click a computer Mouse includes the time to move the mouse $\leq$4 inches (10 cm), align the cursor to one point and click the mouse once or twice. When clicking the mouse twice, the action is quick and short. This is often used when opening a software application. The placement value of the hand to the mouse is a $\mathbf{P_1}$. The Process Time, which often follows the click of the mouse, is not included in the $\mathbf{W_1}$ value.

Example: An automotive technician uses the mouse to click on a customer's account.

Keyboard	
Index Value	*Words*
W_{67}	23
W_{81}	28
W_{96}	33
W_{113}	38
W_{131}	44
W_{152}	51
W_{173}	58
W_{196}	65
W_{220}	73
W_{245}	81
W_{270}	90
W_{300}	99
W_{330}	109

FIGURE 3.61 Extended values table for Keyboard–Words.

W_6 Date

A $\mathbf{W_6}$ is used when typing the date (either in the form 08–17–20, August 17, 2020 or 17 August 2020).

Example: The help desk associate enters the date of the customer issue.

W_{42} Address

The value for Address includes the time to type a four-line address. The four lines may be a contact name, company, address and a line for the city, state/province, country and zip or postal code. The four lines could also be used for the company name, address one, address two and a line for the city, state/province, country and zip or postal code. The placement value for the hand to the keyboard is a $\mathbf{P_1}$.

Example: An associate types a customer's shipping address.

Keypad (K)

Keypad refers to the coordination of the eyes and hand and the finger movements of depressing the keys to register numerals on a numeric keypad or to make the calculation visible on a tape printout. This heading is divided into two parts: Digits and Data, depending on the method used.

Digits

The values for Digits should be used with a *low method level* such as using one finger to depress one key at a time. For example, pressing the numbers on a security keypad would be analyzed with digits. The index value is selected by the number of keys pressed.

Example: A customer presses four buttons to enter a PIN (personal identification number) code on a keypad: $\mathbf{K_3}$.

Data

The values for Data should be used when analyzing a *high method level* such as using all five fingers to depress the keys on a printing calculator or a numeric keypad on a keyboard. The motions are quick and short. For example, a bank associate reconciling receipts would most likely need to be analyzed with the Data values given the high method level. The index value is selected by the number of keys pressed—this includes the numeric and function keys. The extended values table for Keypad–Data is shown in Figure 3.62.

Depending on the type of keypad used, if there is a difference in the Keypad values and those studied, new elements for the equipment being used must be created using the formula in the Development of Elements for Special Tools or Situations section for developing new elements. If the activity is used infrequently, the analyst could analyze the activity with General and Controlled Moves (the process time of the printing calculator will need to be developed using a stopwatch).

Example: An associate presses 15 keys to enter the order numbers to be shipped for the day: $\mathbf{K_{10}}$.

Writing Method Descriptions

Below is the recommended format for a clear and concise method description for the Equipment Use Sequence Model. Additional words may be used to enhance the method description. These could be Action Distances, Body Motions or adjectives. Examples of this structure can be found in the Equipment Use examples listed below and throughout the Equipment Use section.

The recommended method description format for Equipment Use is:

Gain Control Object Equipment Use Activity At Location Aside

Keypad – Data	
Index Value	***Data***
K_{67}	122
K_{81}	146
K_{96}	173
K_{113}	202
K_{131}	235
K_{152}	270
K_{173}	306
K_{196}	345
K_{220}	386
K_{245}	428
K_{270}	474
K_{300}	524
K_{330}	574

FIGURE 3.62 Extended values table for Keypad–Data.

Equipment Use Examples for Keyboard and Keypad

Note: For the examples listed below, all distances are within reach unless specified otherwise.

1. An associate reaches to the mouse to click in the address field. He then puts his hands on a keyboard and types a three-line address.

   Click mouse

   $A_0\ B_0\ G_0\ A_1\ B_0\ P_1\ W_1\ A_0\ B_0\ P_0\ A_0$

   $(1+1+1)\times 10=30$ TMU

   Type address

   $A_0\ B_0\ G_0\ A_1\ B_0\ P_1\ W_{42}\ A_0\ B_0\ P_0\ A_0$

   $(1+1+42)\times 10=440$ TMU
   30 TMU
   440 TMU
   470 TMU Total
2. A manufacturing associate presses five digits on a keypad to enter a secure area.

   Enter 5 digits

   $A_0\ B_0\ G_0\ A_1\ B_0\ P_1\ K_3\ A_0\ B_0\ P_0\ A_0$

   $(1+1+3)\times 10=50$ TMU

Letter/Paper Handling (H)

This parameter covers the use of commonly performed manual activities involved with the arrangement or distribution of documents and paper. The parameter index values refer to the time to move the fingers and/or hands to perform the actions necessary to find, change, prepare or distribute papers. In addition, the **H** parameter applies to the handling of paper that results in changes in shape, direction or position at some point (i.e., leaf through paper, tap edges, etc.). The values under the **H** parameter apply to the particular activity of the letter or paper. The equipment used in these cases is normally manipulated by the hands. Since you do not have to gain control of the hands or 'aside' the hands, the Get and Aside phases of the sequence model will often contain zeros. The Put phase allows for putting the hand to the equipment. The Letter/Paper Handling parameter is broken down into six main categories:

1. Operations
2. Jog or Tap
3. Staple
4. Stamp
5. Leaf Through Paper
6. Filing

The placement value of the empty hand or tool for the **H** parameter is $\mathbf{P_1}$.

Operations

The activities performed under this heading include the use of the hand to manipulate paper or envelopes. The placement value for all parameter variants is $\mathbf{P_1}$.

H_3 Open Envelope

The Open Envelope value includes the motions necessary to open an envelope using a knife, letter opener or fingers. This includes time to fully insert the instrument used to open and opening the envelope. The placement value, as with all the **H** values, is a $\mathbf{P_1}$. The actions to get, put and aside the opener must be analyzed in the other phases within the Equipment Use Sequence Model.

Example: A legal associate uses a letter opener to open an envelope.

H_6 Interleaf

The value for Interleaf includes the time to lift a sheet or sheets of paper with one hand while simultaneously reaching for the second or divider sheet with the other hand, grasping the divider sheet and inserting it beneath the sheet that was lifted. The time to lower the top sheet of paper is also included in the $\mathbf{H_6}$, Interleaf value. The placement value for the hand is a $\mathbf{P_1}$. Figure 3.63 provides an example of Interleaf.

Example: A production associate in a copying facility puts a divider sheet between pages.

H_{10} Seal Envelope

The value for Seal Envelope includes time to get an envelope flap slightly wet by using a wetting device or pulling back the protective strip of a self-sticking envelope to seal the envelope. When using a wetting device, the time includes lifting the envelope flat and putting the flap on the wetting device or putting the wetting device to the flap, using the wetting device and sealing the flap. If pulling back an adhesive strip, the $\mathbf{H_{10}}$ value includes the time to open the flap, pull back the adhesive strip, aside the adhesive strip and then fold the flap down and seal the envelope. The Seal Envelope value does not include the time to gain control or aside the envelope or the wetting device.

Example: A billing associate removes the protective strip to seal the envelope of an invoice.

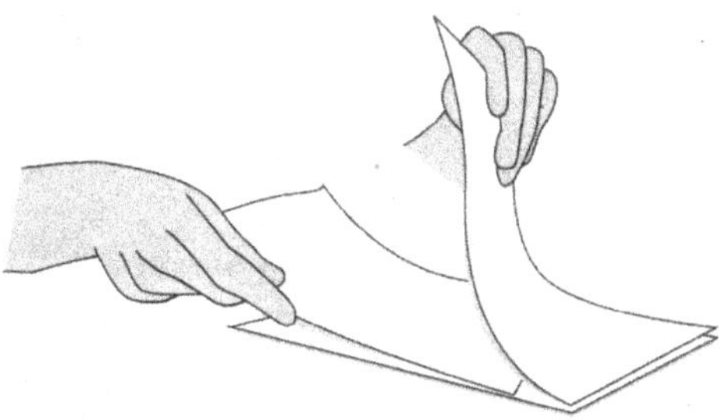

FIGURE 3.63 Example of Interleaf.

H_{16} Fold and Crease

The index value for Fold and Crease includes all of the movements of the hand and fingers to fold a piece of paper into three equal parts and to run the fingers along the length of the folds to crease it as shown in Figure 3.64.

Example: An administrative assistant folds and creases a notification letter.

Jog or Tap

The index values for Jog or Tap include the movements necessary to strike (jog) the sides of sheets of paper on a hard surface to align edges (Figure 3.65) or the use of the hands to tap the edges of pieces of paper for alignment purposes (Figure 3.66). The index value is chosen by the number of jogs or taps. One jog or tap is a complete up and down motion. The placement of the hand is normally a $\mathbf{P_1}$.

Example: A bank associate grasps a stack of forms and then taps them three times to align the edges: $\mathbf{H_3}$.

Staple

Staple analyzes the use of either a hand-operated or electrically powered instrument to attach a metal staple to join the papers together. Index values are based on the various parameter definitions. The placement value for Staple is normally a $\mathbf{P_1}$.

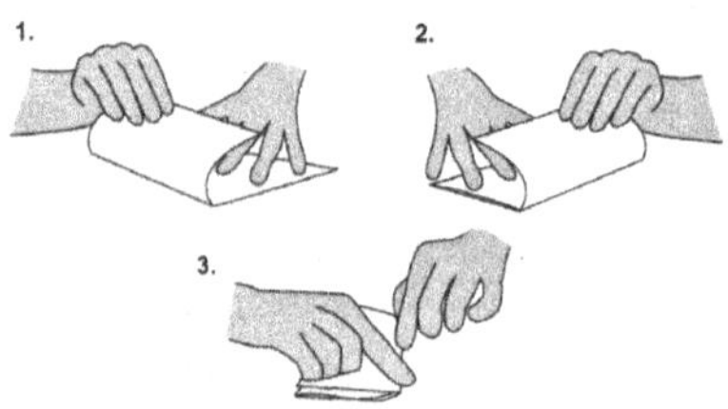

FIGURE 3.64 Example of Fold and Crease.

FIGURE 3.65 Example of jogging paper.

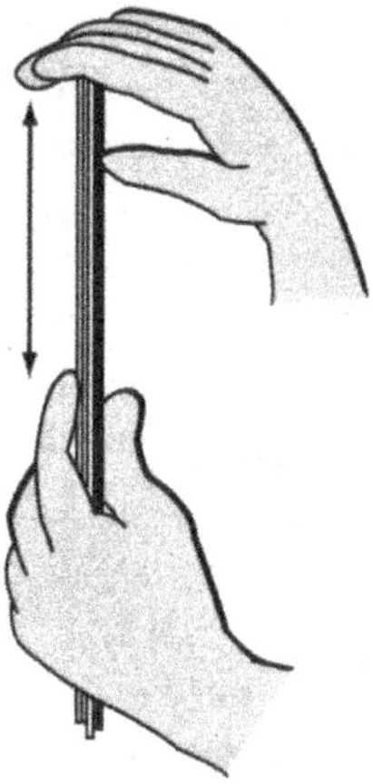

FIGURE 3.66 Example of a hand tapping paper.

H_1 Electric Stapler

Covers the time to staple the paper after it is put into an electric stapler. Getting and asiding the paper are not included in the $\mathbf{H_1}$ value and must be analyzed in the Get and Aside phases of the sequence model.

Example: A pharmacy associate staples a prescription information sheet to a bag.

H_3 Hole Punch

Covers the finger and hand movements of pressing down the handle of a hole punch and releasing it. This value does not include the time to pick up the paper or the hole punch or to aside the paper.

Example: Associate uses hole punch on report to prepare for client meeting.

H_3 Hand Stapler

Covers the finger and hand movements of pressing down on the handle of the stapler and releasing it. This value does not include the time to pick up or aside the paper or stapler.

Example: A retail associate staples a returned item 'credit' receipt to the original receipt.

H_3 Remove Staple

Covers all of the hand and finger actions used in order to remove a staple from paper. This value does not include the time to pick up or aside the paper or staple remover. The placement value to remove a staple is $\mathbf{P_1}$.

Example: Associate removes a staple from a set of papers in preparation to scan the papers.

Stamp

The index values for Stamp cover the use of a self-inking stamp. The values for Stamp include the time to move the stamp to the location, make the stamp impression and lift the stamp. The values for Stamp assume that inspection of the stamp marking is internal to lifting the stamp. If a separate inspection is necessary, an additional Tool Use Sequence Model is required. The index value is chosen by the number of impressions made.

Example: A postal service employee uses the 'Priority Mail' stamp on a box: $\mathbf{H}_6$.

Leaf Through Paper

The activity to quickly 'thumb' through pages in a book or papers in a stack can be analyzed with this parameter variant. The data presented is to be used when leafing through papers is performed using one of the following techniques:

1. Using the fingers, grasp the corner of paper with one hand and pull it back to be grasped and held by the other hand. Repeat this cycle.
2. Using the fingers, grasp the corner of a paper with one hand and pull it back. While pulling the paper back, reach for the next corner with the other hand and pull it back. Repeat this cycle. To the analyst, this operation will appear as alternating short reaches and moves of the fingers.

Note: Each reach (left or right) is considered one action when leafing through papers.

The data on the Equipment Use data card refers to the number of leafing actions observed. An action is defined as the displacement of one page or one group of pages. To assign an index value, simply count the number of actions observed. The placement value will only be shown for the initial control. Time to subsequently gain control of each sheet is included in the Leaf Through Paper value. Leaf Through Paper values are for very short leafing actions. If the leafing action is greater than 12 inches (30 cm), this data will not apply.

Example: Associate leafs through documents with eight actions to confirm each document has been signed: $\mathbf{H}_{10}$.

Filing

Filing refers to the activity of placing an item (a file, paper or group of papers) in or removing an item from a specific location. The placement value in all cases for Filing is a $\mathbf{P}_1$ for the initial hand placement to the file or drawer. The location is determined by an alphabetic or numeric sequence. Filing is broken up into four categories:

1. Select (select file from an open drawer)
2. Open/Close Select (open drawer, select file and close drawer)
3. File (return file to an open drawer)
4. Open/Close File (open drawer, return file and close drawer)

Select

The values for Select include the time for the finger, hand or arm actions necessary to obtain a particular file from an alphabetically or numerically ordered file container and remove it (lifting it high enough to clear the remaining files). The values include the time to reach into the open file drawer and while reading the file titles, 'thumb' or 'flip' through several files until the appropriate file is discovered. The number of thumbing or flipping actions used is the primary variable in the activity. The index value is chosen by the number of actions used to push files aside by the operator during the selection process. A $\mathbf{P_1}$ needs to be assigned to the initial hand placement to the file. Time to put the file aside is not included in the Filing values and should be analyzed in the Aside phase of the sequence model.

Example: A legal clerk reaches into an open desk drawer to select a file. The clerk uses six actions to thumb through files, removes the needed file and asides it to the top of the desk.

Select file from open drawer and aside to desk.

$\mathbf{A_0\ B_0\ G_0\ A_1\ B_0\ P_1\ H_{16}\ A_1\ B_0\ P_1\ A_0}$ **200 TMU**

Open/Close Select

The values for Open/Close (O/C) Select include the time to select a file as noted above under Select plus the activity to open and subsequently close the file drawer. The values are based on a standard filing cabinet. The index value is chosen based on the number of thumbing or flipping actions to select the file. A $\mathbf{P_1}$ needs to be assigned to the initial hand placement to the drawer.

Example: An administrative assistant walks eight steps to a filing cabinet, opens the top drawer and flips through files with three flipping actions and selects one file. He then removes the file, closes the drawer, walks back to the desk, sits down and puts the file on top of the desk.

Select file from closed drawer and put on desk.

$\mathbf{A_0\ B_0\ G_0\ A_{16}\ B_0\ P_1\ H_{24}\ A_{16}\ B_{10}\ P_1\ A_0}$ **680 TMU**

File

The index values for file include the finger, hand or arm actions necessary to place a file in a specific location in a filing cabinet either in alphabetic or numerical order. The values include the time to reach into the open file drawer and while reading the file titles, thumb or flip through several files to determine the exact location for placement and then place the file into the selected location while holding the surrounding files out of the way. File is a two-handed activity where one hand finds the location and the other hand places the file. The index value is chosen by the number of thumbing or flipping actions. A $\mathbf{P_1}$ needs to be assigned to the initial hand placement to the file.

Example: A medical assistant uses six actions to flip through files to find the proper location and then places a patient file in the correct spot.

File document with 6 actions

$\mathbf{A_0\ B_0\ G_0\ A_1\ B_0\ P_1\ H_{24}\ A_0\ B_0\ P_0\ A_0}$ **260 TMU**

Open/Close File

The index values for Open/Close (O/C) File include the time to file an item plus the activity to open and subsequently close a standard filing cabinet drawer. The index values are chosen based on the number of thumbing or flipping actions to find the file location. A $\mathbf{P_1}$ needs to be assigned to the initial hand placement to the drawer.

Example: An administrative assistant grasps a document, stands from his chair and walks four steps to a filing cabinet. He then bends down to open the bottom drawer to place the document in its proper location with six flipping actions.

Grasp document and file in closed drawer with 6 actions.

$\mathbf{A_1\ B_0\ G_1\ A_6\ B_{16}\ P_1\ H_{32}\ A_0\ B_0\ P_0\ A_0}$ **570 TMU**

Letter/Paper Handling Examples

Note: For the examples listed below, all distances are within reach unless specified otherwise.

1. Warehouse associate walks three steps to grasp a 'perishable' stamp from the work table and then walks six steps to use the stamp on the side of a box. The associate asides the stamp back at the work table.

   Grasp stamp from work table, stamp box and aside stamp

   $\mathbf{A_6\ B_0\ G_1\ A_{10}\ B_0\ P_1\ H_6\ A_{10}\ B_0\ P_1\ A_0}$

   $(6+1+10+1+6+10+1)\times 10=350$ TMU

2. An administrative assistant picks up a divider page and puts it on top of a section in a report. She repeats the process four more times.

   Interleaf 5 divider pages

   $\mathbf{A_0\ B_0\ G_0\ A_1\ B_0\ P_1\ H_6\ A_0\ B_0\ P_0\ A_0\ 5}$

   $(1+1+6)\times 5\times 10=400$ TMU

3. An administrative assistant picks up a staple remover from a desk and puts it to papers already in hand. The assistant then removes the staple from the papers, drops the staple over the trash can and then asides the paper to the desk.

   Put staple remover to papers, remove staple and toss into trash can

   $\mathbf{A_1\ B_0\ G_1\ A_1\ B_0\ P_1\ H_3\ A_1\ B_0\ P_0\ A_0}$

   $(1+1+1+1+3+1)\times 10=80$ TMU

   Aside paper

   $\mathbf{A_0\ B_0\ G_0\ A_1\ B_0\ P_1\ A_0}$

$(1+1)\times 10=20$ TMU

80 TMU
20 TMU
100 TMU

THE MANUAL CRANE SEQUENCE MODEL

As stated in the introduction, the four sequence models covering the manual handling of objects constitute the BasicMOST Work Measurement System. These sequence models, General Move and Controlled Move in particular, can be used to measure the handling of heavy objects with lifting or moving equipment as well. For simplification, special sequence models were developed to cover equipment handling. (See also Chapter 5—MaxiMOST.)

The values on the data card for using a manual crane are based on a representative sample of equipment found in industry. Therefore, the data is valid for most situations. However, before applying the data, it is recommended that individual parameter values be reviewed and adjusted to specific methods if necessary.

The Manual Crane Sequence Model deals with the movement of objects using a manually traversed crane. The sequence model is appropriate for a crane that may resemble either a jib crane or an overhead bridge crane (Figure 3.67), as long as the crane is moved laterally and longitudinally by hand, not under power. The Manual Crane Sequence Model is best applied when there are several cranes in use or one crane is being used for multiple activities. If one type of crane is being used (e.g., one crane always moving 10 feet or 3 m), it is more practical to use General and Controlled Moves to analyze the activity. The process time will need to be developed using a stopwatch, process specifications or engineering calculations.

Like General Move, Controlled Move, Tool Use and Equipment Use, the index values for Manual Crane are added together for one sequence model and the total is multiplied by 10 to convert to TMU.

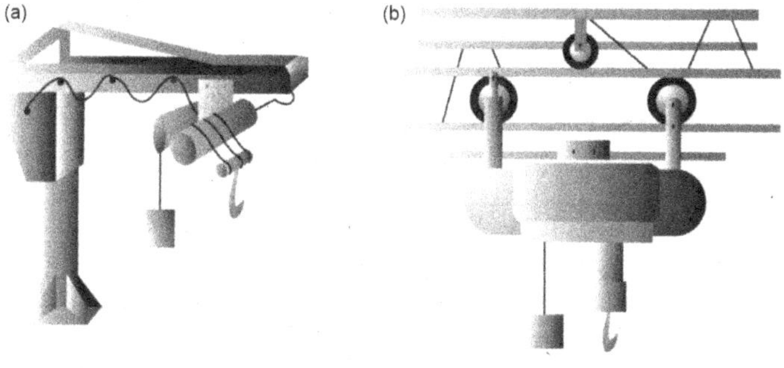

FIGURE 3.67 Manually traversed cranes; jib crane (a) and overhead crane (b).

The Sequence Model

The movement of an object with a manual crane is described by the following sequence model:

A T K F V L V P T A

where:
A = Action Distance
T = Transport Unloaded
K = Hook-up and Unhook
F = Free Object (from surroundings, pallet, fixture, etc.)
V = Vertical Move
L = Loaded Move
P = Placement

As with all sequence models in MOST, all manual operations can be identified with a certain sequence of events that repeats from cycle-to-cycle, regardless of the description, size or name of the object being moved. Unlike the four sequence models in BasicMOST, the Manual Crane Sequence Model is not broken into phases. The analyst follows the movement with the parameters in order of the sequence model.

1. The operator moves to the crane (Action Distance).
2. The empty crane is transported to the location of the object to be moved (Transport).
3. The object is hooked up and freed from its surroundings (Hook-up, Free).
4. The object is raised vertically using the crane (Vertical Move).
5. The crane is moved, with the load, to the placement location (Loaded Move).
6. The object is lowered vertically (Vertical Move).
7. The object is placed in a new location (Placement).
8. The object is released from crane (Unhook).
9. The empty crane is transported to a rest position (Transport).
10. The operator returns to the original location (Action Distance).

Figure 3.68 illustrates the sequence of events that occur when an object is moved with a manual crane.

Parameter Definitions

A Action Distance

This is defined in the General Move section and is indexed by the distance (in steps) the operator walks to or from the crane.

T Transport Unloaded

Transport Unloaded includes getting the empty crane and transporting it horizontally to the location of the object to be moved. Note that the movement is a result of the

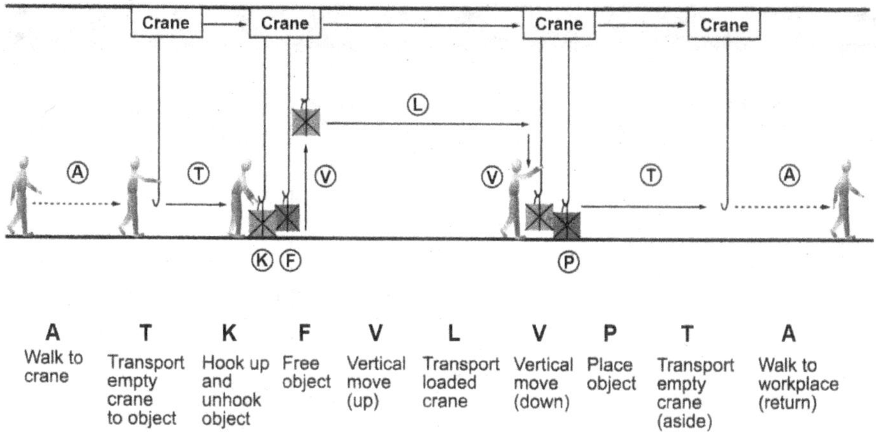

FIGURE 3.68 Illustration of Manual Crane Sequence Model.

operator pulling or pushing the crane from one location to another. Vertical movement of the hook during the Transport Unloaded parameter is an *internal* function.

K Hook-up and Unhook

Hook-up and Unhook includes both connecting and disconnecting the object to and from the holding device.

F Free Object

Free Object includes the actions necessary to work the object free from its surroundings (e.g., container or fixture) and raise the object, at a low speed, 2–3 inches (5–7.5 cm).

V Vertical Move

Vertical Move is the raising or lowering of the object at high speed following the **F** and **L** parameters.

L Loaded Move

Loaded Move covers the horizontal movement of the object with the crane. Note that the movement with a manual crane is a result of the operator pulling or pushing the crane from one location to another.

P Placement

Placement covers the actions in lowering the object the last 2–3 inches (5–7.5 cm) at low speed and placing the object in the desired location.

Parameter Indexing

The Manual Crane data card (Figure 3.69) is divided into six columns. Index values are selected either by the distance involved (the **T, L** and **V** parameters) or by the holding device used or difficulty involved in moving an object (the **F** and **P** parameters).

BasicMOST System			Manual Crane				ATKFVLVPTA
Index x 10	T Transportation up to 2 Tons	L	K Hook-up and Unhook	F Free Object	V Vertical Move	P Placement	Index x 10
	Unloaded	Loaded			Inches (cm)		
	Feet (m)						
3	-	-	-	Without Direction Change	8 (20)	Without Direction Change	3
6	-	-	-	With Single Direction Change	16 (40)	Align with One Hand	6
10	5 (1.5)	5 (1.5)	-	With Double Direction Change	30 (75)	Align with Two Hands	10
16	13 (4)	12 (3.5)	-	With One or More Direction Changes, Care in Handling or Apply Pressure	45 (115)	Align and Place with One Adjustment	16
24	20 (6)	18 (5.5)	Single or Double Hook		60 (150)	Align and Place with Several Adjustments	24
32	30 (9)	26 (8)	Sling			Align and Place with Several Adjustments and Apply Pressure	32
42	40 (12)	35 (10)					42
54	50 (15)	45 (13)					54

FIGURE 3.69 Manual Crane data card. Values are read up to and including. Transportation times for the **T** and **L** parameters must be validated before application of the Manual Crane Sequence Model.

Action Distance (A)

Choose the index value by the distance the operator walks to get to or move away from the crane. Select the values from the Action Distance column on the General Move data card (Figure 3.1).

Transport Unloaded (T)

Select the proper index value by the distance (feet, meters) the operator moves the empty crane to or from the object moved.

Loaded Move (L)

Select the proper index value by the distance (feet, meters) the operator moves the loaded crane.

Hook-up and Unhook (K)

Choose the proper index value by the holding device used. The parameter begins at the point at which the transport of the empty crane ends and is complete when the object is fastened to the crane hook or sling. The parameter also includes time to remove the holding device. Note that getting the hook or slings to the workplace will be analyzed separately with a General Move. The values for Hook-up and Unhook include:

$\mathbf{K_{24}}$ Single or Double Hook
$\mathbf{K_{32}}$ Sling

Free Object (F)

Choose the proper index value by the difficulty involved in freeing the object, raising the object 2–3 inches (5–7.5 cm) and positioning such that the next action will be an unobstructed vertical move. This parameter includes all actions necessary to position the load so that the next activity will be an unrestricted vertical move. The values for Free Object include:

$\mathbf{F_3}$ Without Direction Change
$\mathbf{F_6}$ With Single Direction Change
$\mathbf{F_{10}}$ With Double Direction Change
$\mathbf{F_{16}}$ With One or More Direction Changes, Care in Handling or Apply Pressure

Vertical Move (V)

Select the proper index value by the distance (inches, centimeters) the object is raised or lowered. The hook is raised after the object is freed and lowered after the loaded crane is moved to the placement location.

Note: If the hook is raised or lowered during the transportation of the crane, the time is covered by the **T** or **L** parameters.

Placement (P)

Choose the proper index value by the difficulty involved in lowering the object the last 2–3 inches (5–7.5 cm) and placing it in the desired location. Index values are based on the degree of difficulty affecting placement and include the following:

P_3 Without Direction Change

The object is simply lowered into position without any additional manual guidance from the operator.

P_6 Align with One Hand

While lowering the last 2–3 inches (5–7.5 cm), the operator reaches out with one hand and steers or swings the load into position.

P_{10} Align with Two Hands

During the placement activity, the operator must release the controls and steer or swing the object into position using two hands.

P_{16} Align and Place with One Adjustment

To position an object, the operator must steer or swing the object in addition to making one directional adjustment (longitudinally, laterally or vertically).

P_{24} Align and Place with Several Adjustments

To position an object, an operator must steer or swing the object in addition to making several directional adjustments (longitudinally, laterally and/or vertically).

P_{32} Align and Place with Several Adjustments and Apply Pressure

To position an object, an operator must steer or swing the object and also make several directional adjustments (longitudinally, laterally and/or vertically) in addition to exercising care in handling or applying pressure. A pause or hesitation must also be observed at the point of placement to indicate the application of heavy pressure required to seat the object or an obvious slow motion is observed in placing the object carefully.

Manual Crane Data Card Backup Information

The data provided on the data card is to be treated as sample data only. The methods represented on the data card must be *verified* and the vertical speeds (process time) must be *validated* for the particular cranes in question. Guidelines for validating the process times using a crane can be found in the Application of the BasicMOST Work Measurement System section.

Methods to be verified are the Hook-up and Unhook (**K**) and Placement (**P**) subactivities. Backup data for methods other than specified on the data card can be developed and placed on the data card according to the procedure outlined in the Development of Elements for Special Tools or Situations section. Equipment data to be verified and validated are Loaded and Unloaded transportation speeds and Vertical speeds (**T, L** and **V**).

The transportation time per traveled distance and the corresponding index values can be calculated using the following formula:

$$t = c + (s \times n)$$

where: t = time (TMU)
c = fixed manual time (TMU), including grasping of control and crane acceleration and deceleration times
s = crane horizontal speed (TMU/foot or meter)
n = distance variable (number of feet or meters moved)

Writing Method Descriptions

Below is the recommended format for a clear and concise method description for the Manual Crane Sequence Model. Examples of this structure can be found following the example listed below.

The recommended method description format for Manual Crane is:

Transport	Object	Holding Device	To Location	Placement

Manual Crane Example

1. A machine operator walks 10 feet (3 m) to a crane and manually transports it to a fixture (66 pounds, 30 kg) located 7 feet (2 m) away. The fixture, which is lying by itself on a pallet, is hooked up to the crane with a single hook and moved 14 feet (4.5 m) to a workbench 3 feet (1m) higher than the pallet. The fixture is then lowered

4 inches (10 cm) and placed on top of the workbench. The operator transports the empty crane 3 feet (1 m) away and returns to the workbench.

Transport fixture from pallet to workbench using jib crane with single hook and transport empty crane aside

$\mathbf{A_6\ T_{16}\ K_{24}\ F_3\ V_{16}\ L_{24}\ V_3\ P_3\ T_{10}\ A_3}$
$(6+16+24+3+16+24+3+3+10+3)\times 10=1080$ TMU

DEVELOPMENT OF ELEMENTS FOR SPECIAL TOOLS OR SITUATIONS

Another important feature of the BasicMOST System is the provision for developing elements for unique cases. An example may be for tools not included in the BasicMOST System.

The Tool Use and Equipment Use data cards were designed to provide accurate parameter values for a wide range of common tools and activities found throughout industry. Although the majority of tools can be analyzed using the data from the Tool Use data cards (Figures 3.20, 3.21 and 3.22) either directly or by comparison, special tools used in an operation may not be covered by any of the Tool Action categories. If the tool is infrequently used, the General and Controlled Move Sequence Models can be used to measure the activity. If the tool is frequently used, however, it may be desirable to develop special Tool Action elements specifically for the tool. This same process can be used for equipment not listed on the Equipment Use data card.

Three alternatives are available to the analyst for describing the use of those tools or equipment *not found* on the data cards:

1. Identify the method used, compare it with existing data and select an appropriate index value from a similar Tool Action method. (It is always the *method of using a tool,* not the name of the tool that determines the parameter value.)
2. Make a detailed BasicMOST analysis using a combination of General and Controlled Moves.
3. For frequently used tools, develop an element with index values based on a MiniMOST, MTM-1 or MTM-2 analysis using the Element Development Procedure.

Alternative 1: Compare Method and Use Existing Data

Sometimes, a special tool will resemble another tool in appearance as well as the method employed. A corkscrew, for example, which requires the use of wrist actions, looks very much like a small T-wrench. Therefore, as this alternative suggests, the activity to 'turn' a corkscrew into a cork (e.g., with six wrist actions) can be analyzed using the Fasten or Loosen data for a small T-wrench. Since light pressure is needed to start the corkscrew, a $\mathbf{P_3}$ is required for the tool placement.

$\mathbf{A_1\ B_0\ G_1\ A_1\ B_0\ P_3\ F_{16}\ A_0\ B_0\ P_0\ A_0}$

$(1+1+1+3+16) \times 10 = 220$ TMU

Another example of comparing the method can be found in food preparation. The activity to shake seasoning onto food is similar to wrist taps and would be indexed by the number of shaker taps. In the following case, the associate grasps the shaker, puts it to the dish, taps the shaker with five wrist taps and asides the shaker.

$\mathbf{A_1\ B_0\ G_1\ A_1\ B_0\ P_1\ F_6\ A_1\ B_0\ P_1\ A_0}$

$(1+1+1+1+6+1+1) \times 10 = 120$ TMU

Alternative 2: Analyze the Method Using General and Controlled Moves

If an appropriate index value is not found after comparing a special tool method with the existing data, the activity can be analyzed using General and Controlled Moves. For example, the method of using a crank-operated hand drill does not seem to fit any of the tools listed in the Tool Use data cards. However, a detailed BasicMOST analysis can be made by breaking down the complete drilling activity into its basic sub-activities. The analysis for using a hand drill to make a hole in a wooden block with eight revolutions of the crank handle would require three sequence models as follows:

1. Grasp and place hand drill to a mark on the block:

   $\mathbf{A_1\ B_0\ G_1\ A_1\ B_0\ P_3\ A_0}$ **60 TMU**

2. Grasp handle and drill hole with eight cranking actions:

   $\mathbf{A_1\ B_0\ G_1\ M_{16}\ X_0\ I_0\ A_0}$ **180 TMU**

3. Disengage and put hand drill aside:

   $\mathbf{A_0\ B_0\ G_3\ A_1\ B_0\ P_1\ A_0}$ **50 TMU**

Note: This alternative is for *infrequently used* tools because of the amount of analysis effort involved.

Alternative 3: Develop Elements for the Tool

One of the most useful features of the MOST Work Measurement Technique is the provision for the development of elements for special tools or sub-activities. This feature is particularly applicable when a *frequently used* tool (or applicable method) is not found in the Tool Action or Use Equipment data. The element development procedure first requires that the method be analyzed using MiniMOST, MTM-1 or MTM-2. Index values are then assigned to the element according to the BasicMOST time interval table for the tool.

Consider, for example, an assembly operation in which a spiral screwdriver is frequently used. The MiniMOST analysis for this activity might be:

1. Turn spiral screwdriver ≤ 10 inches (25 cm) for power stroke:

   $\mathbf{A_0\ B_0\ G_0\ M_{10}\ X_0\ I_0\ A_0}$ **10 TMU**

2. Return stroke:

$$A_0\ B_0\ G_0\ M_{10}\ X_0\ I_0\ A_0\quad 10\ \text{TMU}$$

3. Seat screwdriver for final tightening:

$$A_0\ B_0\ G_0\ M_{16}\ X_0\ I_0\ A_0\quad 16\ \text{TMU}$$

The formula used to develop new elements is:

$$y = mx + c$$

where: y = maximum time per tool action in TMU
m = TMU per unit
x = number of tool actions
c = constant time

For the example above, the formula would be written:

$$y = 20x + 16$$

Using the formula above, but now solving for x, one can determine the maximum number of tool actions for each index value. The maximum interval limits are assigned to y and the solutions for the x value are rounded down to the nearest whole number. The formula to solve for x would then be:

$$x = (y - c)/m \quad \text{or} \quad x = (y - 16)/20$$

where: y = total maximum time to fasten screws (use upper limits of index value ranges)
c = constant for using screwdriver (16 TMU for final tightening)
m = time per tool action (20 TMU for each stroke)
x = number of tool actions

Taking the upper limit values from the table in Appendix A, Figure A.3, the data table for a spiral screwdriver is shown in Figure 3.70.

The steps to develop elements for a tool or situation not on the data card using the element development procedure are:

1. Perform MiniMOST, MTM-1 or MTM-2 analysis.
2. Apply algebraic formula: $y = mx + c$.
3. Solve formula for x: $x = (y - c)/m$.
4. Develop supplementary index value table.

Figure 3.71 represents the simplified supplementary index value table for a spiral screwdriver.

Index Value	Upper (y value) interval limit	Number of actions (x value) x = (y-16)/20	Number of actions x (truncated)
1	17	0.05	0
3	42	1.30	1
6	77	3.05	3
10	126	5.50	5
16	196	9.00	9
24	277	13.05	13
32	366	17.50	17
42	476	23.00	23
54	601	29.25	29

FIGURE 3.70 Data table for spiral screwdriver.

Spiral Screwdriver	
Index Value	***Number of Tool Actions***
F_3	1
F_6	3
F_{10}	5
F_{16}	9

FIGURE 3.71 Supplementary index values for a spiral screwdriver. Values are read up to and including.

If the spiral screwdriver was used to fasten a screw with four tool actions, the BasicMOST analyst could now use one Tool Use Sequence Model and the table (Figure 3.71) that has been developed. The analysis would appear as:

$$\mathbf{A_1\ B_0\ G_1\ A_1\ B_0\ P_3\ F_{10}\ A_1\ B_0\ P_1\ A_0}$$

$$(1+1+1+3+10+1+1)\times 10=180\ \text{TMU}$$

The preceding situation dealt with the development of elements for a spiral screwdriver based on a detailed MiniMOST backup analysis. Other situations that may lend themselves to MiniMOST backup analyses could be polishing, grinding, painting, gluing or any other activity involving a short process time (i.e., using power tools or office machines). Elements should be developed for these situations when they occur frequently enough to justify the time taken to develop such elements and when consistency of application is required.

To determine new elements, the method, the unit of the variable and frequencies should be specified. The proper analyses should be performed and the results should be entered into the formula. For example, the *method* for polishing might be based on push or pull (Controlled Move) with resistance, the *unit* per square foot (0.1 m^2)

and the *frequency* of 20 strokes per square foot (0.1 m^2). This would be calculated and a supplementary data table for polishing per square foot would be developed. To use the data, values from this table could then be applied to the Tool Use Sequence Model and placed under the Surface Treat (**S**) parameter.

APPLICATION OF THE BASICMOST WORK MEASUREMENT SYSTEM

THE BASICMOST ANALYSIS

The application of MOST is consistent among all three systems. There are several key factors to a MOST analysis which are listed below and detailed in Chapter 2. Analysts need to consider the following factors and the content included in this section when developing a BasicMOST analysis:

- Identifying the best method
- Practical analysis procedures
- Analyst consistency
- General application rules for all MOST Systems

METHOD LEVELS AND SIMULTANEOUS MOTIONS

When analyzing activities, it's important to understand the method level used by the operator during the activity and the documentation needed for possible method improvements. Understanding the method level used and how it should be documented can assist the analyst with identifying improvement opportunities.

Method level refers to the degree of coordination between the right and left hands during two-handed work. A *high method level* occurs when a large percentage of manual and body motions are performed simultaneously. Obviously, it is desirable to have as much work as possible performed at high method levels because of the reduced time to accomplish a given amount of work.

The method level at which an activity is performed is determined by its occurrence frequency, that is, the practice opportunity available to the operator. The more often the activity occurs, the greater the operator's opportunity to improve the method level. If the activity is seldom performed, the short learning period prevents any development of simultaneous skills. For example, with mass production and large batch-size operations, which allow ample training and practice opportunity, one would expect to find operators using a high percentage of simultaneous motions. On the other hand, job shop and setup activities will most likely be performed with few simultaneous motions. Therefore, method level depends to a large extent on the type of work being performed. Three different method levels are defined for the application of BasicMOST.

1. *High method level* includes all possible simultaneous motions with the right and left hands. The analysis and time for the limiting (longest) hand is allowed. If the analysis for both hands is shown, the time value for the limited hand is not included in the total. As shown in Figure 3.72,

Step	Method Description	Seq. Model	Freq.	Simo	TMU	Total
1	RH GRASP LIGHT OBJECT AND PUT ASIDE	A1 B0 G1 A1 B0 P1 A0	1.000	N	40	40
2	LH GRASP LIGHT OBJECT AND PUT ASIDE	A1 B0 G1 A1 B0 P1 A0	1.000	Y	40	0

FIGURE 3.72 High method level.

Step	Method Description	Seq. Model	Freq.	Simo	TMU	Total
1	RH GRASP LIGHT OBJECT AND PUT ASIDE	A1 B0 G1 A1 B0 P1 A0	1.000	N	40	40
2	LH GRASP LIGHT OBJECT AND PUT ASIDE	A1 B0 G1 A1 B0 P1 A0	1.000	N	40	40

FIGURE 3.73 Low method level.

the activity performed by the left hand (LH) occurs simultaneously with the activity performed by the right hand (RH). This means the LH time is 'limited out' by the activity performed in step 1 and the TMU is not counted in the total:

2. *Low method level* involves no simultaneous motions. Figure 3.73 shows that the LH and RH perform an activity with no simultaneous motions. The analysis time for both hands must be allowed.
3. *Intermediate method level* refers to a method performed partially with simultaneous motions. For example, the Action Distance 'Within Reach' to two objects may be performed simultaneously with both hands but gaining control and placing two objects simultaneously may not be possible. In the BasicMOST analysis, the appropriate parameters are documented to show that they are performed simultaneously to another action and the associated time should be excluded from the sequence model calculation. In the activity shown in Figure 3.74, a portion of the sequence model for the LH (the reach to get the object) is performed simultaneously with the reach of the RH. In this case, the $\mathbf{A}_1$ is shown in brackets and is not included in the time calculation because it is 'limited' by another activity.

Method Level and Simultaneous Motion Examples

The activity 'Grasp two pins and place in assembly' is analyzed using three different method levels. Each hand grasps a pin and places it in an assembly with adjustments. In Figure 3.75, the high method level will only include the time for one hand. The time for both hands is included in the low method level and only the Put phase time is allowed for the intermediate method level.

As the example shows, there is a variation in the total time between method levels. Therefore, one of the analyst's most important considerations in a work measurement situation is to represent the correct method level in the analysis. This relationship between method and time should always be emphasized in BasicMOST analysis work based on the theory that the greater the practice opportunity for the operator, the higher the method level. It is not required that the analyst break out two-handed work in a BasicMOST analysis; however, it is important to know the method level used to accurately write and document each method step.

Step	Method Description	Seq. Model	Freq.	Simo	TMU	Total
1	RH GRASP LIGHT OBJECT AND PUT ASIDE	A1 B0 G1 A1 B0 P1 A0	1.000	N	40	40
2	LH GRASP LIGHT OBJECT AND PUT ASIDE	<A1 > B0 G1 A1 B0 P1 A0	1.000	N	30	30

FIGURE 3.74 Intermediate method level.

Method Description	Seq. Model	Freq.	Simo	TMU	Total
HIGH METHOD LEVEL - BOTH HANDS WORK SIMULTANEOUSLY					
RH GRASP LIGHT OBJECT AND PLACE WITH ADJUSTMENTS	A1 B0 G1 A1 B0 P3 A0	1.000	N	60	60
LH GRASP LIGHT OBJECT AND PLACE WITH ADJUSTMENTS	A1 B0 G1 A1 B0 P3 A0	1.000	Y	60	0
*					
LOW METHOD LEVEL - BOTH HANDS WORK SEPARATELY					
RH GRASP LIGHT OBJECT AND PLACE WITH ADJUSTMENTS	A1 B0 G1 A1 B0 P3 A0	1.000	N	60	60
LH GRASP LIGHT OBJECT AND PLACE WITH ADJUSTMENTS	A1 B0 G1 A1 B0 P3 A0	1.000	N	60	60
*					
INTERMEDIATE METHOD LEVEL - ONLY THE GET PHASE OCCURS SIMULTANEOUSLY					
RH GRASP LIGHT OBJECT AND PLACE WITH ADJUSTMENTS	A1 B0 G1 A1 B0 P3 A0	1.000	N	60	60
LH GRASP LIGHT OBJECT AND PLACE WITH ADJUSTMENTS	< A1 B0 G1 > A1 B0 P3 A0	1.000	N	40	40

FIGURE 3.75 Examples of high, intermediate, and low method level.

BasicMOST Analysis Documentation

The BasicMOST Analysis form is shown in Figure 3.76 to provide the user with a history of BasicMOST. Today, most companies use software applications, such as StandardsPro, to develop and maintain engineered standards. MOST has progressed from a manual system to a fully integrated digital solution. Figure 3.77 shows a Method and TMU report for the activity to *Cut box and remove floor unit from cardboard box* using the StandardsPro software application. Whether the manual form or a software application is used to document a BasicMOST analysis, it is important to include essential information. One of the key features of the MOST Work Measurement Technique is the predefined sequence models. Using a manual system or a software application, the use of the predefined sequence models is critical for the analyst to maintain applicator accuracy and consistency.

Regardless of the documentation method, the information listed below should be included in the analysis.

1. *Identification and Header*

   This section of the analysis should include all key information for ease of searching and updating the analysis.

   - Title or description

     The title often begins with a verb of the activity and may include the object, product/equipment, tools and a work area description. An example title could be:

     Cut box and remove floor unit from cardboard box.

   - Activity

     The activity should be a verb that indicates the overall context and/or main goal of the actions which are included in the analysis. The activity for the example to *Cut box and remove floor unit from cardboard box* would be 'Cut.'

- Unit of measure
  Used to designate what the activity is based on. Examples of units of measure are per unit, per part, per box, per customer, per occurrence per pallet, etc.
- Operator, applicator or safety instructions

MAYNARD

BasicMOST® Analysis

DATE

ANALYST

PAGE ____ OF ____

DESCRIPTION

ACTIVITY • OBJECT • IN/ON/FOR • PRODUCT/EQUIPMENT • USING/WITH • TOOL • TO/AT • WORK AREA

UNIT OF MEASURE

INSTRUCTIONS ☐ APPLICATOR (A) ☐ OPERATOR (O) ☐ SAFETY (S)

TMU FROM PREVIOUS PAGE:

STEP NO.	METHOD STEP DESCRIPTION	SEQUENCE MODELS	FF	FR	SIMO TO	TMU
1		A B G A B P A	( )			
		A B G M X I A	( )			
		A B G A B P A B P A	( )			
2		A B G A B P A	( )			
		A B G M X I A	( )			
		A B G A B P A B P A	( )			
3		A B G A B P A	( )			
		A B G M X I A	( )			
		A B G A B P A B P A	( )			
4		A B G A B P A	( )			
		A B G M X I A	( )			
		A B G A B P A B P A	( )			
5		A B G A B P A	( )			
		A B G M X I A	( )			
		A B G A B P A B P A	( )			
6		A B G A B P A	( )			
		A B G M X I A	( )			
		A B G A B P A B P A	( )			
7		A B G A B P A	( )			
		A B G M X I A	( )			
		A B G A B P A B P A	( )			
8		A B G A B P A	( )			
		A B G M X I A	( )			
		A B G A B P A B P A	( )			
9		A B G A B P A	( )			
		A B G M X I A	( )			
		A B G A B P A B P A	( )			
10		A B G A B P A	( )			
		A B G M X I A	( )			
		A B G A B P A B P A	( )			
		A T K F V L V P T A	( )			
		A T K F V L V P T A	( )			

TOTAL TIME:	HOURS:________	MINUTES:________	SECONDS:________	TMU:________

FIGURE 3.76 BasicMOST Analysis form.

Instructions can be added to clarify key points in the analysis. Examples of instructions are listed below:

Operator Instructions—Check for quality on Step 2 before adding additional part.

Applicator Instructions—Checking for quality is internal to moving the part.

Safety Instructions—Wear safety glasses while welding parts.

- Date of analysis
- Identification of analyst
- Total TMU

  The total time for the activity is shown by simply adding all of the numbers in the TMU column for each method step. That number is then noted in the Total Time section of the form or even better, calculated automatically in a software solution. Most electronic systems automatically convert the TMU to the organization's choice of time in seconds, minutes or hours.

2. *Method Steps and Sequence Models*

   This section of the analysis should include the step number, method step description, sequence model, including documentation of any frequencies or simultaneous motions, and the TMU for each method step.

- Method steps

  The amount of information placed in the method description section is usually a function of its eventual use, that is, the description can be used for detailed operator instructions or for an outline of the manual work for time computation only. Each method step has *only one* corresponding sequence model so the method description should be phrased in terms of moving an object or using a tool as to easily identify the sequence model used for that activity.

- Sequence model analysis

  Regardless of the system used to document a BasicMOST analysis, it should include the predefined sequence models for General Move, Controlled Move, Tool Use and Equipment Use. This helps to ensure applicator accuracy and consistency. After the index values to the selected sequence model are applied, any frequencies if they occur in the method step or if the method step is performed simultaneously to another activity are documented.

Frequencies and partial frequencies were discussed earlier in the chapter and are used when one or more parameters of a sequence model occurs more or less than once. The default for any system is a frequency of one (1). As shown in Figure 3.77 for the Cut box and remove floor unit from cardboard box, a '4' is noted in the 'Freq' column for Step 3 to show that the activity to 'Pull back flaps' occurs four times.

Partial frequencies, where a portion of the sequence model occurs more than once, can be documented as shown in Figure 3.77. In step 2 of Figure 3.77, a partial frequency occurs when cutting three sides of the box. The parameters that occur

Sub-Operation - Method / TMU Report
H. B. Maynard and Co.

Sub-Op ID:	70873	**Status:**	Private
Title:	CUT BOX AND REMOVE FLOOR UNIT FROM CARDBOARD BOX		
Activity:	CUT	**Sub-Activity:**	
Object:	BOX	**Prod/Equip:**	
Size/Capacity:		**Tool:**	
WA Number:		**WA Origin:**	
5-S Level:	SOME EFFORT	**Other:**	
Unit of Measure:	OCCURRENCE	**OFG:**	2
WorkArea ID:		**Starting Oper:**	
Starting Location:		**Total TMU:**	920

Operator Instr:

Applicator Instr:

This sub-operation includes the time to cut the exterior of a cardboard box and remove a floor unit from the box.

Safety Instr:

Applicator:	MCS		
Create Date:	3/25/2020	**Effect Date:**	3/23/2020

Method Steps

Step	Method Description	Freq.	Simo	TMU	Total
1	GET BULKY BOX FROM COUNTER AND PUT ON FLOOR A1 B0 G3 A3 B6 P1 A0	1.000	N	140	140
2	GRASP KNIFE, SLICE TOP AND BOTH SIDES OF TAPE ON BOX A1 B0 G1 A0 B0 (P1 A1 C3) A1 B0 P1 A0 (3.00)	1.000	N	190	190
3	CONTACT BOX AND PUSH EACH FLAP OPEN A1 B0 G1 M3 X0 I0 A0	4.000	N	50	200
4	GET PART FROM BOX AND PUT ON COUNTER A1 B6 G3 A3 B0 P1 A0	1.000	N	140	140
5	GRASP BOX AND PUSH TO FLATTEN A1 B0 G1 M3 X0 I0 A0	3.000	N	50	150
6	GET BULKY BOX AND TOSS IN BIN A1 B0 G3 A6 B0 P0 A0	1.000	N	100	100

FIGURE 3.77 Method and TMU report.

more than once ($\mathbf{P_1}$ $\mathbf{A_1}$ $\mathbf{C_3}$) are shown in parentheses with the partial frequency of '(3.00)' noted at the end of the sequence model.

The 'Simo' column in an analysis is used to document that a method step or a portion of the method step occurs at the same time as another step. In Figure 3.78, Step 4 is simultaneous to Step 3. A '0' is listed in the Total field for Step 4 so that time is not included in the total time. If the index values are equal for the simultaneous actions, select which method steps should be identified as the limited step and document accordingly.

Step	Method Description	Seq. Model	Freq.	Simo	TMU	Total
1	GRASP PART AND PLACE TO MACHINE	A3 B0 G1 A3 B0 P3 A0	1.000	N	100	100
2	PRESS BUTTON TO START MACHINE	A1 B0 G1 M1 X0 I0 A0	1.000	N	30	30
3	PTIME - 45 SECONDS	PROCESS TIME 45.000 SECONDS	1.000	N	1,250	1,250
4	GRASP RAG AND WIPE 2 SQ FT AREA AND ASIDE CLOTH	A3 B0 G1 A1 B0 P1 S16 A1 B0 P1 A0	1.000	Y	240	0

FIGURE 3.78 Documentation of a simultaneous activity.

If a portion of a method step is simultaneous to another, the proper use of the Simo notation is to indicate the method step and parameters to which the activities are simultaneous. From its inception, BasicMOST has used a simple coding system to document simultaneous motions. Since the General Move and Controlled Move Sequence Models consist of seven parameters, they are numbered as follows:

A	**B**	**G**	**A**	**B**	**P**	**A**	
1	2	3	4	5	6	7	(parameter number)

The Tool Use Sequence Model and Equipment Use Sequence Model are numbered in a similar manner:

A	**B**	**G**	**A**	**B**	**P**	*	**A**	**B**	**P**	**A**	
1	2	3	4	5	6	7	8	9	10	11	(parameter number)

As an example, if the Get phase of the second method step is simultaneous to the Get phase of step one, then the code for a Simo activity for the second method step would read 1:1–3. The **A B G** parameters of Step 2 would be documented to show the Simo activity and not counted in the total for that method step. Figure 3.79 shows that the Get phase for Step 2 is simo to the Get phase in Step 1.

The time for each method step is then calculated by adding the index values, applying any frequencies as needed and then multiplying by 10 to get the time value for the sequence model in TMU.

In software applications, the use of the coding system can be noted in the method description or in the applicator instructions as shown in Figure 3.80.

Updating a BasicMOST analysis is very simple using software applications which feature the functionality to update one sub-operation, determine the impact of a method change on the overall time and even complete a mass update of the new time to automatically update all sub-operations that include a particular method step.

The ease with which BasicMOST analyses can be updated and/or new methods determined is one of the greatest features of the MOST Work Measurement Technique. It makes simulation and comparison easy.

Step	Method Description	Seq. Model	Freq.	Simo	TMU	Total
1	GRASP PART AND PUT TO WORK TABLE	A1 B0 G1 A1 B0 P1 A0	1.000	N	40	40
2	GRASP COMPONENT AND SNAP ONTO PART	< A1 B0 G1 > A1 B0 P3 A0	1.000	N	40	40

FIGURE 3.79 Partial Simo notation in BasicMOST.

Operator Instr:	
Applicator Instr:	Step 2 is simo: 1:1-3
Safety Instr:	

FIGURE 3.80 Use of the Simo coding system in the Applicator Instructions in an analysis.

Summary of the BasicMOST Analysis

A BasicMOST analysis is completed by confirming the following is included:

1. Identification and Header Information
    - Title or description
    - Activity
    - Unit of measure
    - Operator, applicator or safety instructions
    - Date of analysis
    - Identification of analyst
    - Total TMU
2. Method Steps and Sequence Models
    - Method steps and method step descriptions
    - Sequence model analysis
    - Select *one* sequence model for each method step
        - Apply the correct index value for each parameter within each sequence model.
        - Note the documentation for PF, FR or Simo activities as needed.
        - Add parameter index values together, apply frequencies as needed and multiply by the BasicMOST factor of 10 to get the time for the sequence model in TMU.
    - Add all method step totals to get a total TMU

Note: A MOST analysis will provide the normal time (manual time+allowance time). Allowances need to be added to generate an engineered standard.

Validation of Process Times

It will be necessary to validate elements that are based on process times such as power tools and manual cranes. Also, if new elements involving process times are being developed, these elements must be validated for different types of equipment. In all cases, the validation should be carried out to ensure that the desired level of accuracy will be achieved. The analyst compares the index value on the data card with its allowed deviation range to the process time for the selected equipment determined by stopwatch time study.

The steps required to perform the validation are as follows:

1. Review the specification and method used for the existing equipment.
2. Establish criteria for the time study based on the characteristics and method for the selected equipment.
3. Conduct and compile time study.
4. Compare time study results to existing index values.
5. Determine if the current data card can be applied.

6. If necessary, develop required elements and a supplementary data card for the selected equipment according to the principles described earlier in this chapter.
7. Document the validation process for future use.

Because it is impractical to cover the wide variety of available and potential future equipment on data cards, it will be necessary to validate all process times in order to achieve the desired level of accuracy and consistency when using MOST.

BasicMOST Summary

The MOST Work Measurement Technique uses a structured approach to measure work based on the movement of objects. There is a consistent approach the analyst should always use prior to analyzing an activity with MOST. It begins with determining the starting and stopping points of the activity to be analyzed and ends with a total time for the activity. The BasicMOST Analysis Process (Figure 3.81) depicts this procedure and can be used to lead the analyst through the basic thought processes and decisions that need to be considered in order to arrive at a thorough and consistently applied BasicMOST analysis. Follow the process to make the proper

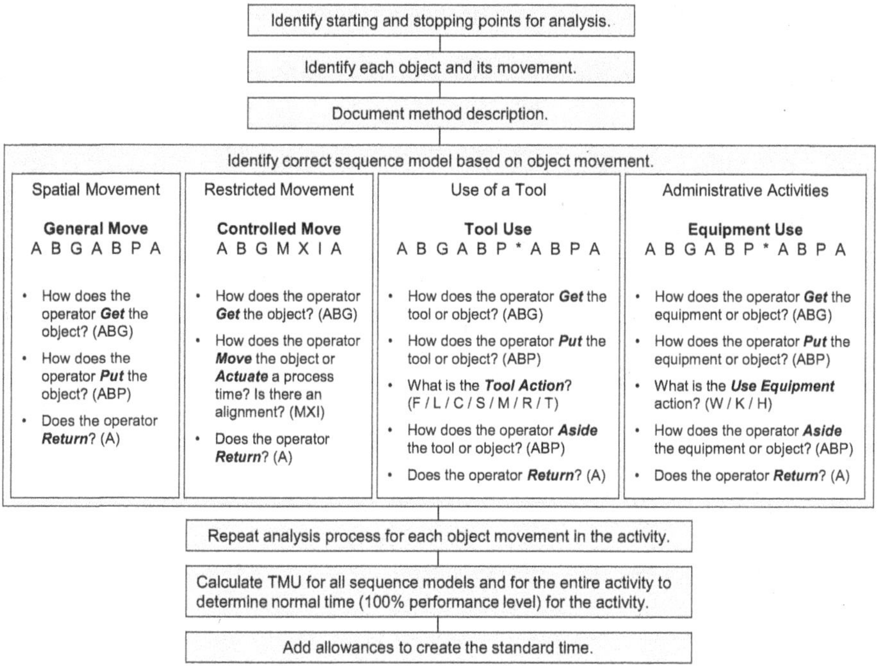

FIGURE 3.81 BasicMOST Analysis Process.

decisions to complete an analysis of an activity. The BasicMOST Analysis Process includes the four main sequence models of BasicMOST.

It must be remembered that measured times produced with any MOST System represent a performance level of 100%. That is, the performance of an average skilled worker following a prescribed method, working under average work conditions at a normal pace and with capable supervision. The total time value for an activity analyzed with BasicMOST produces a normal time without allowances. Typically, the allowances as a percentage of normal time are applied as a final step to establish a standard time.

Following the analysis process and answering the questions is key to the effective application of BasicMOST. The answers will help the analyst to:

- Determine the correct sequence model to be used.
- Determine the index value for each parameter (sub-activity).
- Avoid overlooking any other objects being moved or analyzing any unnecessary activity.
- Apply BasicMOST consistently.

4 The MiniMOST System

MiniMOST was created with the goal of developing a version of the MOST Work Measurement Technique based on the MOST concept and format that can be applied to *identical cycles* (typically of short duration) with a high level of accuracy.

MiniMOST satisfies the more rigorous accuracy requirements associated with short-cycle and highly-repetitive operations. These types of operations are performed following an *identical* or almost identical motion pattern from cycle-to-cycle. MiniMOST is a more detailed system that takes more time to apply than BasicMOST and should therefore be applied only to activities that have been determined to be short-cycle and identically repeated. Guidelines for when to select MiniMOST as the appropriate work measurement system are included in Chapter 2.

Following the same philosophy of the BasicMOST System, MiniMOST was designed to replace the more 'detailed' systems such as MTM-1 and Work Factor. It consists of two sequence models which represent the two ways of moving an object from one location to another:

1. *General Move Sequence Model*—Used to analyze the spatial movement of an object freely through the air.
2. *Controlled Move Sequence Model*—Used to analyze the movement of an object when it remains in contact with a surface or is attached to another object during the movement.

The methodology behind the MOST Work Measurement Technique is that the system can be used for any type of manual work. The sequence models in MiniMOST provide a structure to analyze different types of work. Technology is advancing rapidly and can impact how work is performed. MOST, however, can still be used to analyze any manual motions affected by technology. Through proper certification training, it will become clearer how to measure new technologies with MOST.

The two sequence models of MiniMOST are presented below.

Activity	Sequence Model	Parameter	
General Move	A B G A B P A	**A**	Action Distance
		B	Body Motion
		G	Gain Control
		P	Placement
Controlled Move	A B G M X I A	**M**	Move Controlled
		X	Process Time
		I	Alignment

The index values applied in a MiniMOST analysis follow the same scale as BasicMOST (e.g., 1, 3, 6, 10, 16, 24, etc.), however, the multiplier for the index values is one (1). This means that each index value directly represents the time in TMU (Time Measurement Unit). Since the time interval represented by each index value is very small, MiniMOST provides a highly detailed system for measuring work while maintaining many of the benefits of the BasicMOST System. The MiniMOST measurement level of detail is somewhere between MTM-1 and BasicMOST. Because of the increased level of detail with MiniMOST, the analyst must look at the work to be measured differently than with BasicMOST.

THE GENERAL MOVE SEQUENCE MODEL

General Move deals with the spatial displacement of one or several objects. Under manual control, the object follows an unrestricted path through the air. If the object is in contact with, restricted by or attached to another object during the move, the General Move Sequence Model is not applicable. This type of move will be defined later in the chapter as a Controlled Move activity.

As defined in Chapter 1, MOST measures the movement of objects. One or more objects can be moved with one or both hands. For simplification of the text, when one object is referenced it can mean one or more objects unless it specifically states only one object in the definition.

In MiniMOST, a General Move follows a sequence of sub-activities identified by the following steps:

1. *Reach* with one or two hands a distance to the object, either directly or in conjunction with body motions.
2. *Gain* manual *control* of the object.
3. *Move* the object a distance to the point of placement, either directly or in conjunction with body motions.
4. *Put* the object in a temporary or final position.
5. *Return* the hand to an undefined location.

These five sub-activities form the basis for the activity sequence describing the manual displacement of one or more objects freely through space. This sequence describes the manual events that can occur when moving an object freely through the air and, as in BasicMOST, is known as a *General Move Sequence Model.* The major function of the sequence model is to guide the analyst through an activity of having a structured and standardized analysis format. The predefined sequence model allows for increased analyst consistency and reduces sub-activity omission.

THE SEQUENCE MODEL

The sequence model consists of a series of letters, or parameters, representing the various sub-activities of General Move. The parameters of the General Move Sequence Model identify the sub-activities included in the five-step pattern already indicated:

A B G A B P A

where: **A** = Action Distance
B = Body Motion
G = Gain Control
P = Placement

Phases of the General Move Sequence Model

The displacement of an object through space occurs in three distinct phases as shown by the following General Move Sequence Model breakdown:

Get	Put	Return
A B G	**A B P**	**A**

The first phase, referred to as *Get*, describes the actions to reach the object with body motions (if necessary) and gain control of the object. The **A** parameter indicates the distance the hand or body travels in order to reach the object. The **B** parameter indicates the need for a body motion during this action. The degree of difficulty encountered in gaining control of the object is described by the **G** parameter.

The *Put* phase of the sequence model describes the actions to move the object to another location. As before, the **A** and **B** parameters indicate the distance the hand or body travels with the object and the need for any body motions during the move before the object is placed. How the object or objects are placed is described by the **P** parameter.

The third phase simply indicates a return or clearing a hand out of the way, often for safety reasons, to permit the next activity. Normally the 'return' of the hand is accounted for in the first Action Distance of a subsequent sequence model.

The MOST analyst should strictly adhere to the three-phase breakdown of the General Move Sequence Model. Following the sequence model and phases provides consistency in application and ease in communication.

Parameter Definitions

A Action Distance

This parameter covers all spatial movement or actions of the fingers, hands and/or feet, either loaded or unloaded (loaded means carrying an object, unloaded means the hands are free).

B Body Motion

This parameter is used to analyze vertical motions of the body or the actions necessary to overcome an obstruction or impairment. The movement of the head to exercise Eye Travel is also part of the Body Motion parameter.

G Gain Control

This parameter covers all manual motions (mainly of the finger, hand or foot) used to obtain complete manual control of an object or objects. The **G** parameter can include one or several short motions to gain full control of the object before moving it to another location.

P Placement

This parameter is used to analyze actions at the final stage of an object's displacement to align, orient and/or engage the object with another object before control of the object is relinquished.

Parameter Indexing

Before assigning index values, the MOST analyst must ask two questions to determine the appropriate sequence model:

1. What is the object being moved?
2. How is the object moved (this question is key to determining the appropriate sequence model to use)?

Then, assuming a General Move:

- What does the operator do to *get* the item (determine index values for **A**, **B** and **G**—first phase)?
- What does the operator do to *put* the item (determine index values for **A**, **B** and **P**—second phase)?
- Does the operator *return* or 'clear' hands (determine index value for the final **A**—third phase)?

Asking these questions is vital to the effective application of MiniMOST. The answers will help the analyst:

- Avoid overlooking any operator activity or analyzing any unnecessary activity.
- Correctly divide a process into method steps and phases.
- Write accurate and clear method descriptions.
- Determine the index value for each parameter (sub-activity).
- Apply MiniMOST consistently.

Indexing each parameter of the General Move Sequence Model is accomplished by observing or visualizing the object movement during each phase of an activity and selecting the appropriate index value from the data card (Figure 4.1). The index value for each parameter is taken from the extreme left or right column of the data card and is manually documented as $\mathbf{A_3}$. Software applications feature a built-in data card and index values can be selected from a drop-down list or input from memory.

Consider the example of an associate grasping a clip from a bin located on a workstation and placing it to a part on a conveyor. To apply MiniMOST, it is necessary to know the distances to reach and place objects. Assume the associate reaches 10 inches (25 cm) to grasp the clip from the bin and then moves the clip 16 inches (40 cm) and places the clip on the part. The sequence model for this activity is:

$\mathbf{A_{10}}\ \ \mathbf{B_0}\ \ \mathbf{G_6}\ \ \mathbf{A_{16}}\ \ \mathbf{B_0}\ \ \mathbf{P_{10}}\ \ \mathbf{A_0}$

Since the associate reaches 10 inches (25 cm), the first **A** parameter in the sequence model is indexed $\mathbf{A_{10}}$ because the bin is more than 8 inches (20 cm) and less than 14 inches (35 cm) away from the associate. Refer to the Action Distance column of the data card in Figure 4.1 for an $\mathbf{A_{10}}$. No Body Motion is needed to reach the clip; therefore, a 0 is assigned to the Body Motion parameter ($\mathbf{B_0}$) and control of the clip is gained with no difficulty which is indexed with a $\mathbf{G_6}$, Grasp. ($\mathbf{G_6}$—Grasp in Gain Control column). The clip is then moved 16 inches (40 cm) away ($\mathbf{A_{16}}$). There is no Body Motion requiring another $\mathbf{B_0}$ in the Put phase. There is some orientation to place the clip which requires a $\mathbf{P_{10}}$ to be indexed in the sequence model. An $\mathbf{A_0}$ is used in the Return phase since there is no return of the hands for safety reasons.

The time to perform this activity is calculated by adding all index values in the sequence model. Remember that the multiplier for MiniMOST is one (1) so the index values can simply be added together for a total TMU: 10+0+6+16+0+10+0=42 TMU. Refer to Chapter 1 for a review of TMU.

MiniMOST System				General Move		ABGABPA	
Index x 1	A Action Distance			B Body Motion	G Gain Control	P Placement	Index x 1
	Hand		Leg				
	Inches (cm)	Degrees	Inches (cm)				
0	≤ 1 (2.5)	≤ 30	-	-	Sweep	Hold Drop	**0**
1	≤ 2 (5)	≤ 60	-	-	-	-	**1**
3	≤ 4 (10)	≤ 120	-	-	Contact (hand or foot)	Toss Set and Retain	**3**
6	≤ 8 (20)	≤ 180	≤ 8 (20) Foot Action	-	Grasp Regrasp	Set Aside Put	**6**
10	≤ 14 (35)		≤ 12 (30)	Eye Travel	Transfer Select	Place with Some Orientation	**10**
16	≤ 24 (60)		≤ 18 (45) 1 Step	-	Select – Small Disengage	Position with Complete Orientation	**16**
24	> 24 (60)		≤ 26 (65)	-	Collect		**24**
32			> 26 (65) 2 Steps	Bend or Arise Sit			**32**
42				Stand			**42**
NOTES	When hand reaches more than 8 inches (20 cm) to an object in a fixed location or to the other hand, use the next lower index value.				1. ENW = Effective Net Weight (max. 2.5 lbs., 1 kg). 2. For an ENW > 2.5 – 10.0 lbs. (1 – 5 kg), use the next higher index value except for 'Grasp.' 3. For Controlled Move, use 40% of weight of object to determine ENW.	1. For accurate positioning, difficult handling or insertion 1/8 in. – 1 in. (3 mm – 2.5 cm), use the next higher index value. 2. For binding or applying pressure, go up two index values per occurrence. 3. For insertions > 1 in. (2.5 cm), use an additional Controlled Move Sequence Model.	NOTES

FIGURE 4.1 General Move Sequence Model data card.

Limiting or Limited

In this chapter, there are many references to 'limiting' or 'limited' time. These terms are often used with two-handed work to define which time can be counted in an analysis and which time should be excluded. The terms are defined as:

Limiting Time—The time allowed (counted) in an analysis. This time is included in the total TMU.
Limited Time—The time not allowed (not counted) in an analysis. This time is not included in the total TMU.

Action Distance (A)

Action Distance covers all spatial movements or actions of the fingers, hands, feet or leg, either loaded or unloaded. This includes turning the hand in the air, leg and foot motions. Action Distance also includes horizontal transportation of the body up to two steps.

Hand—Inches (cm)

The value for a hand Action Distance is determined by the total net distance (inches or centimeters) the hand travels. This distance must be measured. It is necessary to trace the arc the hand travels in measuring the distance. *Do not* use the straight-line distance, since the hand follows an arc.

A_0 ≤ 1 Inch (2.5 cm)

Any displacement of the fingers and/or hands a distance of up to 1 inch (2.5 cm) will result in a zero index value. Since the time for traveling these short distances is minimal, it is included within the Gain Control and Placement parameters.

Examples: Moving the fingers from one key to another on a keyboard is usually an Action Distance of 1 inch (2.5 cm) or less.

A_1–A_{24} ≤ 2 Inches (5 cm) to > 24 Inches (60 cm)

The index values for hand movements as part of Action Distance are based on any displacement of the fingers and/or hands to get or move an object. As with all index values, the value is read 'up to and including.' If the index value is $\mathbf{A_3}$, then the distance measured is greater than 2 inches (5 cm) and less than or equal to 4 inches (10 cm).

$\mathbf{A_1}$ ≤ 2 inches (5 cm)
$\mathbf{A_3}$ ≤ 4 inches (10 cm)
$\mathbf{A_6}$ ≤ 8 inches (20 cm)
$\mathbf{A_{10}}$ ≤ 14 inches (35 cm)
$\mathbf{A_{16}}$ ≤ 24 inches (60 cm)
$\mathbf{A_{24}}$ > 24 inches (60 cm)

Examples: Reach 6 inches (15 cm) to a part jumbled with others in a tote pan: $\mathbf{A_6}$.
Place a bolt in a parts bin located 9 inches away (22.5 cm): $\mathbf{A_{10}}$.

Reach a net distance of 20 inches (50 cm) with both hands to a suitcase on a conveyor: $\mathbf{A_{16}}$.

Reaching to a Fixed Location or to the Other Hand

When reaching to an object in a fixed location a distance of more than 8 inches (20 cm), the next lower index value is applied. Reaching to objects repetitively in a fixed location can be accomplished with no visual attention which significantly reduces the time required for completion. If, however, the hand must make a sharp change of direction in reaching to the object, the time reduction does not occur and the time in the hand column is allowed without adjustment. Reaches to a fixed location frequently occur when reaching to machine controls such as buttons or levers. Reaches to the other hand apply to any object held in the hand or on which the hand is resting. This contingency applies only if the new grasping point is within 3 inches (7.5 cm) of the hand previously in contact with the object.

Note: Occasionally, a reach is observed that does not require visual attention but does not meet the fixed location or other hand location criteria. This may occur when there is unusually high practice in performing the motion such as repeatedly reaching for an item from a shelf. If this is the case, the time should be reduced one index value when the length of the reach is greater than 8 inches (20 cm). When this occurs, make sure that the absence of visual control is possible because of the nature of the work, not due to unusual skill or coordination of the individual operator. The rule for reducing the time applies only to reaches performed by the hand. It does not apply to placing an object or to any action of the foot or leg. This adjustment, when required, is made *only* to the first **A** parameter in the sequence model.

The criteria for reducing the Action Distance value is listed below. All criteria must be met to reduce the Action Distance index value.

1. Movement is a Hand Action Distance.
2. Reach is to gain control of an object or location (first **A** in the Get phase) a net distance that exceeds 8 inches (20 cm).
3. No visual control is required as one of the following is occurring:
   a. Object is in a fixed location.
   b. Object is in the other hand and the grasping point is within 3 inches (7.5 cm) of the hand currently in contact with the object.

Examples: Reach 10 inches (25 cm) to press the start button on a machine (fixed location): $\mathbf{A_6}$.

With the right hand, reach 16 inches (40 cm) to a thin rod held in the left hand. The new grasping point is 2 inches (5 cm) from the left hand's current grip on the rod: $\mathbf{A_{10}}$.

Hand—Degrees

In addition to linear movements, the Action Distance parameter also covers rotational movements of the hand. This refers to the rotational movement of an empty or loaded hand about the long axis of the forearm. The rotation of the hand would be estimated to the nearest 30 degrees for the 0 and 1 index values or the nearest 60 degrees for the 3 and 6 index values. Estimate the rotation from the thumb knuckle or the base knuckle of the little finger.

$\mathbf{A_0}$ ≤ 30 degrees
$\mathbf{A_1}$ ≤ 60 degrees
$\mathbf{A_3}$ ≤ 120 degrees
$\mathbf{A_6}$ ≤ 180 degrees

Examples: Rotate wearable on arm approximately 90 degrees to read picking location: $\mathbf{A_3}$.
Rotate part about 50 degrees before putting in bin: $\mathbf{A_1}$.

When a linear Action Distance occurs simultaneously with a rotational Action Distance, both index values must be identified and the greater time value or the limiting action is allowed.

Example: Turn a work order 90 degrees upright while moving it 12 inches (30 cm) to a bin:
90 degrees: $\mathbf{A_3}$
12 inches (30 cm): $\mathbf{A_{10}}$
Allow the time for the $\mathbf{A_{10}}$ in the analysis.

Note: The data is for rotation of the hand with or without an object while moving freely through the air. If rotating a dial or other attached object, use the Move Controlled (**M**) parameter in the Controlled Move Sequence Model.

Leg Action Distance

Leg Action Distance applies to Action Distances of the leg or foot. These actions are for displacements of the leg or foot, not for leg and foot actions that transport the body. Leg actions are pivoted at the knee or hip and are measured at the ankle. Measurements are taken to the nearest inch (cm) and the appropriate index value is applied.

This type of Action Distance is commonly limited by movements made by other body members (i.e., fingers and hands). Leg Action Distance values are as follows:

$\mathbf{A_6}$ ≤ 8 inches (20 cm)
$\mathbf{A_{10}}$ ≤ 12 inches (30 cm)
$\mathbf{A_{16}}$ ≤ 18 inches (45 cm)
$\mathbf{A_{24}}$ ≤ 26 inches (65 cm)
$\mathbf{A_{32}}$ > 26 inches (65 cm)

Foot actions move the ball of the foot, with either the heel or the instep acting as a pivot point for the foot. Foot actions are by their nature quite short and the index value for a foot action is 6. An $\mathbf{A_6}$ would be allowed for an Action Distance performed by the foot where the heel is used as the pivot point to locate the foot to a floor actuating device as shown in Figure 4.2.

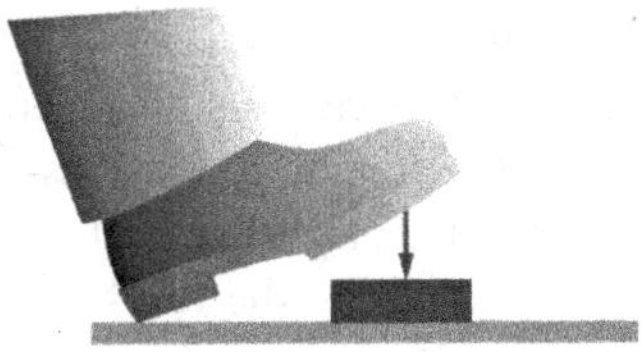

FIGURE 4.2 Action Distance performed by the foot.

Examples: While seated, reach 14 inches (35 cm) to contact a foot-operated pedal: $\mathbf{A_{16}}$.
With both feet on the ground, raise the right foot 24 inches (60 cm) and put it on a pumping device: $\mathbf{A_{24}}$.
Keeping heel on the floor, reach to a sewing machine floor pedal: $\mathbf{A_6}$.

Sometimes there is confusion about when to allow a step as opposed to a leg action. To make this decision, determine the primary purpose of the action. If the primary purpose is to *locate the foot,* such as on a pedal, allow a leg action of the appropriate distance. If the primary purpose is to *locate the body,* such as moving the body to read a gauge located on a machine, allow a step. The trunk of the body is shifted or displaced during a step.

Steps

Steps refer to horizontal displacement of the body. For convenience, they are shown in the Leg Action Distance column.

A_{16} 1 Step

The trunk of the body is shifted or displaced by walking, sidestepping or turning the body around taking one step (the foot hits the floor once).

Example: Take one step to put a document in a bin.

A_{32} 2 Steps

The trunk of the body is shifted or displaced by walking, sidestepping or turning the body around. The number of steps is equal to the number of times the foot hits the floor (twice).

Because of the repetitiveness of the activities that are analyzed with MiniMOST, the need for values for more than two steps seldom arises. If it does occur in practice, the layout should be reviewed and the distance shortened. However, in the rare activities that call for more than two steps, the procedure should be to use the one-step data and apply a partial frequency for the number of steps needed.

Example: Take three steps to get a part

$\mathbf{(A_{16})\ \ B_x\ \ G_x\ \ A_x\ \ B_x\ \ P_x\ \ A_x\ \ (3)}$ **48 TMU**

If walking more than two steps occurs frequently, BasicMOST is usually the preferred system.

Body Assistance to Hand Action

When the operator takes only one step, caution must be taken by the analyst to determine whether to allow the step or to treat it as body assistance. Normally, a step that displaces the body to the side less than 12 inches (30 cm) or turns the body less than 45 degrees for the purposes of extending the reach is considered body assistance. This body assistance does not have to be analyzed separately as a step because the time will be included in the Action Distance for the hand. Longer steps that must be completed prior to the next action in the operation are usually time limiting and must be recorded in the analysis to account for the time.

Body assistance is broken down into two categories:

1. Lateral assistance
2. Rotational assistance

Lateral assistance is the bending or leaning of the trunk in the direction of travel to bring the hand closer to its destination. This tends to shorten the distance the hand must travel. When the operator uses lateral assistance, the net distance is determined by subtracting the lateral assist distance from the total distance moved to determine the index value.

Example: From a standing position, reach 18 inches (45 cm) to a small component in a bin. The shoulder moves 6 inches (15 cm) while reaching for the component. The net distance is 12 inches (30 cm) and the index value is an $\mathbf{A}_{10}$.

The second form of body assistance is rotational assistance. Rotational assistance is the rotating or pivoting of the torso about the vertical axis of the body. The distance moved by the hand as a result of rotational assistance typically follows a 4 to 1 ratio, meaning that the hand travels four times the distance of the body.

Example: An operator places a part and then reaches to the side 16 inches (40 cm). The shoulder moves 2 inches (5 cm) while reaching. The net action distance is 8 inches (20 cm). The rotational distance of 8 inches (2 inches×4) or 20 cm (5 cm×4) is subtracted from the total distance. The index value would be $\mathbf{A}_{6}$.

Note: Body assistance motions should be eliminated, if possible, through method improvements.

Movement of the Hand with Steps

When steps are involved in getting or placing an object, one or both hands usually begin the required movement as the steps are being taken. When the step is complete, the hand will have typically moved 5 inches (12.5 cm) closer to the object; therefore, a 5 inch (12.5 cm) Action Distance is already included in the step data. When an Action Distance for one or two steps is included in a MiniMOST analysis, any additional distance the hand moves after the conclusion of the step is analyzed as a separate hand Action Distance.

Example 1: Take a side step and then reach 9 inches (22.5 cm) over an obstruction to get a part ('reach' begins at the conclusion of the step). To clearly note the activities occurring, the analyst needs to show both Action Distances in two sequence models.

Sidestep: $\mathbf{A_{16}}$
Reach 9 inches (22.5 cm): $\mathbf{A_{10}}$

The analysis would be shown with two sequence models:

$\mathbf{A_{16}\ B_x\ G_x\ A_x\ B_x\ P_x\ A_x}$
$\mathbf{A_{10}\ B_x\ G_x\ A_x\ B_x\ P_x\ A_x}$

Example 2: Grasp a part 25 inches (62.5 cm) from the edge of a table two steps away (reach begins during the second step).

2 steps: $\mathbf{A_{32}}$
20 inches (50 cm) of the reach: $\mathbf{A_{16}}$

$\mathbf{A_{32}\ B_x\ G_x\ A_x\ B_x\ P_x\ A_x}$
$\mathbf{A_{16}\ B_x\ G_x\ A_x\ B_x\ P_x\ A_x}$

Final A

The Final **A** in a sequence model may be used for only two activities:

1. Disengage greater than 5 inches (12.5 cm). This application will be clearer once the section on Gain Control has been reviewed and the rules of Disengage defined.
2. Hand Action Distance to an undefined location for safety purposes or to permit the next activity.

Examples: Place an object on a balance and move hand aside 10 inches (25 cm) to permit the movement of the balance: $\mathbf{A_{10}}$.
Light liquor with match in saucepan and move hand aside 20 inches (50 cm) to avoid burning hand: $\mathbf{A_{16}}$.
After loading part into press, operator retracts hands 8 inches (20 cm) to clear light curtain permitting the press to cycle: $\mathbf{A_6}$.

Body Motion (B)

Body Motion refers primarily to vertical motions of the body to overcome an obstruction or impairment. Body Motion also includes time for Eye Travel.

B_{10} Eye Travel

Eye Travel, when required, may be allowed as a Body Motion. *Eye Travel* is the basic eye motion used to shift the axis of vision from one location to another. Eye Travel rarely occurs as a limiting motion and is allowed only when the next manual motion depends upon its completion. To assign a $\mathbf{B_{10}}$ for Eye Travel, a necessary, recognizable pause must occur. This pause occurs only when the items requiring attention are not within the 'area of normal vision'; that is, the distance between the items is greater than one-quarter the perpendicular distance from the eyes. For objects 16

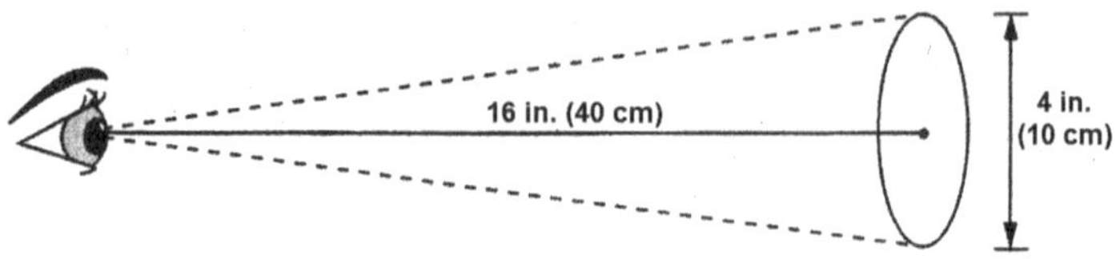

FIGURE 4.3 Area of Normal Vision.

inches (40 cm) from the eyes, the area of normal vision is a circular area 4 inches (10 cm) in diameter (see Figure 4.3).

Example: An operator places two pins, one held in each hand, into an assembly. The pin locations on the assembly are located 16 inches (40 cm) from the operator's eyes. The placement of the pins is 8 inches (20 cm) apart. The operator places the first pin into the designated area on the assembly. Eye Travel is then needed to place the second pin in the second area on the assembly.

B_{32} Bend or Arise

From an erect standing position, the body is lowered to allow the hands to reach below the knees *or* the subsequent return to an upright position. It is not necessary for the hands to actually reach below the knees, only that the body is lowered sufficiently to allow the reach. This value is only for bend *or* arise, not both bend and arise.

Note: Care must be exercised to distinguish between body assistance and Body Motion. Leaning, which does not lower the shoulders enough to permit the hands to reach below the knees, is body assistance. The time for this body movement is covered in the Action Distance parameter by measuring the hand movement. However, when the body is lowered far enough to permit the hands to reach below the knees, Body Motion rather than an Action Distance is allowed.

Often, when a Body Motion occurs, any accompanying Action Distance is completed during the Body Motion and only the Body Motion is allowed. However, there are times when both the Body Motion and the Action Distance must be accounted for. This is because the Action Distance occurs before or after the Body Motion due to some obstruction in the workplace, or it may be necessary to assist an operator in maintaining balance.

Example: Bend to bottom shelf and then reach 10 inches (25 cm) to an object on the back of the shelf:

$\mathbf{A_{10}\ \ B_{32}\ \ G_x\ \ A_x\ \ B_x\ \ P_x\ \ A_x}$

The Action Distance had to take place after the Body Motion was completed.

Example: With a 15 pound (7 kg) object held by both hands, bring the object 12 inches (30 cm) nearer to the body for balance and arise from a bending position. In this case, the Action Distance took place prior to the start of the Body Motion.

$\mathbf{A_x\ \ B_x\ \ G_x\ \ A_{10}\ \ B_{32}\ \ P_x\ \ A_x}$

When these types of combinations occur, care must be taken to allow only the length of Action Distance that *must* be done before or after the Body Motion. The time for Action Distances performed during a Body Motion are considered a limited activity and not included in the total time.

B_{32} Sit

The value for Sit includes lowering the body to a seated position on a chair and does not include any adjustments to the chair. The Sit value is not used very much with MiniMOST applications. If sitting occurs frequently, the BasicMOST System most likely should be used.

Example: Sit on a stool in work area.

B_{42} Stand

The value for Stand includes raising the body out of a chair to an erect position without any adjustments to the chair. The Stand value is not used very much with MiniMOST applications. If standing occurs frequently, the BasicMOST System most likely should be used.

Examples: Stand from a bench.

Gain Control (G)

The Gain Control parameter covers all motions of the fingers, hand or foot required to obtain control of one or more objects. This includes making contact with an object and closing the hand on an object. The index value selected is based on the motions necessary for obtaining control due to the nature of the object, its surroundings and its size. The **G** parameter includes time for up to 1 inch (2.5 cm) hand movement prior to or in combination with the grasp of one or more objects.

When several objects are grouped together or arranged in such a way that they can be picked up as one object such as a stack of tiles or tray of washers, one Gain Control value will be applied. Picking up several objects separately requires multiple Action Distances and Gain Control values.

G_0 Sweep

The object is obtained without interrupting the flow of the Action Distance—the hand does not stop. Closing the fingers around the object occurs internally to the Action Distance; therefore, no hesitation or pause is seen. An object obtained in this manner is of minimal weight and size and can be located by itself on the surface with no interference at the grasping point.

Example: With an open hand, sweep the scrap wire remains from a work station.

G_3 Contact

Control is gained simply by touching the object with the fingers, hand or foot as shown in Figure 4.4. In the illustration, the hand is brought to the ruler, rests on the ruler and ultimately pulls the ruler away with no closing of the fingers. Also, when the hand is already in contact with the object, but one or more fingers close around

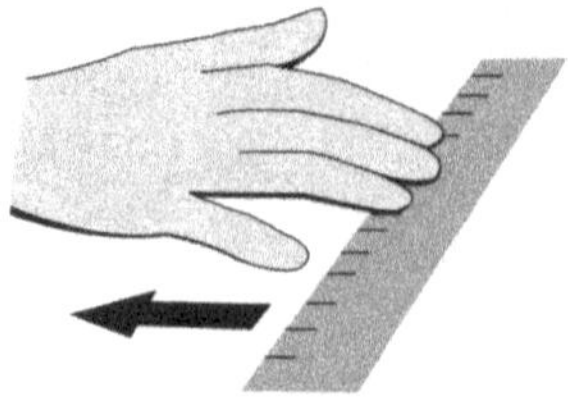

FIGURE 4.4 Example of Contact motion.

the object to fully make contact, use a G_3 Contact grasp. In practical work situations, however, this type of grasp is often limited by an Action Distance movement. Analysts should watch for these types of limited actions when analyzing work.

Examples: Contact power button, light switch or foot pedal.
Close thumb on dollar bill that has been pulled from the cash register.

G_6 Grasp

This is the most common case of gaining control. A Grasp is a simple pickup with the fingers closing around the object prior to the next action. The object can be any size, can be lying close against a flat surface or by itself.

The value for Grasp allows the last 1 inch (2.5 cm) of hand motion to the object, closing the fingers on the object and, when required, a short scraping or digging in of the fingers or a slight recoil of no more than 1 inch (2.5 cm). A Grasp motion is illustrated in Figure 4.5.

Examples: Grasp light switch component.
Grasp soft drink can.
Using both hands, grasp light box from conveyor.
Grasp test tube.

G_6 Regrasp

Regrasp differs from the other forms of Gain Control in that the object is already controlled by the operator but the regrasp motion is short and the grip is shifted to improve or change the control. While control of the object is maintained, the grasp is shifted slightly for the purpose of improving control or bringing the object into position for use. Regrasp is characterized by two or three short finger actions and can occur repetitively when a major repositioning is required. Shifting one finger with one motion to a new location on an object already under control is *not* a Regrasp, but qualifies as a G_3, Contact.

Because of restrictions that may occur with the initial gain control, it is common for a Regrasp to immediately follow a Grasp. Therefore, analysts should watch for these adjustments to train their eyes to detect this motion.

Example: Grasp stylus from holder and then use fingers to regrasp the stylus to prepare for use.

A_x B_x G_6 A_x B_x P_x A_x 6 TMU
A_x B_x G_6 A_x B_x P_x A_x 6 TMU

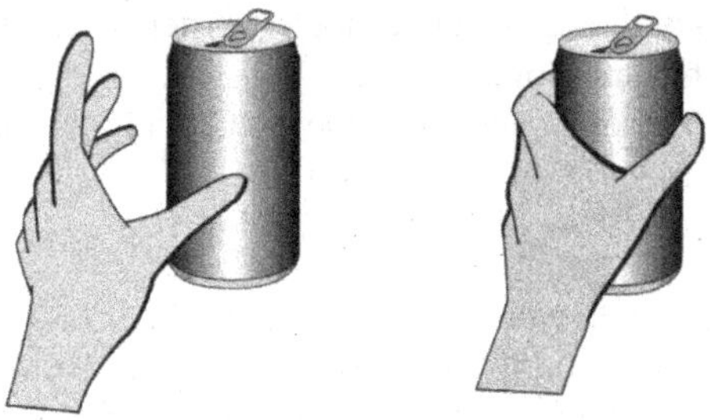

FIGURE 4.5 Example of Grasp motion.

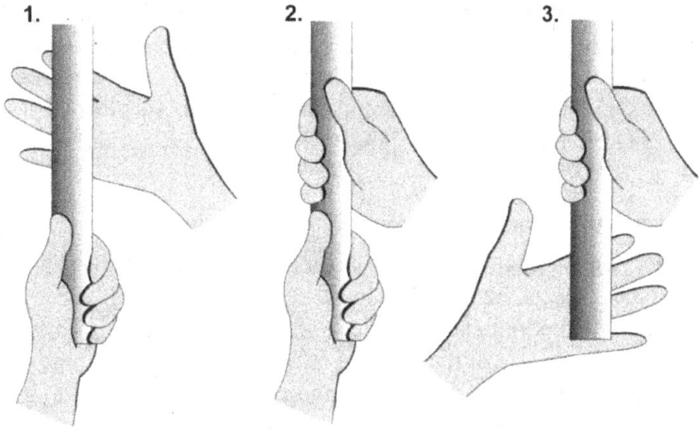

FIGURE 4.6 Series of hand motions used in a Transfer Grasp.

Regrasps can frequently occur during an Action Distance while transporting the object and are normally limited. The action of the regrasp should be documented and shown as simultaneous to the Action Distance.

Example: Grasp stylus from holder and use fingers to regrasp the stylus while moving the stylus to a screen.

$$\mathbf{A_x \; B_x \; G_6 \; A_x \; B_x \; P_x \; A_x \quad 6\ TMU}$$
$$\mathbf{A_x \; B_x \; \textcircled{G_6} \; A_x \; B_x \; P_x \; A_x \quad 0\ TMU}$$

G_{10} Transfer

Control of the object is exchanged from one hand to the other. This includes the brief holding period required by the fingers of both hands before release of the giving hand occurs. This hesitation or pause highlights a Transfer grasp.

The layout should be arranged in a way that Transfer grasps are reduced to a minimum. Figure 4.6 shows the series of hand motions during a Transfer grasp.

A Transfer takes place only when one hand is closed on the object and then the other hand opened. For instance, picking up a nut lying in the palm of the left hand by closing the fingers of the right hand is not a transfer. There was no opening of the left hand to relinquish control of the object.

Example: Transfer a component from the right hand to the left hand.

G_{10} Select

Normally, Select occurs when the object is in a closed or limited area and the grasp is accomplished after overcoming some restriction or impairment encountered at the grasping point due to the surroundings. Short motions are involved to locate the fingers around an object jumbled with other objects or to 'roll out' a cylindrical object to separate it from others. The objects may be jumbled, or if cylindrical, restricted on the bottom and one side.

A restriction or impairment is necessary for a Gain Control to have the Select value. Getting a baseball from a basket of balls does not require a Select Gain Control. Here the object is quite large, there is no restriction or hesitation and Gain Control can be accomplished with a $\mathbf{G_6}$ Grasp.

Examples: Obtain a 1 1/2 inch (40 mm) nut from a bin filled with nuts.
Get a 1/4 inch (6 mm) dowel from a neat row of other dowels.

G_{16} Select–Small

The criteria for Select–Small is the same as for Select in that the object is restricted or impaired by other objects. As with Select, the objects may be jumbled, or if cylindrical, restricted on the bottom and one side. The distinguishing factor between Select and Select–Small is the size of the object. For Select–Small to be chosen, the size of the object needs to be very small and care and precision are needed to gain control. A hesitation or pause will be noticed for the Select–Small value.

Examples: Get small flat washer located in bin with others.
Get 3/32 inch (2 mm) diameter plastic tube from stack of tubes.

G_{16} Disengage

Disengage is the application of muscular force needed to free the object from its surroundings. This parameter variant is characterized by the application of pressure to overcome resistance followed by the sudden movement and recoil of the object greater than 1 inch (2.5 cm) up to and including 5 inches (12.5 cm). Recoil of the object must follow an unrestricted path through the air (not to be confused with unseating a lever, crank or other attached or restricted device).

The Disengage index value includes time to bring the hand the last 1 inch (2.5 cm) to the object, gain control of the object, build up muscular force to free the object and recover from the recoil when the object breaks free. Occasionally, additional rocking or twisting motions may be required to free the object from its surroundings. These additional activities are not included in the Disengage value and must be separately analyzed, usually with the Controlled Move Sequence Model.

Examples: Remove tightly fitted cap from cylinder.
Remove electric plug from socket.

Should the recoil in freeing an object exceed 5 inches (12.5 cm), consider whether method improvements might reduce the recoil. A recoil of more than 5 inches (12.5 cm) suggests that BasicMOST would be the preferred work measurement system. If MiniMOST is still being used in the analysis, the additional recoil distance of greater than 5 inches (12.5 cm) can be measured and included in the final **A** of the sequence model.

G_{24} Collect

Gaining control of more than one object may be accomplished with $\mathbf{G_{24}}$, Collect. The objects may be jumbled together in a pile or lying closely together on a surface. If jumbled, control of several objects is achieved by reaching into the pile with the hand and bringing up a handful. When lying on a surface, the objects may be brought together with the hand and fingers and picked up as one object. The index value for Collect includes the time to grasp the desired number of objects, move to close the hand around the objects and regrasp for improved control.

Examples: Get a handful of adapters from a bin.
Collect two rubber plugs lying on the top of a work surface.

Consideration of Effective Net Weight

The preceding discussions of index value selection for Gain Control (G) are based on an object of nominal weight not more than 2.5 pounds (~1 kg). For objects with an Effective Net Weight (ENW) of greater than 2.5–10 pounds (1–5 kg), the next higher index value should be used. This rule applies to all Gain Control activities *except* Grasp ($\mathbf{G_6}$) for which no consideration of weight is necessary and no adjustment to the index value is necessary. The determination of the ENW of an object depends upon the way the object is being moved.

Weight Consideration for the General Move Sequence Model. If an object is being moved freely through the air, the actual weight of the object is allocated to each body member performing the work based on the portion of the weight supported by that body member. If the object is being moved by one hand, the actual weight of the object is the ENW. If two hands are moving the object, the ENW is equal to the weight of the object divided by two if the weight is evenly distributed. Examples of ENW for a lead pipe lying against a neat row of stacked pipes are shown below.

Weight of Object	Number of Hands	ENW*
8 pounds (4 kg)	One	8/1 = 8 pounds (4 kg)
10 pounds (5 kg)	Two	10/2 = 5 pounds (2 kg)
5 pounds (2 kg)	Two	5/2 = 2.5 pounds (1 kg)

* Assuming that the weight is distributed equally to each hand.

Therefore, when moving an object through space, the actual weight *per hand* must exceed 2.5 pounds (1 kg) before an adjustment to the G parameter is considered for weight purposes. Occasionally an operator will support most of an object's weight in one hand while the other hand guides the object to its destination. This is often seen in the use of power tools. An operator may initially grasp a heavy power tool with the left hand and transfer it to the right hand. The right hand then bears the weight of the tool, but the left hand may guide the tool to the proper location for use as the right hand activates the tool. The ENW is calculated for one hand only—in this case, it is the right hand which is bearing the weight of the tool.

To measure the handling of an object with an ENW of more than 10 pounds (5 kg), BasicMOST is the preferred analysis system.

Placement (P)

Placement refers to actions occurring at the final stage of an object's displacement to align, orient and/or engage the object with another. The time for a 1 inch (2.5 cm) move prior to making contact with a surface in combination with the placement of the object is included in the placement value except in the values for Drop and Hold.

The Placement parameter includes sub-activities for placing objects to a general location and for a more precise placement.

P_0 Hold/Indefinite Location

A part is retained in space where its location is unimportant. This can occur as a preliminary step to another motion or to clear the part from an area. In many cases, this $\mathbf{P_0}$ to an indefinite location is followed by a pause or waiting time.

Examples: Pick up part with left hand and hold while positioning a part with the right hand.

Move a part clear of a machine and hold prior to operating the machine.

P_0 Drop

No deceleration or placing motions occur; the object is released with the hand in motion and the hand continues in motion into the next action.

Examples: Drop item in chute and continue to grasp the next item.

Drop scrap in tote and continue to grasp tool.

P_3 Toss

The object is tossed or thrown with the hand stopping or reversing direction prior to the next action.

Examples: Toss scrap pieces into scrap bin.

Toss finished assembly onto conveyor.

Note: The distinction between Drop and Toss is whether the hand must stop or reverse directions. For Toss, the hand stops or reverses direction. For Drop, the hand continues.

P_3 Set and Retain

The object is moved to a location, usually on a surface, and remains under control for subsequent work.

Note: If the retained object was subsequently released with no further Placement or Controlled Move of the object, the correct value would be $\mathbf{P_6}$ (Set Aside) to account for the release.

Examples: Set base on bench and hold while assembling an additional part.
Set bin on workstation and hold in place with left hand while grasping contents inside bin with right hand.

P_6 Set Aside

The object is moved to a stop, a general location or an exact location with a radial tolerance greater than 3/8 inch (10 mm). Precise and predetermined placement is not required and the object is typically released after the placing action. The value for $\mathbf{P_6}$, Set Aside also includes time to move the object to a preliminary location and then slide the hand or object up to 1 inch (2.5 cm) to a secondary location. The object may be retained if setting the object aside and then sliding it 1 inch (2.5 cm). For actions where the object must be slid more than 1 inch (2.5 cm) after placement, the object should be placed with the appropriate **P** value and any sliding motion analyzed with the **M** parameter in the Controlled Move Sequence Model.

Examples: Set aside completed part on conveyor.
Set part aside and slide 1 inch (2.5 cm) to a stop.
Put stylus to tablet in a general location and make a checkmark.

Precise Placement

Precise placement involves locating an object or point on an object to a *precise* and *predetermined* destination. These placement values include all incremental motions necessary to locate an object in a predetermined destination and to put the object in or on the destination. The time for alignment (linear and tilting), orientation, contact at the destination and insertion is also included in the precise placement values. Alignment is always present in these precise placements.

Alignment includes linear adjustments to bring the object to the desired location with the required accuracy plus any tilting of the object that may be required. Figure 4.7 provides a summary of the rules for precise placement and the subsequent adjustments to the index values.

P_6 Put

An object or point on an object is put to a predetermined location. Time is included for linear and tilting alignment and for making contact with the destination. Insertions of 1/8 inch (3 mm) or less are considered a part of making contact. There is no significant orientation* required because the object can be placed in more than 10 ways about the contact axis. The tolerance does not demand a high degree of accuracy. A radial clearance from 5/32 inch (4 mm) to 3/8 inch (10 mm) is present with this placement. Control of the object may be retained or relinquished.

Examples: Put a bolt into a cavity and insert 1/8 inch (3 mm) with a tolerance of 1/4 inch (6 mm).
Put a round part into a hole 1/16 inch (2 mm) deep with a loose tolerance.

---

* *Orientation* refers to the rotation of the object about its axis, alignment and contact in order to properly engage the object with another. It takes into account the shape of the object at the surface of contact or insertion.

Precise Placement			
	P_6 - Put	P_{10} - Place	P_{16} - Position
Orientation	Not significant	≤ 90 degrees	> 90 and ≤ 180 degrees
	More than 10 ways	2 to 10 ways	Only 1 way
Accuracy	Radial tolerance of 5/32 to 3/8 in. (4 to 10 mm)		
Insertion	Up to 1/8 in. (3 mm)		
	Additional Adjustments		
Accuracy < 5/32 in. (4 mm)	+1 index value		
Insertion > 1/8 in. (3 mm) up to 1 in. (2.5 cm)	+1 index value		
Difficult Handling	+1 index value		
Binding	+2 index values		
Applying Pressure	+2 index values		

FIGURE 4.7 Precise placement summary.

P_{10} Place with Some Orientation

An object or a point on the object is placed at a predetermined location on a surface. In addition to allowing time for a simple placement of the object, time is allowed for a rotational (≤ 90 degrees) adjustment of the object about the contact axis. The criterion is that the object is rotated about its axis regardless of the manual motion made to accomplish that rotation. Objects needing this type of classification are those that could be placed in 2 to 10 possible ways about the axis and that have not been pre-oriented. That is, the object must be rotated at the point of placement. Again, an insertion of up to 1/8 inch (3 mm) is considered part of bringing the object into contact and tolerances are loose enough with a radial clearance from 5/32 inch (4 mm) to 3/8 inch (10 mm) so that a high degree of accuracy is not required. Release of the object may or may not occur.

Examples: Place a metal band through a slot in a bracket with an insertion of 1/8 inch (3 mm).

Place a square block into a square hole with a 3/8 inch (10 mm) tolerance and an insertion of no more than 1/8 inch (3 mm).

P_{16} Position with Complete Orientation

An object or a point on the object is positioned at a predetermined location on the surface. In addition to allowing time for a simple position, time is allowed for a rotational adjustment (> 90 degrees, ≤ 180 degrees) of the object about the contact axis. Objects needing this type of classification are those that can be placed in one and only one possible way around their axis and that have not been pre-oriented. The rotational adjustment must occur during the Placement activity. Again, the manual motion made to accomplish the rotation is not the deciding factor; it is essential only that the object require rotation. Insertions of up to 1/8 inch (3 mm) are considered part of bringing the object to the surface. This placement does not require a high degree of accuracy with a radial clearance from 5/32 inch (4 mm) to 3/8 inch (10 mm). Release of the object may or may not occur.

As noted previously, the orientation values apply only at the point of placement. In many cases, the object may be pre-positioned or pre-oriented either prior to or during the Action Distance. This pre-orientation normally reduces the orientation required at the point of placement. As a result, the Placement index value is applied accordingly.

Examples: Position USB cable into port.
Position key in lock.

Adjustments to the Values for Precise Placement

Accuracy

The previous discussions define the precise placement of an object or point on the object that can be accomplished without slowdown, tension or corrective motions because the tolerances involved are loose. However, if an object's positioning demands a more exact placement, the next higher index value should be selected (see Precise Placement summary in Figure 4.7). The increased precision may be observed as a sequential adjustment or as light pressure. The tolerance associated with this accurate positioning is a radial clearance of less than 5/32 inch (4 mm) at the plane of initial insertion or at the point of contact if no insertion occurs.

Example: Put computer part into tight slot: $\mathbf{P}_{10}$.

$\mathbf{P}_6$ No orientation
- Go to next higher index value for

$\mathbf{P}_{10}$ Accuracy

Insertion

The three classes of precise placement previously described all allow an insertion of the object up to 1/8 inch (3 mm). If the object or point on the object must be inserted more than 1/8 inch (3 mm), the next higher index value should be selected. By doing so, an insertion of up to 1 inch (2.5 cm) is allowed.

Example: Put cap on end of component: $\mathbf{P}_{16}$.

$\mathbf{P}_6$ No orientation
- Go to next higher index value for

$\mathbf{P}_{10}$ Accuracy
- Go to next higher index value for

$\mathbf{P}_{16}$ Insertion up to 1 inch (2.5 cm)
For an insertion greater than 1 inch (2.5 cm), use an additional Controlled Move Sequence Model (**M** parameter) to account for the remaining distance.

Example: Place object with some orientation with an insertion of 3 inches (7.5 cm): $\mathbf{P}_{16}$ and $\mathbf{M}_6$.

$\mathbf{P_{10}}$ Some orientation

- Go to next higher index value for

$\mathbf{P_{16}}$ Insertion up to 1 inch (2.5 cm)

- Use a Controlled Move Sequence Model for insertion greater than 1 inch (2.5 cm)

$\mathbf{M_6}$ Moved Controlled value for additional insertion (see **M** parameter values under Figure 4.8)

Difficult to Handle

At times during the placing of an object, a regrasp, hesitation or pause is required because the object is difficult to handle or hard to control. This can be due to the nature of the object (flexible items such as yarn, paper and cloth), the type of grip that has to be used or to distance, if the object is grasped a distance from the point of placement. In other cases, the handling difficulty is observed as a shifting of the grip during placing. If difficulty in handling is observed, the next higher index value is selected to account for this additional adjusting motion.

Example: Put small soldering iron to electronic component board: $\mathbf{P_{16}}$.

$\mathbf{P_6}$ Put soldering iron to board

- Go to next higher index value for

$\mathbf{P_{10}}$ Accuracy

- Go to next higher index value for

$\mathbf{P_{16}}$ Difficult to handle: Soldering iron is at considerable distance from point of contact and must be regrasped to accurately complete the process.

Binding

At times when an object is being inserted, a part of the object will catch or snag. This will result in the application of muscular force to overcome the obstruction. This can be seen as the snapping action to seat an object during the placing activity. Little or no movement of the object occurs as the bind is freed. For each observation of a bind, go up *two* index values.

Binding will occur only when dealing with an insertion. This allowance for binding should be applied after all others.

Example: Replace rubber stopper in bottle—one occurrence of binding, insert 1 inch (2.5 cm): $\mathbf{P_{32}}$.

$\mathbf{P_6}$ No orientation

- Go to the next higher index value for

$\mathbf{P_{10}}$ Accuracy

- Go to next higher index value for

$\mathbf{P_{16}}$ Insertion up to 1 inch (2.5 cm)

- Go up two index values for

$\mathbf{P_{32}}$ Binding

Apply Pressure

In some instances, an application of muscular force may occur without insertion. If the application of force occurs as part of placement without insertion, go up to *two* index values for each application of pressure.

Example: Firmly push a drill to drill a hole: $\mathbf{P_{24}}$.

$\mathbf{P_6}$ Put drill
- Go to next higher index value for

$\mathbf{P_{10}}$ Accuracy
- Go up two index values for

$\mathbf{P_{24}}$ Apply pressure

Precise Placement to a Point in Space

On occasion, an object or point on an object is located in a precise manner to a predetermined point in space with no contact made with any surface. When this occurs, allow the precise placement time for the appropriate accuracy and handling. Do not allow any orientation, regardless of whether orientation occurs. All considerations except orientation are the same for placement where contact is made.

Example: Locate an oil can spout over a small turning shaft: $\mathbf{P_{10}}$.

$\mathbf{P_6}$ Locate oil can
- Go to next higher index value for

$\mathbf{P_{10}}$ Accuracy

Is a Precise Placement Value Required?

A precise placement value is required when the object or point is brought into a *precise* and *predetermined* relationship with another object or point. The placing actions require high care and visual attention. However, not all placing activities requiring high care and visual attention are precise placements. For instance, placing an egg in a wire basket requires high care and visual attention to avoid breaking the egg, but a Set Aside value is assigned because the egg may be placed anywhere in the basket.

Parameter Frequencies

Frequency

Frequency is an entire sequence occurring more than once. If an activity occurs more than once, the frequency will be specified in a MiniMOST analysis and the time for the activity multiplied by the frequency indicated. The time calculation for a frequency activity is as follows:

1. Add all index values for the entire sequence model.
2. Multiply this total by the number of frequency occurrences.

Since the multiplier for MiniMOST is one (1), the total after Step 2 is the TMU total.

Example: Select a small part from a parts bin 8 inches (20 cm) away and put in bag 4 inches (10 cm) away. Continue until 15 parts are in the bag.

A_6 B_0 G_{16} A_3 B_0 P_6 A_0 15

GET	A_6	Reach to part
	B_0	No body motion
	G_{16}	Select small part
PUT	A_3	Move to place part
	B_0	No body motion
	P_6	Put part in bag
RETURN	A_0	No return

The time calculation steps for the example above are as follows:

1. $6+16+3+6=31$
2. $31\times15=465$ TMU

Partial Frequency

Partial frequency is one or more parameters within a sequence model occurring more than once such as individually placing several objects from a handful. This activity is shown in the sequence model by placing parentheses around the parameters that are repeated and documenting the number of occurrences as a partial frequency in a MiniMOST analysis, also within parentheses. The time calculation for partial frequency is shown below:

1. Add all index values for the parameters within parentheses.
2. Multiply this value by the number of occurrences (the number noted as a partial frequency at the end of the sequence model).
3. Add this total to the remaining parameter index values for the total TMU.

Example: Collect three washers from a bin 4 inches (10 cm) away, hold and then put each washer onto three bolts 8 inches (20 cm) away one at a time. The bolts are 1 inch (2.5 cm) apart.

A_3 B_0 G_{24} A_6 B_0 (P_6) A_0 (3)

GET	A_3	Reach to washers
	B_0	No body motion
	G_{24}	Collect washers
PUT	A_6	Move to put washers
	B_0	No body motion
	P_6	Put washers on bolt
RETURN	A_0	No return

As indicated, only one parameter in the example is repeated three times. The operator collects the washers (**G_{24}**) and then puts each washer over a bolt (**P_6**) one at a time.
The time calculation is:

1. 6×3=18
2. 3+24+6+18=51 TMU

Note: More than one set of parentheses may be used in a sequence model provided the same frequency applies to all parameters within parentheses.

Frequency with a Partial Frequency

It is possible that a partial frequency activity and frequency can occur together. Using the example above where an associate collects three washers and puts each one over a bolt, but now the whole activity to collect and put the washers is completed four times, the following analysis would be used:

A_3 B_0 G_{24} A_6 B_0 (P_6) A_0 (3) 4

The calculation for this type of activity is as follows:

1. Add all index values for the parameters within parentheses.
2. Multiply this value by the number of partial frequency occurrences (the number in parentheses noting partial frequency).
3. Add this total to the remaining parameter index values.
4. Multiple this total by the activity frequency (number at end of sequence model, not in parentheses) for the total TMU.
   1. (**P_6**)=6
   2. 6×3=18
   3. 3+24+6+18=51
   4. 51×4=204 TMU

Note: A frequency or partial frequency can be a whole number, decimal or fraction. For example, a box of parts is put on a conveyor each time it gets filled. The box holds 10 parts. Moving the box then only happens once out of 10 times.

Writing Method Descriptions

One of the advantages of MOST is using a standard sequence model to accurately determine time values. Another advantage is the method description that accompanies each sequence model can be written to consistently and clearly define the activity. It is recommended that the analyst follow a prescribed sentence structure and use consistent wording when writing method descriptions. This will provide other analysts and future readers of the analysis a clear understanding of the process. This information is especially important in MiniMOST because of the level of detail needed.

Below is the recommended format for a clear and concise General Move method description. Additional words for Action Distance, Body Motion or Precise Placement modifiers are optional and may be used to enhance the method description. Often, the Action Distance is stated in the method description for clarification because of the level of detail needed in MiniMOST.

The recommended sentence structure for General Move is as follows:

Gain Control   Object   ⟨From Location⟩   Placement   To Location

⟨If the From Location is apparent, it is not necessary to indicate it in the method description.⟩

General Move Examples

1. Reach 8 inches (20 cm) to grasp an outlet cover lying on the table, move 10 inches (25 cm) and set aside on the table.

   Grasp outlet cover 8 inches (20 cm) away and set aside to table 10 inches (25 cm)

   $\mathbf{A_6 \quad B_0 \quad G_6 \quad A_{10} \quad B_0 \quad P_6 \quad A_0}$ **28 TMU**

2. Reach 5 inches (12.5 cm) to a rod, gain control and put in a bin 16 inches (40 cm) away, regrasping it while moving.

   Grasp rod 5 inches (12.5 cm) away and put in bin 16 inches (40 cm)

   $\mathbf{A_6 \quad B_0 \quad G_6 \quad A_{16} \quad B_0 \quad P_6 \quad A_0}$ **34 TMU**

   Note: The regrasp is limited by the Action Distance to move the rod.

3. An assembly worker reaches 10 inches (25 cm) to collect two washers jumbled with other washers. The worker then moves the washers 8 inches (20 cm) and puts them onto two bolts with 1/8 inch (3 mm) radial clearance. The second bolt is 5 inches (12.5 cm) from the first and the washer is moved down 1/2 inch (12 mm) and released.

   Collect 2 washers 10 inches (25 cm) from bin and put to two bolts with accuracy and insertion

   $\mathbf{A_{10} \quad B_0 \quad G_{24} \quad (A_6\, B_0\, P_{16}) \quad A_0 \quad (2)}$ **78 TMU**

4. A staff member in a post office picks up a stack of bundled mail 3 inches (7.5 cm) away and sets it aside in a mail slot 6 inches (15 cm) away.

   Grasp mail 3 inches (7.5 cm) and sets aside 6 inches (15 cm)

   $\mathbf{A_3 \quad B_0 \quad G_6 \quad A_6 \quad B_0 \quad P_6 \quad A_0}$ **21 TMU**

5. An accounts payable clerk tosses a ball of paper 10 inches (25 cm) into a trash can.

   Toss paper 10 inches (25 cm) into trash can

   $\mathbf{A_0 \quad B_0 \quad G_0 \quad A_{10} \quad B_0 \quad P_3 \quad A_0}$ **13 TMU**

THE CONTROLLED MOVE SEQUENCE MODEL

Controlled Move describes the manual displacement of an object over a controlled path. That is, movement is restricted in at least one direction by contact with or an attachment to another object or the nature of the work requires the object be deliberately moved on a specific path. A Controlled Move is performed under one of three conditions:

1. The object or device is restrained by its attachment to another object such as a button, lever, door or crank.
2. It is controlled during the move by the contact it makes with the surface of another object, such as pushing a box across a table; or
3. The object must be moved on a controlled path to accomplish the activity such as folding a cloth, coiling a rope, winding a spool, moving a balanced item or to avoid a hazard, such as electricity, sharp edges or running machinery.

Similar to General Move, Controlled Move follows a fixed sequence of sub-activities identified by the following steps:

1. *Reach* with one or two hands a distance to one or more objects, either directly or in conjunction with body motions.
2. *Gain* manual *control* of the object.
3. *Move* the object over a controlled path.
4. Allow time for a *process* to occur.
5. *Align* the object following the Move Controlled or at the conclusion of the Process Time.
6. *Return* the hand to an undefined location.

These six sub-activities form the basis for the activity sequence describing the manual displacement of one or more objects over a controlled path.

THE SEQUENCE MODEL

The sequence model is a series of letters or parameters representing the various sub-activities of Controlled Move and is listed below. The **A**, **B** and **G** parameters shown below were introduced with the General Move Sequence Model. The three new parameters of **M**, **X** and **I** will be presented in this section.

A B G M X I A

where:
A = Action Distance
B = Body Motion
G = Gain Control
M = Move Controlled
X = Process Time
I = Alignment

Phases of the Sequence Model

Similar to the General Move Sequence Model, object movement within the Controlled Move Sequence Model is broken down into three phases.

Get	Move or Actuate	Return
A B G	M X I	A

The Get and Return phases of Controlled Move carry the same parameters as in the General Move Sequence Model and therefore describe the same sub-activities. The fundamental difference between these two sequence models is the activity immediately following the **G** parameter. This Move or Actuate phase describes the actions either to simply *move* an object over a controlled path or to *actuate* a control device—often to initiate a machine process. Normally, 'Move' implies that the **M** and possibly the **I** parameter of the sequence model is involved. A Move, for example, would occur when opening a tool cabinet door or sliding a box across a table. 'Actuate' usually applies to situations involving the **M** and **X** parameters. Engaging the clutch on a machine or flipping an electrical switch to start a process are examples of Actuate. For either situation of Move or Actuate, any or all of the parameters in the sequence model could be used.

MiniMOST System			Controlled Move					ABGMXIA	
	M Move Controlled					**X Process Time**	**I Alignment**		
Index x 1	Push / Pull / Turn			Crank			To Point or Line		Index x 1
	Hand		Foot or Leg	≤ 5 in. (12.5 cm)	≤ 20 in. (50 cm)	TMU	Within Area of Normal Vision	Outside Area of Normal Vision	
	Inches (cm)	Degrees	Inches (cm)	Revolutions					
1	-	-	-	-	-	1.7	-	-	**1**
3	≤ 1 (2.5) Button	-	-	-	-	4.2	-	-	**3**
6	≤ 4 (10)	≤ 90	-	-	-	7.7	Check or Inspect 1 Point	-	**6**
10	≤ 10 (25)	≤ 180	≤ 10 (25)	-	-	12.6	1 Point Accurate 2 Points	-	**10**
16	≤ 18 (45) Seat or Unseat		≤ 16 (40) Foot with Pressure	1	-	19.6	2 Points Accurate	Check or Inspect	**16**
24	≤ 30 (75)		≤ 22 (55)	-	1	27.7		2 Points	**24**
32			≤ 30 (75)	2	-	36.6		2 Points Accurate	**32**
42				3	2	47.6			**42**
54				4	3	60.1			**54**

FIGURE 4.8 Controlled Move Sequence Model data card.

Parameter Definitions

The three new parameters introduced in Controlled Move are shown on the Controlled Move data card in Figure 4.8. The **A, B** and **G** parameters were discussed with the General Move Sequence Model and remain unchanged.

M Move Controlled

This parameter is used to analyze all manually guided movements or actions of an object over a controlled path.

X Process Time

This parameter is used to account for the time for work controlled by electronic or mechanical devices or machines, not by manual actions.

I Alignment

This parameter is used to analyze manual actions following the Move Controlled or at the conclusion of Process Time to achieve the alignment of objects.

Parameter Indexing

Move Controlled (M)

Move Controlled covers all manually guided movements or actions of an object over a controlled path. That is, movement of the object is restricted in at least one direction by contact with or attachment to another object. Index values for the **M** parameter are listed under two separate categories on the Controlled Move data card. The most frequently occurring parameter variants of Move Controlled (**M**) are under the Push/Pull/Turn heading. The Crank category applies to a special type of Controlled Move dealing with cranks, handwheels or other devices requiring a circular cranking motion.

Push/Pull/Turn

The object or device may be hinged or pivoted at some point (e.g., door, lever or dial) or restricted due to guides, slots, friction from surface or restricted by other special circumstances requiring movement over a controlled path such as using optical scanning devices or restricted by other components of the surroundings as described by the following **M** parameter variants.

Hand–Inches (cm)

The index values for Hand–Inches movements as part of Move Controlled are based on the displacement of an object over a controlled path using the hand or fingers. Distance is measured to the nearest 1 inch (2.5 cm) as described in the Action Distance section. As with all index values, the value is read 'up to and including.' If the index value is $\mathbf{M_6}$, then the distance measured is greater than 1 inch (2.5 cm) and less than or equal to 4 inches (10 cm).

M_3 $\leq$ 1 inch (2.5 cm)
M_6 $\leq$ 4 inches (10 cm)
M_{10} $\leq$ 10 inches (25 cm)
M_{16} $\leq$ 18 inches (45 cm)
M_{24} $\leq$ 30 inches (75 cm)

Examples: Push connector 3/4 inch (20 mm) closer to part: **M_3**.
Move lever on machine 5 inches (12.5 cm): **M_{10}**.

M_3 Button

A button is actuated by a short pressing action of the fingers, hand or foot.
Examples: Press power button on computer.
Press the 'Enter' button to initiate process time.

M_{16} Seat or Unseat

This parameter variant covers the application of muscular force with little or no movement to 'seat' or 'unseat' an object or, if necessary, the short manual actions employed to latch or unlatch the object. In the *unseating* of an object, the recoil must follow a restricted path (not to be confused with a **G_{16}** Disengage, which allows for the recoil of an object freely through the air).
Examples: Unseat lever before pulling it down.
Push stapler to staple pages together.

Hand–Degrees

The object is displaced over a controlled path with the hand rotating greater than 15 degrees about the long axis of the forearm. Rotations are estimated using the thumb knuckle or the base knuckle of the little finger as a reference point and the appropriate index value selected. Measure the distance the hand or fingers move to rotate the device and select the proper value from the distance column. If a rotation less than 15 degrees occurs, it should be analyzed as a linear movement using the Move Controlled Hand–Inches data. Index values are read 'up to and including.' If the index value is **M_6**, then the distance measured is greater than 15 degrees but less than or equal to 90 degrees.

M_6 $\leq$ 90 degrees
M_{10} $\leq$ 180 degrees

Examples: Turn dial on machinery approximately 120 degrees: **M_{10}**.
Fasten screw with screwdriver with one turn less than 90 degrees (performed as a wrist turn): **M_6**.

Foot or Leg Motion

Object is displaced over a controlled path using the *foot* or *leg*. Distance is measured as discussed in the Action Distance section. Leg distances are measured at the ankle. Foot distances are always less than 10 inches (25 cm). If the index value is **M_{16}**, then the object is displaced with the leg or foot over a controlled path greater than 10 inches (25 cm) but less than or equal to 16 inches (40 cm).

$\mathbf{M_{10}}$ ≤ 10 inches (25 cm)
$\mathbf{M_{16}}$ ≤ 16 inches (40 cm)
$\mathbf{M_{24}}$ ≤ 22 inches (55 cm)
$\mathbf{M_{32}}$ ≤ 30 inches (75 cm)

Examples: Depress a machine pedal 12 inches (30 cm) with the foot: $\mathbf{M_{16}}$.
Depress a clutch pedal 8 inches (20 cm). Action hinged at hip: $\mathbf{M_{10}}$.

M_{16} Foot with Pressure

A foot motion when muscular force is needed to overcome friction or resistance due to the nature of the surroundings. This includes time to overcome the resistance and to complete the motion.

Note: There is no index value provided for overcoming resistance when the leg is used. This is because of the strength of the leg. The leg commonly exerts great force and does not require added time to overcome the resistance normally encountered in industrial settings. If great strain is observed, consider modifying the process or using BasicMOST.

Example: Push foot pedal with 35 pounds (16 kg) resistance.

Crank

With the forearm pivoting at the elbow, the object or device is moved in a circular or nearly circular path by the fingers, hand or forearm. In MiniMOST, crank applies only to those specific motions with the hand following a circular path, pivoting at the wrist and/or the forearm pivoted at the elbow with the upper arm essentially fixed. Crank values are not used for any activity in which there is significant motion of the elbow.

Index values are based on the diameter of the crank, the method of cranking (intermittent or continuous) and the number of cranking revolutions rounded to the nearest whole number. For less than half a revolution, use the Move Controlled Push or Pull index values.

Caution should be exercised to apply the crank values appropriately. Crank is determined by the motions employed, *not* the device being used. If the elbow is displaced, crank is not the motion being used. Figure 4.9 illustrates two examples of cranking motions.

Figure 4.10 is *not* a crank motion. Here the elbow moves with the action pivoted at the shoulder. Turning the wheel is analyzed as a series of push and pull motions.

Continuous Cranking

The object is moved in a circular path without pause or interruptions between revolutions. It begins with the object at rest, allows for the number of revolutions needed and

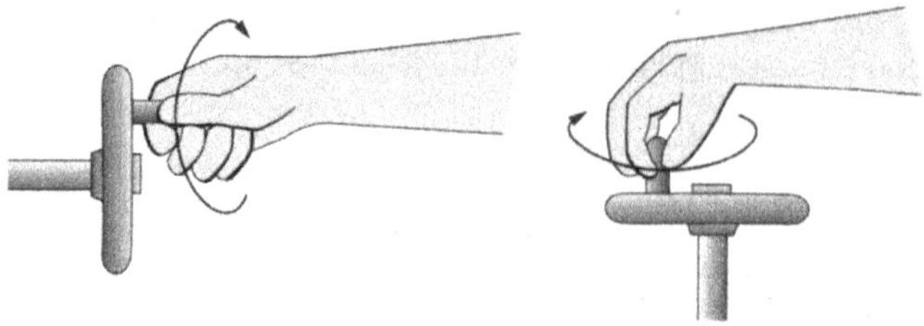

FIGURE 4.9 Crank motion examples.

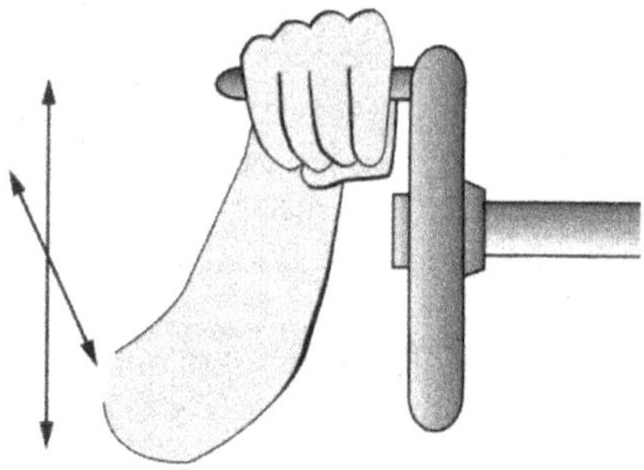

FIGURE 4.10 Example of a non-cranking motion.

ends when the cranking stops completely. The index value for continuous cranking is selected based on the crank diameter and the total number of revolutions rounded to the nearest whole number.

Crank diameter less than or equal to 5 inches (12.5 cm):

$\mathbf{M_{32}}$ Two revolutions
$\mathbf{M_{42}}$ Three revolutions
$\mathbf{M_{54}}$ Four revolutions

Crank diameter greater than 5 inches (12.5 cm), up to and including 20 inches (50 cm):

$\mathbf{M_{42}}$ Two revolutions
$\mathbf{M_{54}}$ Three revolutions

Examples: Continuously rotate a towel dispenser with two revolutions using a 2 inch (5 cm) crank handle: $\mathbf{M_{32}}$.
Continuously rotate a manual drum with three revolutions using a 7 inch (17.5 cm) crank handle: $\mathbf{M_{54}}$.

Intermittent Cranking

The object is moved in a circular path with noticeable pauses occurring between revolutions. The cranking begins with the object at rest, allows for one revolution and ends when the pause occurs before the next revolution. For intermittent cranking, the index value per revolution is selected based on the crank diameter and then a frequency is applied for multiple cranking actions.

Crank diameter less than or equal to 5 inches (12.5 cm):

M_{16} One revolution

Crank diameter greater than 5 inches (12.5 cm) up to and including 20 inches (50 cm):

M_{24} One revolution

Examples: Intermittently crank three revolutions with a 12 inch (30 cm) crank diameter:

M_{24} **(3)** With a partial frequency of three applied to the **M** value

Intermittently crank a handwheel two revolutions with a 3 inch (7.5 cm) diameter:

M_{16} **(2)** With a partial frequency of two applied to the **M** value

Effective Net Weight in the Controlled Move Sequence Model

If an object is being pulled, pushed or slid across a horizontal surface, the body member or members doing the work do not support all the object weight and must supply only enough force to overcome friction. This amount of force depends on surface texture and composition, but it has been calculated that the ENW of an object being slid on a solid surface is approximately 40% of its spatial ENW. Therefore, when sliding an object, the ENW must exceed 2.5 pounds (1 kg) before an adjustment to the **G** parameter is considered for weight purposes. When an object such as a lever requires muscular force to overcome resistance, a spring scale can be employed to determine the amount of force needed.

Example: Contact a 15 pound (7 kg) carton and push it across a table using both hands:

Actual weight: 15 pounds (7 kg)

ENW (spatial): 15/2=7.5 pounds (3.5 kg)

ENW (sliding): 40% of 7.5=3.0 pounds (1.5 kg)

For general application of this theory, the following table is provided. The weights below can be used as a guideline to help the analyst determine if an object is subject to the theory of ENW. All of the weights below have an Effective Net Weight greater than 2.5–10 pounds (1–5 kg) and would need an adjustment to the **G** value, except when using a Grasp.

Type of Move	Number of Hands	Actual Weight of Object Pounds (kg)
General Move (Spatial)	One	>2.5–10 (1–5)
	Two*	>5–20 (2–10)
Controlled Move (Sliding)	One	>6.50–25 (3–12)
	Two[a]	>13–50 (6–24)

* Assuming that the weight is distributed equally to each hand.

See Gain Control in The General Move Sequence Model section for the appropriate adjustment for Effective Net Weight.

Process Time (X)

Process Time is defined as the portion of work that is controlled by electronic or mechanical devices or machines, not by manual actions. The **X** parameter of the Controlled Move Sequence Model is intended to cover process times of relatively short duration. These process times will normally have minor variations and are often difficult to time. Power fluctuations and minor differences in parts can affect the process time. To apply the correct index value to the **X** parameter, the observed or calculated process time is converted to TMU and the appropriate index value is selected.

Longer process times, such as machining times based on feeds and speeds, are normally calculated and entered separately as a process time in the analysis. The actual clock time is *never* placed on the **X** parameter of the sequence model. Only the index value that statistically represents the actual time should be placed in the sequence model.

Example: A punch press cycles for two seconds after the palm buttons are pressed (2 seconds = 55.6 TMU): $\mathbf{X_{54}}$.

Alignment (I)

Alignment includes manual actions following the Move Controlled or at the conclusion of a Process Time to achieve alignment of an object to a point, line or to check for a single characteristic. Index values for Alignment are influenced by the ability (or inability) of the eyes to focus on a point in more than one area at a time.

Area of Normal Vision

The 'area of normal vision' is the average area covered by a single eye focus and is described by a circle 4 inches (10 cm) in diameter at a normal reading distance of about 16 inches (40 cm) from the eyes (Figure 4.3). Within this area of normal vision, the alignment of an object to two points can be performed without any additional 'eye times.' If one of the two points lies outside this area, two separate alignments are required because of the inability of the eyes to focus on both points simultaneously. In fact, an object would first be aligned to one point, the eyes would then shift to allow the alignment to the second point and then the object would finally be adjusted to correct for the minor shifting from the first point. The area of normal vision is therefore the basis for defining most of the Alignment parameter variants.

Index values for aligning an object are selected from the 'Within Area of Normal Vision' column if no eye travel is required to shift the eye to the point of alignment, that is, if one point is ≤ 4 inches (10 cm) from the current focal point of the eyes at a 16 inch (40 cm) perpendicular distance from the eye. A value from the 'Outside Area of Normal Vision' column is used when the eyes must shift more than 4 inches (10 cm) from their current focal point to assist the alignment.

When aligning a second or any sequential point, the alignment time includes movements of less than 1 inch (2.5 cm) of the object being aligned. Movements of 1 inch (2.5 cm) or more require an additional Controlled Move Sequence Model. The alignment values, which follow, are for alignments or checks *within* the area of normal vision. For an alignment or check *outside* the area of normal vision, the **I** parameter index value is increased two index values. Figure 4.11 provides a summary of the Alignment values.

Whenever a Controlled Move involves the Alignment activity, the preceding **M** parameter is used to describe only the distance the object travels (e.g., <4 inches (10 cm)—$\mathbf{M_6}$).

I_6 Check or Inspect

The Check or Inspect values include the eye and mental activities used to determine a single easily recognized physical characteristic of an object. It is a simple binary recognition of a trait; a yes or no decision. Check and Inspect are not included in Figure 4.11 because these values are simply a recognition of a trait, not an actual alignment. If the Check or Inspect is outside the area of normal vision, the index value is $\mathbf{I_{16}}$.

Example: As the part goes by on a conveyor, check (within the area of normal vision) to see that the company logo is printed at the top of the product.

I_6 Align to 1 Point

The Align value of $\mathbf{I_6}$ includes the time to align an object to one point where radial clearance is 5/32 (4 mm) to 1/2 inch (12 mm). The alignment to one point will always be within the area of normal vision.

Example: Align an arrow to an icon on a screen using a computer mouse.

I_{10} Align to 1 Point Accurate

The 1 Point Accurate alignment is characterized by the alignment of one point and by adjustments of the object to move it into a more precise relationship. The radial clearance must be less than 5/32 (4 mm) for the Align Accurate values to apply. The alignment of one point will always be within the area of normal vision.

Example: After sliding indicator, align to set temperature on machine.

Alignment

	I_6	I_{10}		I_{16}
Data Card Description	1 Point	1 Point Accurate	2 Points	2 Points Accurate
Clearance	5/32 – 1/2 in. (4 – 12 mm)	< 5/32 in. (4 mm)	5/32 – 1/2 in. (4 – 12 mm)	< 5/32 in. (4 mm)

Values above are for use **within** the area of normal vision.

Increase by two index values for objects aligned **outside** the area of normal vision. Aligning to one point will always be within the area of normal vision.

FIGURE 4.11 Alignment values summary.

I_{10} Align to 2 Points

Align to 2 Points is characterized by one manual adjustment of the object to move it into a predetermined relationship to one location where radial clearance is from 5/32 (4 mm) to 1/2 inch (12 mm). If the alignment is outside the area of normal vision, the index value is $\mathbf{I_{24}}$.

Example: Fold piece of material 3×4 inches (7.5×10 cm) in half so that the bottom edge of the paper is within 1/2 inch (12 mm) of the top edge.

I_{16} Align to 2 Points Accurate

Alignment to 2 Points Accurate is characterized by adjustments of the object to move it into a precise relationship. A radial clearance of less than 5/32 inch (4 mm) is considered accurate. If the alignment is outside the area of normal vision, the index value is $\mathbf{I_{32}}$.

Example: Align a ruler to two points with accuracy 3 inches (7.5 cm) apart to draw a line.

Selection of Placement (General Move) or Alignment (Controlled Move)

Placement and Alignment are similar in that both involve moving an object to a general or precise and predetermined location. Placement (General Move) provides for locating an object in space, bringing an object in contact with a surface and, when necessary, inserting one object into another. Alignment (Controlled Move) provides only for locating an object on the same surface. For example, placing a part on a workstation is a General Move. Sliding the part along the workstation is a Controlled Move.

The Alignment (**I**) parameter applies only when an alignment follows a Move Controlled. Should an object be moved freely without restrictions and then be 'aligned to two points,' the General Move Placement (**P**) parameter is the appropriate selection. In fact, there is a direct relationship between the Controlled Move and the General Move activities. That relationship is **M:I** as **A:P**. The Alignment (**I**) of an object occurs after the object is moved over a *controlled path* (**M**) and accounts for the time to align and/or orient the object, just as the Placement (**P**) of an object occurs after the *spatial* displacement of an object (**A**) and accounts for the time to orient and/or position the object. Normally, the nature of the movement of the object, spatial or controlled, is the determining factor behind the selection of a sequence model.

Placement (General Move) Followed by Alignment (Controlled Move)

Sometimes after an object is placed it may require an additional alignment. The Controlled Move is used to provide for the alignment. First allow the precise placement (**P** value) and then use a Controlled Move Sequence Model to allow the Alignment (**I** value).

Example: Insert a round pin in a hole [3/64 inch (1 mm) radial clearance] to precisely 2 inches (5 cm). There is a mark on the pin that must be aligned within 13/64 inch (5 mm) of the surface. The alignment (**I**) is to insert the pin to the exact depth of 2 inches (5 cm) (Figure 4.12). Allow $\mathbf{P_{16}}$ for the placement, an $\mathbf{M_3}$ for insertion and an $\mathbf{I_6}$ for aligning the mark.

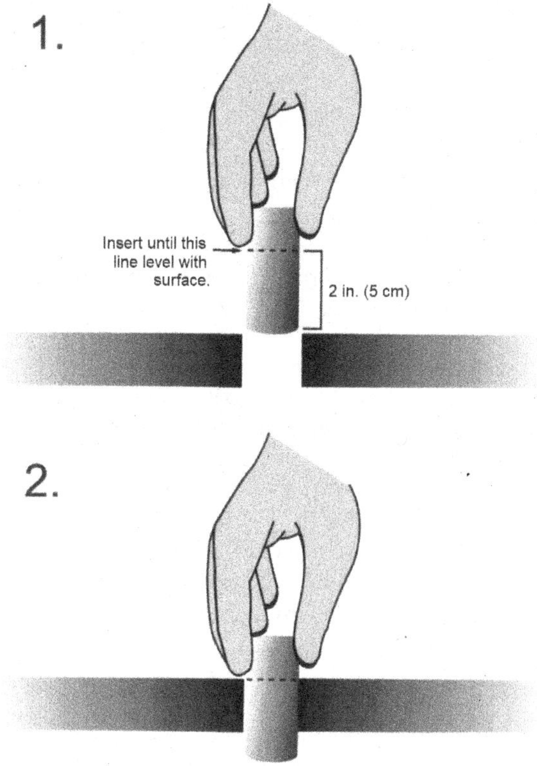

FIGURE 4.12 Alignment to insert pin.

Put pin with accuracy and insertion
$\mathbf{A_x}$ $\mathbf{B_x}$ $\mathbf{G_x}$ $\mathbf{A_x}$ $\mathbf{B_x}$ $\mathbf{P_{16}}$ $\mathbf{A_0}$

Additional insertion and align pin
$\mathbf{A_0}$ $\mathbf{B_0}$ $\mathbf{G_0}$ $\mathbf{M_3}$ $\mathbf{X_0}$ $\mathbf{I_6}$ $\mathbf{A_0}$

Writing Method Descriptions

Below are the recommended minimum requirements for a clear and concise method description for Controlled Move. Additional words for Action Distance and Body Motion are optional and may be used to enhance the method description. Often, the Action Distance is stated in the method description for clarification because of the level of detail needed in MiniMOST.

There are two recommended method description formats for Controlled Move: One for the movement of an object along a controlled path and one for process time.

Gain Control	Object	⟨From Location⟩	Move	To Location
Gain Control	Object	Actuate	At Location	

⟨If the From Location is apparent, it is not necessary to indicate that in the method description.⟩

Controlled Move Examples

1. A worker reaches 18 inches (45 cm) to the rearview mirror in a car and adjusts it by turning it 30 degrees in two directions to confirm it moves correctly.

   Grasp mirror 18 inches (45 cm) in car, turn 2 times 30 degrees each way

   A_{16} B_0 G_6 (M_6) X_0 I_0 A_0 (2) 34 TMU

2. After cutting material, an associate reaches 10 inches (25 cm) to contact scrap and pushes it 12 inches (30 cm) into a waste can.

   Contact scrap 10 inches (25 cm) and push 12 inches (30 cm) into waste can

   A_{10} B_0 G_3 M_{16} X_0 I_0 A_0 29 TMU

3. A machine operator reaches 20 inches (50 cm) to press a button on a machine. The button is in a fixed location. Process time is 0.5 seconds.

   Contact button 20 inches (50 cm) in a fixed location at machine with a process time of 0.5 seconds

   A_{10} B_0 G_3 M_3 X_{16} I_0 A_0 32 TMU

4. An operator releases a part from a fixture by moving a handle 8 inches (20 cm). The handle is grasped 10 inches (25 cm) away.

   Grasp handle 10 inches (25 cm) at fixture and pull 8 inches (20 cm)

   A_{10} B_0 G_6 M_{10} X_0 I_0 A_0 26 TMU

5. An operator reaches 12 inches (30 cm) to a cotter pin, gains control of the pin and puts it into a hole 16 inches (40 cm) away with a 13/64 inch (5 mm) radial clearance. The pin is inserted 2 inches (5 cm) and released.

   Grasp cotter pin 12 inches (30 cm) away and put into hole 16 inches (40 cm) away

   Additional insertion of cotter pin

A_{10}	**B_0**	**G_6**	**A_{16}**	**B_0**	**P_{10}**	**A_0**	**42**
A_0	**B_0**	**G_0**	**M_3**	**X_0**	**I_0**	**A_0**	**3**
							45

APPLICATION OF THE MINIMOST WORK MEASUREMENT SYSTEM

The MiniMOST Analysis

The application of MOST is consistent among all three systems. There are several key factors to a MOST analysis which are listed below and detailed in Chapter 2. Analysts need to consider the following factors and the content included in this section when developing a MiniMOST analysis:

- Identifying the best method
- Practical analysis procedures
- Analyst consistency
- General application rules for all MOST Systems

Since the time interval represented by each index value is very small, MiniMOST provides a highly detailed system for measuring work while maintaining many of the benefits of the BasicMOST System. The desire to analyze very short-cycle identical motion patterns has brought MiniMOST to a level of detail somewhere between MTM-1 and BasicMOST. As a result, analysts must look at the work to be measured differently. When using the MiniMOST System, every analyst must dissect the $\mathbf{A_1}$ (Within Reach) from the BasicMOST System and ascertain more precisely the distance the hand moved in inches (cm) and whether it was rotated during the action. Analysts must now look at the $\mathbf{G_1}$ (Gain Control of Light Object) and determine the type of grasp or grasps employed and the physical surroundings in which the grasp took place. An action that might be analyzed with one parameter in BasicMOST may well appear as one or two complete sequence models in MiniMOST.

The analyst's job becomes more difficult as the cycle gets shorter and the skill level of the operator increases. The sequence models for the *identical* cycles described in this chapter are designed to make the analyst's job as easy as possible while retaining the level of detail demanded by the work being measured.

The sequence models represented in Figures 4.13 and 4.14 provide a comparison of the same activity (PC Board Assembly) using BasicMOST (Figure 4.13) and MiniMOST (Figure 4.14) as shown from the StandardsPro software. The MiniMOST analysis requires greater detail than the BasicMOST analysis. Note that 13 method steps are needed to describe the activities in the MiniMOST analysis and only seven

Step	Method Description	Seq. Model	Freq.	Simo	TMU	Total
1	GRASP PC BOARD PUT ONTO PRESS & HOLD	A1 B0 G1 A1 B0 P1 A0	1.000	N	40	40
2	GRASP TERMINAL & PLACE INTO UPPER FIXTURE	A1 B0 G1 A1 B0 P3 A0	6.000	N	60	360
3	GRASP PRESS HANDLE & PULL LOWERING PIN NEAR BOARD	A1 B0 G1 M1 X0 I0 A0	6.000	N	30	180
4	SLIDE BOARD INTO POSITION TO RECEIVE PIN	A0 B0 G0 M3 X0 I0 A0	6.000	N	30	180
5	SEAT PIN	A0 B0 G0 M3 X0 I0 A0	6.000	N	30	180
6	PUSH PRESS HANDLE	A0 B0 G0 M1 X0 I0 A0	6.000	N	10	60
7	HOLDING PC BOARD INSPECT 6 POINTS AND ASIDE	A0 B0 G0 A1 B0 P0 T10 A1 B0 P1 A0	1.000	N	130	130

FIGURE 4.13 PC Board Assembly analysis using BasicMOST.

Step	Method Description	Seq. Model	Freq.	Simo	TMU	Total
1	LH CONTACT & PULL PC BOARD FROM STACK	A10 B0 G3 M6 X0 I0 A0	1.000	N	19	19
2	LH REGRASP PC BOARD, SET & RETAIN ON PRESS	A0 B0 G6 A10 B0 P3 A0	1.000	N	19	19
3	LH REGRASP PC BOARD & HOLD	A0 B0 G6 A0 B0 P0 A0	1.000	N	6	6
4	RH SELECT-SMALL TERMINAL & PLACE (ACC. & INS.)	A16 B0 G16 A16 B0 P24 A0	6.000	N	72	432
5	RH GRASP PRESS HANDLE & LOWER PIN	A16 B0 G6 M10 X0 I0 A0	6.000	N	32	192
6	LH PUSH PC BOARD & ALIGN TO ONE POINT ACCURATE	A0 B0 G0 M6 X0 I10 A0	6.000	N	16	96
7	RH PULL HANDLE	A0 B0 G0 M10 X0 I0 A0	6.000	N	10	60
8	RH SEAT PIN	A0 B0 G0 M16 X0 I0 A0	6.000	N	16	96
9	RH PUSH HANDLE	A0 B0 G0 M10 X0 I0 A0	6.000	N	10	60
10	LH LIFT BOARD	A0 B0 G0 A6 B0 P0 A0	1.000	N	6	6
11	LH CHECK 6 PINS	A0 B0 G0 M0 X0 I6 A0	6.000	N	6	36
12	LH SET & RETAIN BOARD ON STACK	A0 B0 G0 A16 B0 P3 A0	1.000	N	19	19
13	LH REGRASP & SLIDE BOARD IN PLACE	A0 B0 G6 M6 X0 I0 A0	1.000	N	12	12

FIGURE 4.14 PC Board Assembly analysis using MiniMOST.

are needed for the BasicMOST analysis. Since there is no Tool Use Sequence Model in MiniMOST, an analysis using one sequence model in BasicMOST may require several sequence models in MiniMOST.

MiniMOST Analysis Documentation

A MiniMOST analysis is similar to a BasicMOST analysis but should also reflect the special requirements of the MiniMOST System to show two-handed work. When MiniMOST was developed, two forms were used to provide an analyst with a choice to identify the form that best met their needs (Figures 4.15 and 4.16). The forms are presented to show the history of MiniMOST. Today, most companies use software tools such as StandardsPro to develop and maintain engineered standards. MOST has developed from a manual system to a fully integrated digital solution. Figure 4.17 shows a completed MiniMOST analysis to *Stake 6 Terminals to PC Board Using Arbor Press* using the Method and TMU report from the StandardsPro software application. Whether the manual form or a software application is used to document a MiniMOST analysis, it is important to include essential information. One of the key features of the MOST Work Measurement Technique is the predefined sequence models. Using a manual system or a software application, the use of the predefined sequence models is critical for the analyst to maintain applicator accuracy and consistency.

Regardless of the documentation method, the information listed below should be included in the analysis.

1. *Identification and Header*

   This section of the analysis should include all key information for ease of searching and updating the analysis.

   - Title or description

     The title often begins with a verb of the activity and may include the object, product/equipment, tools and a work area description. An example title could be:

     Stake 6 Terminals to PC Board Using Arbor Press.
   - Activity

     The activity should be a verb that indicates the overall context and/or main goal of the actions which are included in the analysis. The activity for the *Stake 6 Terminals to PC Board Using Arbor Press* sub-operation would be 'Assemble/Disassemble.'
   - Unit of measure

     Used to designate what the activity is based on. Examples of units of measure are per unit, per part, per box, per customer, per occurrence, per pallet, etc.
   - Operator, applicator or safety instructions

     Instructions can be added to clarify key points in the analysis.

     Operator Instructions—Secure wires together tightly before final assembly.

MAYNARD

MiniMOST® Analysis (Vertical)

DATE	
ANALYST	
PAGE	OF

DESCRIPTION

ACTIVITY • OBJECT • IN/ON/FOR • PRODUCT/EQUIPMENT • USING/WITH • TOOL • TO/AT • WORK AREA

UNIT OF MEASURE

INSTRUCTIONS ❒ APPLICATOR (A) ❒ OPERATOR (O) ❒ SAFETY (S)

TMU FROM PREVIOUS PAGE

STEP NO	HAND R/L	METHOD STEP DESCRIPTION	SEQUENCE MODELS	PF	FR	SIMO TO	TMU
1			A B G A B P A	( )			
			A B G M X I A	( )			
2			A B G A B P A	( )			
			A B G M X I A	( )			
3			A B G A B P A	( )			
			A B G M X I A	( )			
4			A B G A B P A	( )			
			A B G M X I A	( )			
5			A B G A B P A	( )			
			A B G M X I A	( )			
6			A B G A B P A	( )			
			A B G M X I A	( )			
7			A B G A B P A	( )			
			A B G M X I A	( )			
8			A B G A B P A	( )			
			A B G M X I A	( )			
9			A B G A B P A	( )			
			A B G M X I A	( )			
10			A B G A B P A	( )			
			A B G M X I A	( )			
11			A B G A B P A	( )			
			A B G M X I A	( )			
12			A B G A B P A	( )			
			A B G M X I A	( )			
13			A B G A B P A	( )			
			A B G M X I A	( )			
14			A B G A B P A	( )			
			A B G M X I A	( )			
15			A B G A B P A	( )			
			A B G M X I A	( )			
16			A B G A B P A	( )			
			A B G M X I A	( )			

TOTAL TIME: HOURS:________ MINUTES:________ SECONDS:________ TMU:________

FIGURE 4.15 MiniMOST Analysis form (vertical).

Applicator Instructions—Eye motion in step four is internal to moving the part.

Safety Instructions—Adjust worktable to correct ergonomic height before assembly.

- Date of analysis
- Identification of analyst

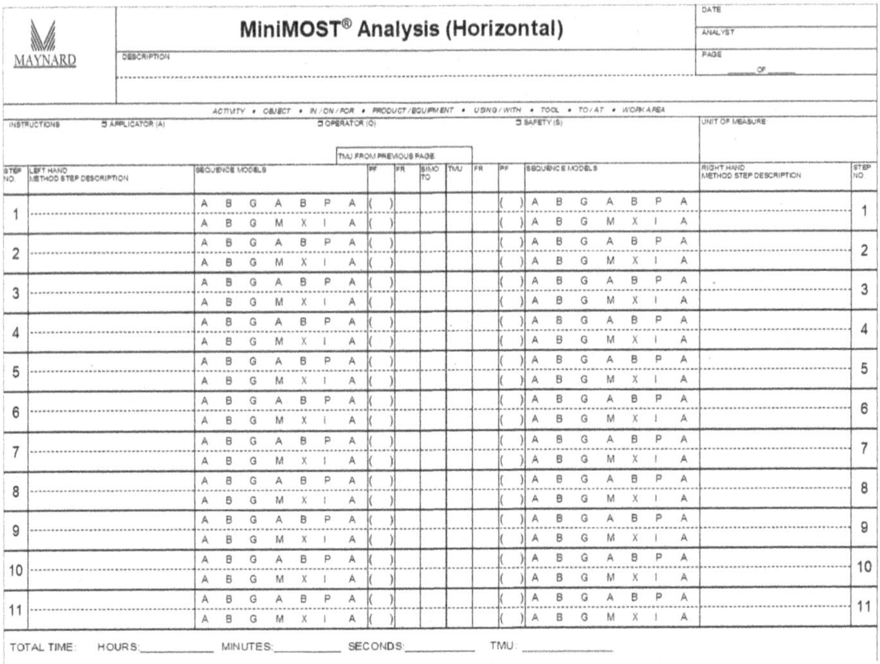

MAYNARD

MiniMOST® Analysis (Horizontal)

DATE | ANALYST | PAGE ___ OF ___

DESCRIPTION

ACTIVITY • OBJECT • IN/ON/FOR • PRODUCT/EQUIPMENT • USING/WITH • TOOL • TO/AT • WORK AREA

INSTRUCTIONS: ☐ APPLICATOR (A) ☐ OPERATOR (O) ☐ SAFETY (S)

UNIT OF MEASURE

TMU FROM PREVIOUS PAGE

STEP NO.	LEFT HAND METHOD STEP DESCRIPTION	SEQUENCE MODELS	PF	FR	SIMO TO	TMU	FR	PF	SEQUENCE MODELS	RIGHT HAND METHOD STEP DESCRIPTION	STEP NO.
1		A B G A B P A / A B G M X I A	( )					( )	A B G A B P A / A B G M X I A		1
2		A B G A B P A / A B G M X I A	( )					( )	A B G A B P A / A B G M X I A		2
3		A B G A B P A / A B G M X I A	( )					( )	A B G A B P A / A B G M X I A		3
4		A B G A B P A / A B G M X I A	( )					( )	A B G A B P A / A B G M X I A		4
5		A B G A B P A / A B G M X I A	( )					( )	A B G A B P A / A B G M X I A		5
6		A B G A B P A / A B G M X I A	( )					( )	A B G A B P A / A B G M X I A		6
7		A B G A B P A / A B G M X I A	( )					( )	A B G A B P A / A B G M X I A		7
8		A B G A B P A / A B G M X I A	( )					( )	A B G A B P A / A B G M X I A		8
9		A B G A B P A / A B G M X I A	( )					( )	A B G A B P A / A B G M X I A		9
10		A B G A B P A / A B G M X I A	( )					( )	A B G A B P A / A B G M X I A		10
11		A B G A B P A / A B G M X I A	( )					( )	A B G A B P A / A B G M X I A		11

TOTAL TIME: HOURS:________ MINUTES:________ SECONDS:________ TMU:________

FIGURE 4.16 MiniMOST Analysis form (horizontal).

- Total TMU

  The total time for the activity is shown by simply adding all of the numbers in the TMU column for each method step. That number is then noted in the Total Time section of the form, or even better, calculated automatically in a software solution. Most electronic systems automatically convert the TMU to the organization's choice of time in seconds, minutes or hours.

2. *Method Steps and Sequence Models*

   This section of the analysis should include the step number, method step description, sequence model, including documentation of any frequencies or simultaneous motions, and the TMU for each method step.

   - Method steps

     The amount of information placed in the method description section is usually a function of its eventual use; that is, the description can be used for detailed operator instructions or for an outline of the manual work for time computation only. Each method step has *only one* corresponding sequence model so the method description should be phrased in terms of moving an object or using a tool as to easily identify the sequence model used for that activity. In MiniMOST, the use of the left hand (LH) or right hand (RH) is often documented in the method step as shown in Figure 4.17.

Sub-Operation - Method / TMU Report
H. B. Maynard and Co.

Sub-Op ID:	70874	**Status:**	Private
Title:	STAKE 6 TERMINALS TO PC BOARD USING ARBOR PRESS		
Activity:	ASSEMBLE/DISASSEMBLE	**Sub-Activity:**	
Object:		**Prod/Equip:**	
Size/Capacity:		**Tool:**	
WA Number:		**WA Origin:**	
5-S Level:	SOME EFFORT	**Other:**	
Unit of Measure:	BOARD	**OFG:**	2
WorkArea ID:		**Starting Oper:**	
Starting Location:		**Total TMU:**	1,053
Operator Instr:			
Applicator Instr:			
Safety Instr:			
Applicator:	MCS		
Create Date:	3/26/2020	**Effect Date:**	3/24/2020

Method Steps

Step	Method Description	Freq.	Simo	TMU	Total
1	LH CONTACT & PULL PC BOARD FROM STACK A10 B0 G3 M6 X0 I0 A0	1.000	N	19	19
2	LH REGRASP PC BOARD, SET & RETAIN ON PRESS A0 B0 G6 A10 B0 P3 A0	1.000	N	19	19
3	LH REGRASP PC BOARD & HOLD A0 B0 G6 A0 B0 P0 A0	1.000	N	6	6
4	RH SELECT-SMALL TERMINAL & PLACE (ACC. & INS.) A16 B0 G16 A16 B0 P24 A0	6.000	N	72	432
5	RH GRASP PRESS HANDLE & LOWER PIN A16 B0 G6 M10 X0 I0 A0	6.000	N	32	192
6	LH PUSH PC BOARD & ALIGN TO ONE POINT ACCURATE A0 B0 G0 M6 X0 I10 A0	6.000	N	16	96
7	RH PULL HANDLE A0 B0 G0 M10 X0 I0 A0	6.000	N	10	60
8	RH SEAT PIN A0 B0 G0 M16 X0 I0 A0	6.000	N	16	96
9	RH PUSH HANDLE A0 B0 G0 M10 X0 I0 A0	6.000	N	10	60
10	LH LIFT BOARD A0 B0 G0 A6 B0 P0 A0	1.000	N	6	6
11	LH CHECK 6 PINS A0 B0 G0 M0 X0 I6 A0	6.000	N	6	36
12	LH SET & RETAIN BOARD ON STACK A0 B0 G0 A16 B0 P3 A0	1.000	N	19	19
13	LH REGRASP & SLIDE BOARD IN PLACE A0 B0 G6 M6 X0 I0 A0	1.000	N	12	12

FIGURE 4.17 Method and TMU report.

- Sequence model analysis

  Regardless of the system used to document a MiniMOST analysis, it should include the predefined sequence models for General Move and Controlled Move. This helps to ensure applicator accuracy and consistency. After the index values to the selected sequence model are applied, any frequencies if they occur in the method step or if the method step is performed simultaneously to another activity are documented.

Frequencies and partial frequencies were discussed earlier in the chapter and are used when one or more parameters of a sequence model occurs more or less than once. The default for any system is a frequency of one (1). As shown in Figure 4.17, a '6' is noted in the 'Freq' column for Steps 4–9 to identify that the tasks occur six times.

For a partial frequency, where a portion of the sequence model occurs more than once, Figure 4.18 shows the parameters that occur more than once (**A_{10} B_0 P_0**) are noted in parentheses with the partial frequency of '(3.00)' noted at the end of the sequence model.

Motion Combinations

Motion combinations occur when two or more motions are performed at the same time by the same body member (hand grasping and then regrasping) or by different body members (right and left hand grasping different objects at the same time). These motion combinations often occur in many industries and are especially prevalent in activities for which MiniMOST is being applied. There are two types of motion combinations: Combined Motions and Simultaneous Motions.

Combined Motions

Combined motions are the motions performed by one body member such as the hand or foot. The analyst's task is to completely document the method and allow index values for the controlling or time-limiting motions. For example, an operator reaches 12 inches (30 cm) to a part on the bench, grasps the part and moves it 10 inches (25 cm) to set and retain it. While moving the part, the operator regrasps the part for a better grip and rotates the part 90 degrees to make assembly easier. Assuming the work is done with the left hand, the analysis is shown in Figure 4.19.

The **A_{10}** for a 10 inch (25 cm) action to place the object in Step 1, **G_6** (regrasp) in Step 2 and the **A_3** (rotate) in Step 3 are combined motions. They are all done with one hand. The **A_{10}** is the controlling or limiting motion and its index value is included in the total time. If using the manual documentation, a circle indicates that a parameter

Step	Method Description	Seq. Model	Freq.	Simo	TMU	Total
1	LH REACH 10 IN COLLECT PARTS AND TOSS SEPARATELY	A10 B0 G24 (A10 B0 P0) A0 (3.00)	1.000	N	64	64

FIGURE 4.18 Partial frequency notation.

Step	Method Description	Seq. Model	Freq.	Simo	TMU	Total
1	LH SET PART TO FRONT OF BENCH, HOLD FOR ASSEMBLY	A10 B0 G6 A10 B0 P3 A0	1.000	N	29	29
2	LH REGRASP PART FOR HOLDING	A0 B0 G6 A0 B0 P0 A0	1.000	Y	6	0
3	LH TURN 90 DEGREES TO RECEIVE NEXT PART	A0 B0 G0 A3 B0 P0 A0	1.000	Y	3	0

FIGURE 4.19 Notation for combined motions.

is limited out by one or more parameters having a greater index value and is not counted in the total TMU. If the total TMU of a sequence model is less than or equal to a simultaneous method step, a circle is placed around the total TMU. Figure 4.19 shows the same situation completed in the StandardsPro application. The notation of the hand is included in the Method Description fields and since Steps 2 and 3 are simultaneous to Step 1, the times in the application area are shaded and a '0' is listed in the Total field so that time is not included in the total time. If the index values are equal for combined motions, select which method steps should be identified as the limited step and document accordingly.

Simultaneous Motions

Motion combinations performed by different body members are referred to as simultaneous motions. When simultaneous motions occur, record the sequence model for each body member (usually the hands) and enter the appropriate index value for each parameter. After completely recording each method description, determine which parameters are performed simultaneously and identify the parameter with the lower index value for each pair of simultaneous parameters and apply and note that the action is simultaneous.

Simo is used to document that a method step or a portion of the method step occurs at the same time as another step. Figure 4.20 shows how the time for a simultaneous activity is noted in the TMU column and the Total is shown as '0' to reflect that time is not included in the total time for the activity. If documenting the analysis manually, the Total TMU should be circled to show that it should not be counted in the analysis time.

Example: With both hands, reach 10 inches (25 cm) and grasp two parts in a bin at the same time and pick up the parts 10 inches (25 cm) and hold. This is an example of simultaneous motions and is analyzed as shown in Figure 4.20.

If a portion of a method step is simultaneous to another, use of the Simo notation is to indicate the method step and parameters to which the activities are simultaneous. Figure 4.21 shows how the Get phase for Step 2 is simo to the Get phase in Step 1. From its inception within the MOST Systems, simo uses a simple coding system. Since the General Move and Controlled Sequence Models consist of seven parameters, they are numbered as follows:

Step	Method Description	Seq. Model	Freq.	Simo	TMU	Total
1	LH REACH 10 IN GRASP PART AND PICKUP 10 IN	A10 B0 G6 A10 B0 P0 A0	1.000	N	26	26
2	RH REACH 10 IN GRASP PART AND PICKUP 10 IN	A10 B0 G6 A10 B0 P0 A0	1.000	Y	26	0

FIGURE 4.20 Notation for simultaneous motions.

Step	Method Description	Seq. Model	Freq.	Simo	TMU	Total
1	LH REACH 10 IN GRASP COMPONENT AND POSITION 10 IN TO PART	A10 B0 G6 A10 B0 P16 A0	1.000	N	42	42
2	RH REACH 10 IN GRASP COMPONENT AND POSITION 10 IN TO PART	< A10 B0 G6 > A10 B0 P16 A0	1.000	N	26	26

FIGURE 4.21 Partial Simo notation in MiniMOST.

A B G A B P A
1 2 3 4 5 6 7 (parameter number)

As an example, if the Get phase of the second method step is simultaneous to the Get phase of Step 1, then the code for a simo activity for the second method step would read 1:1–3. The **A B G** parameters of Step 2 would be documented to show the simo activity and not counted in the total for that method step. The time for each method step is then calculated by adding the index values and applying the frequencies as needed to arrive at the time value for the sequence model in TMU.

In software applications, the use of the coding system can be noted in the method description or in the applicator instructions as shown in Figure 4.22.

Simultaneous Motion Guide

The Simultaneous Motion Guide includes two charts to indicate when activities of various levels of control may and may not be performed simultaneously. The first chart shows the Control Level of Common Activities and is shown in Figure 4.23. The second chart is the Simultaneous Performance Chart and is shown in Figure 4.24. The guide is provided as a reference for the following discussion of control levels.

Operator Instr:	
Applicator Instr:	Step 2 is simo 1:1-3
Safety Instr:	

FIGURE 4.22 Use of the Simo coding system in the Applicator Instructions in an analysis.

Parameter	Control Level		
	Low	*Medium*	*High*
Gain Control (G)	• Sweep • Contact • Grasp • Regrasp • Collect • Fixed Location • Indefinite Location • Other Hand	• Grasp Small Object • Grasp with Care • Disengage • Varying Location • Foot or Leg	• Select • Select - Small
Placement (P)	• Drop • To Other Hand • To Stop	• Toss • Set and Retain • Set Aside • Put	• Place • Position
Move Controlled (M)	• To Other Hand • To Stop	• General Location	• Exact Location
Alignment (I)		• Align • Check	• Align - Accurate

FIGURE 4.23 Control level of common activities.

<table>
<tr><th colspan="2">Control Level</th><th>Low</th><th>Medium</th><th>High</th></tr>
<tr><td colspan="2">Low</td><td>Usually</td><td>Usually</td><td>Usually</td></tr>
<tr><td rowspan="2">Medium</td><td>Within Area of Normal Vision</td><td>Usually</td><td>Usually</td><td>Practice</td></tr>
<tr><td>Outside Area of Normal Vision</td><td>Usually</td><td>Practice</td><td>Rarely</td></tr>
<tr><td colspan="2">High</td><td>Usually</td><td>Practice</td><td>Rarely</td></tr>
</table>

FIGURE 4.24 Simultaneous performance chart.

The *control level* refers to the mental and visual control the operator must exercise to complete the activity or motion parameter. *Low* control level activities require little mental control and little or no visual control. *Medium* control level activities require mental and visual attention during the activity but not at the completion of the activity. *High* control level activities require mental and visual attention during the activity and on completion of the activity.

The Control Level of Common Activities chart shows the control level normally associated with different types of activities (see Figure 4.23). The left column lists the parameters. Common activities within each parameter are listed in the three control level columns by the control they most commonly require. It should be noted that the workplace, parts or tools may impose conditions changing the control level of an activity.

The Simultaneous Performance Chart (Figure 4.24) shows that activities with a low control level can usually be simultaneously performed with an activity of any control level. On occasion, it may not be possible to perform a low control level activity simultaneously with another activity because of some constraint imposed by the parts or workplace. Medium control level activities can usually be performed simultaneously with low control level activities. These activities may or may not be performed simultaneously with other medium or with high control level activities depending on the practice opportunity and whether the activities are within the same area of normal vision. One high control level activity can rarely be performed simultaneously with another high control level activity. Time for both high control level activities should be allowed even though one may occasionally see an unusually skilled or coordinated operator perform them simultaneously.

To use the Simultaneous Motion Guide, the activities are first located in the table showing the Control Level of Common Activities and the control level determined for those activities. Then use the Simultaneous Performance Chart to determine if the activities can be performed simultaneously. Allow the longer time when the activities can be performed simultaneously. Allow both times (index values) when the activities must be performed separately. To make a final determination if two activities can be performed simultaneously or not, the actual method used should be reviewed.

Control Level and Method Level

Control level refers to the mental and visual control the operator must exercise to complete the activity or motion. Method level refers to the degree of coordination between the right and left hands during two-handed work. There are three method levels:

1. High method level
2. Intermediate method level
3. Low method level

The method level at which an activity is performed is determined by its occurrence frequency, that is, the practice opportunity available to the operator. The more often the activity occurs, the greater the operator's opportunity to improve the method level. If the activity is seldom performed, the short learning period prevents any development of simultaneous skills. For example, with mass production and large batch-size operations which allow ample training and practice opportunity, one would expect to find operators using a high percentage of simultaneous motions. On the other hand, job shop and setup activities will most likely be performed with few simultaneous motions. Therefore, method level depends to a large extent on the type of work being performed. The three different method levels are defined below along with the relationship to control level.

1. *High method level* includes all possible simultaneous motions with the right and left hands. The analysis and time for the limiting (longest) hand is allowed. The total time value for the limited hand must be noted to indicate that this value is not included in the total. The activity performed by the left hand (LH) occurs simultaneously with the activity performed by the right hand (RH). Low control activities can often be performed simultaneously with other low or medium control activities and will normally result in a high method level. In Figure 4.25, the time for the right hand is documented to indicate that it is 'limited out' by another activity and is not included in the total.
2. *Low method level* involves no simultaneous motions. High control activities can rarely be performed with other high or medium activities simultaneously and will normally result in a low method level. The example in Figure 4.26 shows the left and right hands performing an activity with no simultaneous motions. The analysis time for both hands must be allowed as shown in Figure 4.26.
3. *Intermediate method level* refers to a method performed partially with simultaneous motions. For example, the Action Distance 'Within Reach' to two objects may be performed simultaneously with both hands but gaining control and placing two objects simultaneously may not be possible. In the MiniMOST analysis, the appropriate parameters are noted to indicate that they are performed simultaneously and the associated time should be excluded from the sequence model calculation. Medium and low control

Step	Method Description	Seq. Model	Freq.	Simo	TMU	Total
1	LH GRASP PART AND SET ASIDE	A3 B0 G6 A3 B0 P6 A0	1.000	N	18	18
2	RH GRASP PART AND SET ASIDE	A3 B0 G6 A3 B0 P6 A0	1.000	Y	18	0

FIGURE 4.25 High method level.

Step	Method Description	Seq. Model	Freq.	Simo	TMU	Total
1	LH SELECT-SMALL PART AND POSITION	A10 B0 G16 A10 B0 P16 A0	1.000	N	52	52
2	RH SELECT-SMALL PART AND POSITION	A10 B0 G16 A10 B0 P16 A0	1.000	N	52	52

FIGURE 4.26 Low method level.

Step	Method Description	Seq. Model	Freq.	Simo	TMU	Total
1	LH SELECT PART AND PLACE IN BOX	A6 B0 G10 A3 B0 P10 A0	1.000	N	29	29
2	RH SELECT PART AND PLACE IN BOX	< A6 > B0 G10 < A3 > B0 P10 A0	1.000	N	20	20

FIGURE 4.27 Intermediate method level.

activities can often be performed simultaneously. Practice opportunity with these control levels will contribute to the intermediate method level. In Figure 4.27, a portion of the sequence model for the right hand (the reach to get and put the object) in Step 2 is performed simultaneously with the reach of the left hand:

In this case, the Action Distance parameters in the Get and Put phases are not included in the time calculation because it is 'limited' by another activity. Refer to Figure 4.24 for the possible combination of control level activities.

Summary of the MiniMOST Analysis

A MiniMOST analysis is completed by confirming the following is included:

1. Identification and Header Information
    - Title or description
    - Activity
    - Unit of measure
    - Operator, applicator or safety instructions
    - Date of analysis
    - Identification of analyst
    - Total TMU

2. Method Steps and Sequence Models
    - Method steps and method step descriptions
    - Sequence model analysis
    - Select *one* sequence model for each method step

- Apply the correct index value for each parameter within each sequence model.
- Note the documentation for PF, FR or Simo activities as needed.
- Add parameter index values together and apply frequencies as needed to get the time for the sequence model in TMU.

• Add all method step totals to get a total TMU.

Note: A MOST analysis will provide the normal time (manual time+allowance time). Allowances need to be added to generate an engineered standard.

Analysis of Activities Involving Tools

There is no sequence model for tool use in MiniMOST. Since the use of a tool can vary from operation to operation, it is necessary to individually analyze each occurrence of tool use with General and Controlled Moves for those activities that require the detailed analysis of MiniMOST.

Example: An operator reaches 12 inches (30 cm) to a screwdriver, grasps the handle, moves the screwdriver 10 inches (25 cm) to a screw and fits the blade to the slot. The operator shifts control of the handle and, rotating the forearm, turns the screw 120 degrees and then opens the fingers and, holding the palm to the handle, rotates the hand to secure a new grip on the handle for four additional 120 degree turns. To tighten the screw, a total of five 120 degree turns are made by rotating the hand about the forearm. The hand is rotated about the forearm for each of the four reaches back. The palm of the hand is always in contact with the screwdriver handle. It requires a force of about 1 1/2 pounds (0.7 kg) for the first three turns and a force of about 3 pounds (1.4 kg) for the last two turns. The example of using the screwdriver is analyzed as shown in Figure 4.28.

This analysis applies only to the example described. Another use of the screwdriver may differ in distance turned, finger moves instead of hand rotations, number of turns, resistance or even the method of placing the screwdriver. Each occurrence of tool use must be separately analyzed.

Step	Method Description	Freq.	Simo	TMU	Total
1	GRASP SCREWDRIVER HANDLE 12 IN (30 CM) AND PLACE 10 IN (25 CM) TO SLOT ACCURATE AND DIFFICULT A10 B0 G6 A10 B0 P24 A0	1.000	N	50	50
2	REGRASP HANDLE AND TURN 120 DEGREES A0 B0 G6 M10 X0 I0 A0	1.000	N	16	16
3	GRASP HANDLE 120 DEGREES AND TURN 120 DEGREES WITH 1.5 POUNDS (0.7 KG) RESISTANCE TWICE A3 B0 G6 M10 X0 I0 A0	2.000	N	19	38
4	GRASP HANDLE 120 DEGREES AND TURN 120 DEGREES WITH 3 POUNDS (1.4 KG) RESISTANCE TWICE A3 B0 G6 M10 X0 I0 A0	2.000	N	19	38
5	PUT SCREWDRIVER ASIDE 10 IN (25 CM) A0 B0 G0 A10 B0 P6 A0	1.000	N	16	16

FIGURE 4.28 Example of using a screwdriver analyzed with MiniMOST.

The Hand Used as a Tool

There are times when the hand is used in a manner similar to the use of a tool. This occurs when an object is rubbed by hand or turned by hand (e.g., turning down a nut). For rubbing, a Controlled Move is used. Sometimes, when an accurate grasp is required, it is necessary to position the hand or fingers. When this occurs, use the Put (**A B P**) phase of the General Move Sequence Model and consider the same variables as in placing an object.

Striking

The General Move Sequence Model is used for striking with the hand or with a tool. The Placement for the blow is almost always a $\mathbf{P_3}$, Set and Retain. Normally the Placement value for the backswing is also a $\mathbf{P_3}$, Set and Retain. At first glance, the backswing for striking may appear to be a $\mathbf{P_0}$, Indefinite Location, but a closer examination will usually reveal that a general location is required at the end of the backswing in order to start a blow that can be properly directed to the target. In a few cases of pounding a general surface requiring very little control of the blow a $\mathbf{P_0}$, Indefinite Location is adequate, but normally the $\mathbf{P_3}$, Set and Retain, is required.

Also, care must be taken to accurately determine the Action Distance in striking, especially for tapping blows delivered with a hammer. The distance the hammer-head moves is often much farther than the hand moves. The measurement must be taken at the hand. Typically, wrist assistance occurs in using a hammer so the Action Distance must be further reduced for the assistance. Often the Action Distance for normal hammering, such as driving a nail, is as little as 1 or 2 inches (2.5–5 cm). The sequence model for each hammer strike with a 2 inch (5 cm) Action Distance is:

$$\mathbf{A_0 \quad B_0 \quad G_0 \quad A_1 \quad B_0 \quad P_3 \quad A_0}$$

This gives a total of 4 TMU per strike. The backswing sequence model is also:

$$\mathbf{A_0 \quad B_0 \quad G_0 \quad A_1 \quad B_0 \quad P_3 \quad A_0}$$

The backswing is also 4 TMU per occurrence.

Development of Special Tool Elements

In BasicMOST, a procedure is available for the development of special elements primarily for tools that may be unique but commonly used in a particular company. Since there is no Tool Use Sequence Model in MiniMOST, there is no need to develop special elements for tools. Nor is it recommended that Tool Use elements be added to the MiniMOST System as presented. The elements available in MiniMOST are adequate to cover all those work measurement situations in which MiniMOST is the suitable system.

MiniMOST Summary

The MOST Work Measurement Technique uses a structured approach to measure work based on the movement of objects. There is a consistent approach the analyst should always use prior to analyzing an activity with MiniMOST. It begins with determining the starting and stopping points of the activity to be analyzed and ends with a total time for the activity. The MiniMOST Analysis Process (Figure 4.29) depicts this procedure and can be used to lead the analyst through the basic thought processes and decisions that need to be considered in order to arrive at a thorough and consistently applied MiniMOST analysis. The MiniMOST Analysis Process guides the analyst through the process to identify whether a move is a General Move or a Controlled Move and then uses the phases to assist the analyst to select the correct index values.

With the use of software applications, updating a MiniMOST analysis is very simple with the functionality to update one sub-operation, determine the impact of a method change on the overall time and even complete a mass update of the new time to automatically update all sub-operations that include a particular method step.

The ease with which MiniMOST analyses can be updated and/or new methods determined is one of the greatest advantages of the MOST Work Measurement Technique. It makes simulation and comparison easy.

It must be remembered that measured times produced with any MOST System represent a performance level of 100%. That is, the performance of an average skilled worker following a prescribed method, working under average work conditions at a

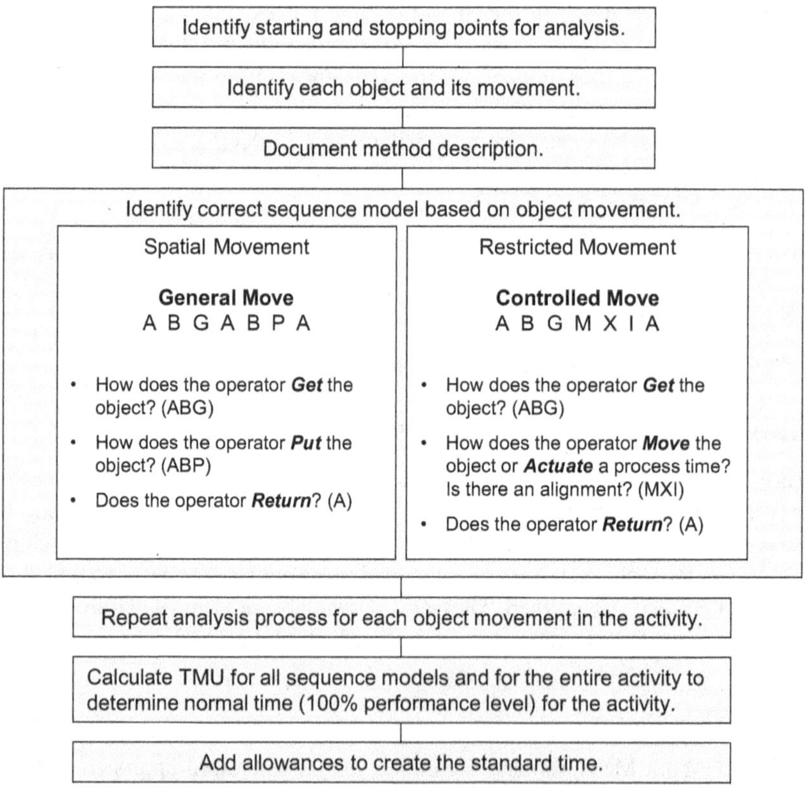

FIGURE 4.29 MiniMOST Analysis Process.

normal pace and with capable supervision. The total time value for an activity analyzed with MiniMOST produces a normal time without allowances. Typically, the allowances as a percentage of normal time are applied as a final step to establish a standard time.

Following the analysis process and answering the questions is key to the effective application of MiniMOST. The answers will help the analyst to:

- Determine the correct sequence model to be used.
- Determine the index value for each parameter (sub-activity).
- Avoid overlooking any other objects being moved or analyzing any unnecessary activity.
- Apply MiniMOST consistently.

5 The MaxiMOST System

In many long-cycle, non-repetitive, non-identical assembly, machining or maintenance operations, the BasicMOST sequence models will likely produce a method description with excessive detail. The variation in the actual method from cycle-to-cycle is often so great that the relatively precise method descriptions of BasicMOST are not required. A more general description of the method is preferred to allow for the variations that occur in the actual method used.

This phenomenon indicates that long-cycle jobs could have a higher-level analysis system applied to them and still produce a meaningful and descriptive method description as well as an accurate analysis in less time. To meet this need, the MOST Work Measurement Technique was expanded to include sequence models designed expressly for the measurement of long-cycle operations. These sequence models produce accurate results, are fast to apply, easy to learn and understand and provide meaningful method descriptions.

MaxiMOST involves the application of larger blocks of time than BasicMOST. The result of using larger blocks of time is that many more combinations of activities can be described. These combinations led to broadening the scope of the MaxiMOST parameters. The data cards for MaxiMOST, therefore, contain a far greater number of entries than the data cards for the other MOST Systems.

However, there are still a number of work activities that were not analyzed and placed on the data cards. Because of this, additional elements may need to be developed. The development of special elements is described in the Developing New Elements section of this chapter.

MaxiMOST can be used in any area where the job is such that significant methods variation occurs from cycle-to-cycle. This variation is a result of the length and low repetition of the cycle. MaxiMOST is commonly used in the following types of activities:

- Heavy assembly
- Welding
- Heavy machining and fabrication
- Long-cycle surface treating such as blasting or coating
- Maintenance
- Setups
- Utility operations

THE SEQUENCE MODELS

As with all sequence models used in MOST, the MaxiMOST sequence models provide for the analysis of the movement of objects. Three primary sequence models are used to analyze long-cycle, non-repetitive, manual activities.

1. *Part Handling Sequence Model*—For the analysis of spatial or restricted movement of parts or objects.
2. *Tool Use Sequence Model*—For the analysis of common hand tools or equipment.
3. *Machine Handling Sequence Model*—For the analysis of operating a machine.

For proper application of the MaxiMOST sequence models, consider the complete activity. The complete activity includes both the *get* and *place* of one or more parts or objects, independent of the number of locations visited to get the parts or objects. This principle applies to the three primary sequence models in MaxiMOST. For example, if an operator walks to a table, gets a flange, moves to a pallet to get a bracket and then carries both back to a workbench where they are set aside, one Part Handling (**A B P**) Sequence Model would be used to analyze this activity.

Two supplementary sequence models are also included in the MaxiMOST System:

- *Powered Crane Sequence Model*—For the analysis of the movement of one or more objects with the aid of an overhead bridge crane.
- *Powered Truck Sequence Model*—For the analysis of the movement of one or more objects with the aid of a powered wheeled truck.

The methodology behind the MOST Work Measurement Technique is that the system can be used for any type of manual work. The sequence models in MaxiMOST provide a structure to analyze different types of work. Technology is advancing rapidly and can impact how work is performed. MOST, however, can still be used to analyze any manual motions affected by technology. Through proper certification training, it will become clearer how to measure new technologies with MOST.

The MaxiMOST sequence models are shown below:

Activity	Sequence Model	Parameter	
Part Handling	**A B P**	**A**	Action Distance
		B	Body Motion
		P	Get and Place Parts
Tool Use	**A B T**	**T**	Get, Use and Aside Tool
Machine Handling	**A B M**	**M**	Operate Machine or Equipment
Powered Crane	**A T K T P T A**	**T**	Transport
		K	Hook-up and Unhook
		P	Place Object
Powered Truck	**A S T L T L T A**	**S**	Start and Stop
		T	Transport
		L	Load or Unload

The index values applied in a MaxiMOST analysis follow the same scale as BasicMOST (e.g., 1, 3, 6, 10, 16, 24, etc.), however, the multiplier for the index values is one-hundred (100). Since the time interval represented by each index value is large,

MaxiMOST can account for non-repetitive, highly-variable activities while maintaining many of the benefits of the BasicMOST System. Because of the decreased level of detail with MaxiMOST, the analyst must look at the work to be measured differently than with BasicMOST.

Parameter Indexing

Before assigning index values, the MOST analyst must ask two questions to determine the appropriate sequence model:

1. What is the object being moved?
2. How is the object moved (this question is key to determining the appropriate sequence model to use)?

Then, assuming a Part Handling Move:

- What is the total object movement to get and place the item and then return?

Asking these questions is vital to the effective application of MaxiMOST. The answers will help the analyst:

- Avoid overlooking any operator activity or analyzing any unnecessary activity.
- Correctly divide a process into method steps.
- Determine Action Distance as a round trip.
- Write accurate and clear method descriptions.
- Determine the index value for each parameter (sub-activity).
- Apply MaxiMOST consistently.

Indexing each parameter of the Part Handling Sequence Model is accomplished by observing or visualizing the total object movement during each phase of an activity and selecting the appropriate index value from the data card (Figure 5.4). The index value for each parameter is identified from the data card or a software application and is manually documented as $\mathbf{A_3}$. Software applications feature a built-in data card, and index values can be selected from a drop-down list or input from memory.

Consider the example of an associate walking four steps to get a part from a bin and then walking eight steps to clip the part onto a side panel. Since MaxiMOST includes the 'round trip,' the steps taken to get the part and the steps to put the part to the panel are included in one Action Distance value. The sequence model for this activity is:

$\mathbf{A_3} \quad \mathbf{B_0} \quad \mathbf{P_1}$

Referring to the Action Distance data card (Figure 5.1), since there is a total of 12 steps, an $\mathbf{A_3}$ is applied. No Body Motion is needed, and the part is attached by placing it with light pressure resulting in a $\mathbf{P_1}$ (Figure 5.4).

MaxiMOST System				A - Action Distance and B - Body Motion					
	A Action Distance	B Body Motion							
Index x 100	Steps	Vertical Motions	Pass Through Openings	Combined Body Motions	Ladder		Obstructed Ladder		Index x 100
					Light Load	Heavy Load	Light Load	Heavy Load	
					Rungs				
0	1 - 2	-	-	-	-	-	-	-	**0**
1	3 - 9	1 or 2 Bends Kneel Sit or Stand Climb On or Off	Door or Hatch	Bend and Sit Stand and Bend	-	-	-	-	**1**
3	10 - 23	Sit and Stand Crawl Creep 2 Kneels Climb On and Off 3 - 6 Bends Climb-Object On-Floor	2 Doors or Hatches Mechanical Door Manhole	Bends and Sit Stand and Bends Sit and Stand and Bends Bends and Climb Bends and Door or Hatch Hatch and On-Floor Bends and Climb On and Off	10	2	5	-	**3**
6	24 - 42	Flat-Crawl	Obstructed-Manhole		25	8	19	6	**6**
10	43 - 70		2 Obstructed-Manholes		45	16	40	14	**10**
16	71 - 108					28		26	**16**

FIGURE 5.1 Action Distance and Body Motion data card. Values are read up to and including.

The time to perform this activity is calculated by adding all index values in the sequence model. The multiplier for MaxiMOST is one-hundred (100) so the index values can be added together to achieve a total in TMU: $(3+0+1)\,4\times100=400$ TMU. Refer to Chapter 1 for a review of Time Measurement Units (TMU).

Action Distance and Body Motion

The Part Handling, Tool Use and Machine Handling Sequence Models all contain **A** and **B** parameters. The definitions for Action Distance and Body Motions are explained below and apply to all of the sequence models in MaxiMOST.

Parameter Definitions

A Action Distance

This parameter covers all spatial movements or actions (mainly horizontal) of the fingers, hands and/or feet during a move from one location to another and returning to the location that occurs within the sequence. Action Distance also covers the walking to or from the location of transportation equipment (e.g., crane or truck).

B Body Motion

This parameter is used to analyze either the vertical motions of the body or the actions necessary to overcome obstructions or impairments to body movement occurring within the sequence.

Action Distance (A)

The Action Distance (**A**) parameter is used for the analysis of the horizontal movement of an operator with or without a load from one work area to another. Action Distance includes the horizontal movement of the trunk of the body caused by taking steps. Steps can be taken forward, to the side or to turn the body around. To correctly use Action Distance, the analyst counts the total number of steps taken and selects the index value directly from the data card.

Action Distance in MaxiMOST is described in a single Action Distance parameter in each of the sequence models. This is different from the BasicMOST sequence models in which a separate Action Distance precedes each action (e.g., Get, Put and Tool Action) of the sequence model. The BasicMOST Action Distance allows time for a one-way action. MaxiMOST Action Distance allows time for a complete round trip to get and place an object as well as the Action Distance for the operator to return. (For examples of operator movement, see Figure 5.2.) Although these values

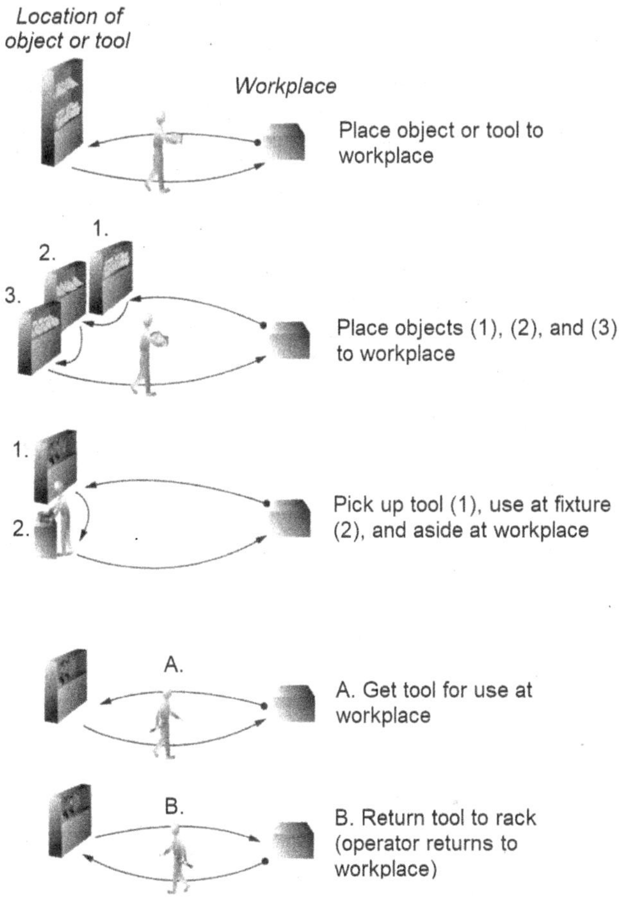

FIGURE 5.2 Examples of operator movement.

generally refer to the horizontal movement of the body, they also apply to walking up or down normally inclined stairs. Index values are given in terms of steps or distance.

A_0

The Action Distance data card (Figure 5.1) shows an index value of zero for distances of up to two steps. The time to travel up to two steps is included in the Part Handling, Tool Use and Machine Handling parameters. This is recognized in the Action Distance (**A**) parameter by assigning an index value of zero to distances that require two steps or less.

Example: Take one side step to press a button and return.

A_1–A_{16} Within Work Area Walking

Normally, the index values up to 16 are used when the Action Distance occurs within a defined work area, when walking is obstructed, when a heavy load is carried or any time the operator's steps must be reduced in length ≤ 2 1/2 feet (0.75 m) per step, such as when a trip includes multiple stops and changes of direction. The Action Distance index values up to 16 are determined according to the number of steps taken. *Step* refers to the number of times each foot hits the floor.

The work *area* is the area in which the worker is primarily engaged, for example, 'assembly line station #24.' A work area most typically is composed of many 'workplaces,' for example, a workbench, a truck chassis or a tub containing parts.

Examples: Walk through welding area, stepping over hoses and cables with 10 steps: $\mathbf{A_3}$.
Walk around end of workbench to pallet with six steps: $\mathbf{A_1}$.

A_1–A_{196} Straight Path Walking

When an operator is walking in a relatively straight path at full stride, the Action Distance index values found in Figure 5.3 should be used. The distance is selected in feet or meters. The index values of $\mathbf{A_1}$–$\mathbf{A_{196}}$ from Figure 5.3 are used when the operator is walking, either unloaded or with a light load only, at full stride. Walking longer distances normally occurs when walking between work areas, for example, when an operator walks from assembly line station #24 to the supervisor's office.

Since workers should not be carrying a heavy load for longer distances, the index values for full stride walking do not apply.

Example: Walk a total of 440 feet (134 m) to the foreman's office, get a job packet and return to the warehouse: $\mathbf{A_{54}}$.

Body Motion (B)

Body Motion refers to either vertical motions of the body or the actions necessary to overcome an obstruction or impairment to body movement. The Body Motion index values allow time for moving or positioning the body. Body Motions are more common in application areas for MaxiMOST than they are with the other MOST Systems.

Action Distance		
Index Value	*Feet*	*Meters*
A_1	24	7
A_3	60	18
A_6	120	36
A_{10}	190	60
A_{16}	300	90
A_{24}	420	130
A_{32}	550	170
A_{42}	720	220
A_{54}	920	280
A_{67}	1120	340
A_{81}	1350	410
A_{96}	1590	490
A_{113}	1860	570
A_{131}	2160	670
A_{152}	2480	770
A_{173}	2820	870
A_{196}	3170	980

FIGURE 5.3 Action Distance values for straight path (full stride) walking.

In addition to the Body Motions, such as bend and kneel, values for climbing on or off platforms or large objects, climbing a ladder and passing through openings have been developed and are included on the Body Motion data card (see Figure 5.1). These Body Motions account for time to overcome an obstacle to the operator's progress during an Action Distance. Note that walking up and down normally inclined stairs is analyzed with the Action Distance parameter, but climbing a ladder always requires grasping the ladder, handrail or rung for balance and is analyzed using the Body Motion parameter.

Additional activities associated with these Body Motions, such as opening or closing a door or hatch or turning toward or away from an obstruction that is climbed over, are included in the Body Motion index values.

Note: When determining the Action Distance with a Body Motion, do not include the steps on or off an obstruction, the steps through a door or hatch or turning to or away from an obstacle that is climbed on or off. These steps are included in the Body Motion index value. Body Motions are broken up into four sections:

1. Vertical Motions
2. Pass Through Openings
3. Combined Body Motions
4. Ladder or Obstructed Ladder

Many of the index values in Body Motions are simply multiple activities of a basic Body Motion or a combination of Body Motions. Therefore, the general definitions of the main body motions are explained below and the detailed body motions with index values are listed later in this section.

Bend and Arise

From an erect standing position, the trunk of the body is lowered by bending from the waist and/or knees to allow the hands to reach below the knees and subsequently return to an upright position. It is not necessary, however, for the hands to *actually* reach below the knees, only that the body be lowered sufficiently to allow the reach. Characteristics of a Bend are bending from the waist with the knees stiff, stooping down by bending at the knees or kneeling down on one knee.

Kneel

The index values for Kneel apply to kneeling onto both knees and arising to an upright position.

Sit or Stand

Sit or Stand is used when the body is lowered onto a seat from an erect position or stands from a seat with or without a series of several hand, foot and body motions to move a chair or stool into a position. All of the motions to manipulate the chair and body are included in the Sit or Stand Body Motion. If the chair or stool is stationary and several foot and body motions are necessary either to situate the body comfortably in the seat or to climb on or off the stool, the Sit or Stand value would also apply.

Climb On or Off

Climb On or Off covers climbing on or off a work platform or any raised surface (approximately 3 feet or 1 m high) using a series of hand and body motions to lift or lower the body. Climbing onto a platform is accomplished by first placing one hand on the edge and then lifting the knee to the platform. By placing the other hand on the platform and bending forward, the weight of the body is shifted, allowing the other knee to be lifted onto the platform. The activity is completed by arising from both knees. Climbing off the platform consists of the same actions but performed in the reverse order. This Body Motion covers Climb On *or* Climb Off, not both.

Door or Hatch

Passing through a door or hatch normally consists of reaching for and turning the handle, opening the door, walking through the door and subsequently closing the door. This value will apply to virtually all hinged, double, sliding or swinging doors or a hatch.

The three or four steps required to pass through the doorway or hatch are included in the **B** value as well as bending, if required, to pass through a low opening having an obstructed step. These steps should not be added to the Action Distance or subtracted from it.

The following list contains all of the Body Motion data card activities (Figure 5.1) and their definitions.

1. Vertical Motions

Body Motion	Definition	Index
1 or 2 Bends	Bend, stoop or kneel on one knee and arise; bend with hand that may go below knees up to two bends and arise.	B_1
Kneel	Kneel onto both knees and arise.	B_1
Sit or Stand	Sit down on stool or chair with or without adjusting the position of the stool or chair **or** stand up from stool or chair with or without positioning the stool or chair.	B_1
Climb On or Off	Climb on **or** off a platform approximately 3 feet (1 m) high.	B_1
Sit and Stand	Sit on a stool or chair and arise or arise from a stool or chair **and** sit on the same stool or chair or a different one.	B_3
Crawl	Kneel and crawl on hands and knees a distance ≤ 11 feet (3 m) and subsequently arise.	B_3
Creep	Lie down on a wheeled creeper, pull in and out a total distance ≤ 13 feet (4 m) and arise from creeper.	B_3
2 Kneels	Kneel on both knees and arise. This activity is performed twice.	B_3
Climb On and Off	Climb on **and** off platform.	B_3
3–6 Bends	Bend and arise three to six times.	B_3
Climb-Object	Climb up or down three steps each 18 to 24 inches (45 to 60 cm) high and then bend or position the body carefully on the object before working, or go up or down two steps and then climb onto a platform with knees in lying, sitting or kneeling position on the object or climb off of the object the same way.	B_3
On-Floor	Lie down on the floor and arise into a standing position.	B_3
Flat-Crawl	Lie down and crawl on the stomach a distance ≤ 12 feet (3.5 m) and arise into a standing position.	B_6

2. Pass Through Openings

Body Motion	Definition	Index
Door or Hatch	Pass through hinged or swinging door by opening and closing; includes the time to take the three or four steps to go through the door **or** pass through a small, low opening requiring bending. The bottom of the opening may be raised, requiring an obstructed step to get through.	B_1
2 Doors or Hatches	Pass through two doors, opening and closing each **or** pass through two hatches.	B_3
Mechanical Door	Pass through mechanically operated door and wait for the door to open and close.	B_3
Manhole	Pass through a vertical manhole at floor level or a horizontal manhole requiring a climb of three rungs to get up or down.	B_3
Obstructed-Manhole	Pass through an obstructed and tight vertical or horizontal manhole.	B_6
2 Obstructed-Manholes	Pass through an obstructed and tight vertical or horizontal manhole twice.	B_{10}

3. Combined Body Motions

Body Motion	Definition	Index
Bend and Sit	Bend and arise **and** sit on a stool or chair at another location.	B_1
Stand and Bend	Arise from a stool or chair **and** bend and arise at another location.	B_1
Bends and Sit	Bend and arise two to three times **and** sit on a stool or chair at another location.	B_3
Stand and Bends	Arise from a stool or chair and bend **and** arise two to three times at other locations.	B_3
Sit and Stand and Bends	Sit **and** stand from a stool or chair **and** bend and arise two to three times at other locations.	B_3
Bends and Climb	Bend and arise two to three times **and** at another location, climb on or off a platform.	B_3
Bends and Door or Hatch	Bend and arise two to three times **and** pass through a door **or** bend and arise two to three times **and** pass through a hatch.	B_3
Hatch and On-Floor	Pass through a hatch **and** lie down on the floor.	B_3
Bends and Climb On and Off	Bend and arise two to three times **and** at another location climb on and off a platform. The activity is performed twice.	B_3

4. Ladder or Obstructed Ladder

Body Motion	Ladder—Definition
Light Load	Climb up or down a ladder more than two rungs without a load or with a light load. The index value is determined by the number of rungs. The characteristic of a light load is hand over hand climbing with no significant pause between rungs of the ladder.
Heavy Load	Climb up or down a ladder more than two rungs with a heavy load. The index value is determined by the number of rungs. The characteristic of a heavy load is the coming to rest of both feet on each rung prior to taking the next step.

Body Motion	Obstructed Ladder—Definition
Light Load	Climb up or down a ladder more than two rungs with obstructed access at upper end (e.g., bend, kneel or crawl under rail) with or without a light load.
Heavy Load	Climb up or down a ladder more than two rungs with obstructed access at upper end, with a heavy load and both feet resting on each rung.

Note: For climbing less than two rungs on a ladder, use the value for Climb On or Off under Vertical Body Motions.

THE PART HANDLING SEQUENCE MODEL

Part Handling represents the movement of objects freely through the air or the restricted movement of an object. The movement is for one or more parts or objects to a general or specific location.

One or more objects can be moved with one or both hands. For simplification of the text, when one object is referenced it can mean one or more objects unless it specifically states only one object in the definition. Normally, the operator moves a distance to get the part or object and moves a distance to place the part or object. Part Handling follows a fixed sequence of sub-activities identified by the following steps:

1. *Reach* with one or two hands a distance to an object either directly or in conjunction with body motions or steps.
2. *Gain* manual *control* of the object.
3. *Move* the object a distance to the point of placement, either directly or in conjunction with body motions or steps.
4. *Place* the object in a temporary or final position.
5. *Return* to the original workplace with steps, if necessary.

The Part Handling Sequence Model accounts for the total walking distance and the total body motions required for the moving of a part and is used for situations such as:

- Get and place part by hand.
- Push or pull objects over a distance.
- Get several parts at different locations and place those parts at another location.
- Exchange a workpiece in a machine.
- Position a hand tie-wrap around cables.
- Place a handful of fasteners and a wrench on bench.

The Sequence Model

The sequence model consists of a series of letters (parameters) representing each of the various sub-activities of Part Handling. The **A** and **B** parameters shown below were introduced earlier in the chapter. The new parameter for Part Handling **(P)** will be presented in this section.

A B P

where: **A** = Action Distance
B = Body Motion
P = Part Handling

Parameter Definitions

P Part Handling

This parameter is used to analyze the gain control and placement of one or more objects to a general or specific location that can be 'final' or from which further handling of the part can be made.

Parameter Indexing

The **P** value in Part Handling includes the time to Gain Control *and* Place one or more objects and is broken down into two data cards:

1. *General Move*—For the spatial movement of objects (Figure 5.4).
2. *Controlled Move*—For the movement of objects along a controlled path (Figure 5.5).

Part Handling activity words are used to describe the activity being done. This provides a more consistent description of the method since analysts use the same words. The words are shown on the data card and provide a guide as to which index values may be used for the activity. The activities are grouped according to difficulty

MaxiMOST System				General Move					P – Part Handling		
Index x 100	Handle Parts					Handle Parts with Adjustments					Index x 100
	Pickup or Hold and Move	Collect and Move	Put	Place	Position	Pickup or Hold and Move	Collect and Move	Put	Place	Position	
	Small	Small, Light or Medium Size and Weight			Medium or Heavy Weight	Small	Small, Light or Medium Size and Weight			Medium or Heavy Weight	
	Actions			Objects		Actions			Objects		
1	2	2	3	2	1	2	-	1	-	-	**1**
3	17	8	10	6	3	13	6	7	4	2	**3**
6		17	18	11	6		15	16	10	5	**6**
10				18	10					10	**10**

FIGURE 5.4 Part Handling data card—General Move.

MaxiMOST System						Controlled Move							P – Part Handling		
Index x 100	Move						Move with Adjustments						Line Handling		Index x 100
	≤ 12 in. (30 cm)				> 12 in. (30 cm)		≤ 12 in. (30 cm)				> 12 in. (30 cm)		Type of Handling		
	Situate	Manipulate	Shove	Draw	Push or Pull	Slide	Situate	Manipulate	Shove	Draw	Push or Pull	Slide	Straight	Into Hand or On Floor	
	Heavy, Large and Bulky	Heavy, Large and Bulky	Small, Light or Medium Size and Weight	Medium or Heavy Weight	Light Weight Cart, Conveyor Table	Medium or Heavy Weight Rails, Conveyor Table	Heavy, Large and Bulky	Heavy, Large and Bulky	Small, Light or Medium Size and Weight	Medium or Heavy Weight	Light Weight Cart, Conveyor Table	Medium or Heavy Weight Rails, Conveyor Table			
	Objects		Actions		Distance in Feet (m)		Objects		Actions		Distance in Feet (m)		Tugs	Coils	
1	-	-	6	1	8 (2.5)	3 (1)	-	-	-	-	-	-	2	1	**1**
3	2	1	19	7	29 (9)	13 (4)	1	-	14	5	21 (6.5)	10 (3)	8	3	**3**
6	4	2		16	58 (17.5)	28 (8.5)	3	2		14	50 (15)	24 (7.5)	16	7	**6**
10	7	4			99 (30)	48 (14.5)	7	3			91 (28)	45 (14)	28	13	**10**
16	12	6			157 (48)	78 (24)	11	6			150 (46)	74 (22.5)	44	21	**16**

FIGURE 5.5 Part Handling data card—Controlled Move.

observed in handling the object and then by the nature of the object itself. The index values for Part Handling are determined by the following criteria:

1. Nature of the object
2. Difficulty of handling
3. Number of actions or objects

Nature of Object

The first criterion that can be used to determine the **P** value is to consider the nature of the part being handled. The terms used on the Part Handling data cards to describe the nature of the part are defined as follows:

Small or light: A small or light part may be held in one hand while working with another object in the same hand.

Medium: A part of medium weight and size cannot be held in one hand while working with another object in the same hand.

Heavy: A heavy part is recognized by the hesitation or pause exhibited by the operator when gaining control of the part (usually with both hands).

Large or bulky: A large or bulky object requires several regrasps when gaining control of the object or intermediate moves when placing it.

Difficulty of Handling

The second criterion for determining the index value to be assigned for the **P** parameter is the difficulty in gaining control and in placing the part. The index value for **P** is never chosen by the nature of the object alone. The nature of the object was explained previously. Although the nature of the object may be an influence in gaining control and in placement, it is the difficulty of both that determines the value chosen for **P**.

Number of Actions or Objects

After the nature of the object and difficulty in handling have been identified, it is then necessary to determine the number of actions or objects to select the correct index value.

Number of Actions is used for small, light objects when one action may result in the handling of a handful of objects. Small, light objects may also be handled one at a time. When only one object is handled at a time, the number of actions is equal to the number of objects.

Number of Objects is used for objects that cannot be handled more than one at a time.

Part Handling–General Move (P)

The Part Handling–General Move data card (Figure 5.4) is used for moving objects along a spatial path. If the object is moved and then slid or is pushed along a controlled path, the Part Handling–Controlled Move data card should be used. The activities for General Move are broken up into two categories: Handle Parts and Handle Parts with Adjustments.

Handle Parts

Pickup

The index values for Pickup include only the time to get a part or object and hold; no placement occurs. The index value is determined by the number of grasping actions. Typical part or object characteristics are small, light objects where the gain control is easily accomplished either by simply grasping or collecting the objects.

Examples: Pick up a handful of washers from tote pan: $\mathbf{P_1}$.
Reach around machine to pick up part with two grasping actions: $\mathbf{P_1}$.

Hold and Move

The index values for Hold and Move include only the time to place a part or object or a handful of parts or objects that are already in the hand. The index value is determined by the number of placing actions. Typical part or object characteristics are small, light parts or parts of medium size and weight where the placing activity occurs with either a simple asiding motion to a general location or where adjustments, light pressure or two distinct placements are observed to move the object to a more specific location.

Examples: Hold and move bracket to workbench: $\mathbf{P_1}$.
Hold and move three rubber grommets to location in firewall: $\mathbf{P_3}$.

Collect and Move

The index values for Collect and Move include getting and placing a part or object into the other hand or a container being held in the other hand and putting the collection of parts or the container aside. A sliding action of ≤ 12 inches (30 cm) of the container may occur as part of this index value. The index value is determined by the number of collecting actions. Typical part or object characteristics are small, light parts or parts of medium size and weight where the placing activity occurs with either a simple asiding motion to a general location or where adjustments, light pressure or two distinct placements are observed to move the object to a more specific location.

Example: With three collecting actions, move seven bolts to a pan being held in the other hand and then put the pan on a counter: $\mathbf{P_3}$.

Put

The index values for Put include getting and placing a part or object to a general location. The index value is determined by the number of placing actions. Typical part or object characteristics are small, light parts or parts of medium size and weight where the gain control is easily obtained, and the placement is a simple asiding motion to a general location. No further movement or insertion of the part is covered.

Examples: Put base from conveyor to bench for assembly: $\mathbf{P_1}$.
Put completed small tire assembly on pallet: $\mathbf{P_1}$.

Place

The index values for Place include getting and placing a part or object at a specific location where adjustments, light pressure or two distinct placements are observed. An insertion of ≤ 12 inches (30 cm) may occur as part of this index value. The index values are determined by the number of objects placed. Typical part or object characteristics are small- or medium-weight parts where the gain control is easily obtained.

Examples: Place one part into fixture: $\mathbf{P_1}$.
Place three parts onto a workbench: $\mathbf{P_3}$.

Position

The index values for Position include getting and placing a part or object at a specific location where care or precision, heavy pressure, intermediate moves or blind or obstructed access occurs at the point of placement. An insertion of ≤ 12 inches (30 cm) may occur as part of this index value. The index value is determined by the number of objects positioned. Typical part or object characteristics are medium- or heavy-weight parts where the gain control is more difficult and may include a hesitation or pause before complete control of the object is obtained.

Examples: Position casting into milling fixture: $\mathbf{P_1}$.
Position two medium-weight molds into machinery: $\mathbf{P_3}$.

Handle Parts with Adjustments

Handle Parts with Adjustments is similar to Handle Parts in that the difficulty observed in handling the object and the nature of the object are considered in identifying the index value (Figure 5.4). The only difference between the two categories is the added adjustments.

Handle Parts with Adjustments may be used when preliminary or subsequent adjustments are required to prepare for, or as a result of, placing the part. These adjustments may be performed on the part or on the surroundings of the part. 'With Adjustments' does not refer to adjustments in the initial placement of the part, but to preliminary adjustments required to prepare for the placement and to supplemental adjustments occurring after the placement of up to 4 seconds in duration.

Adjustments may include adjustments to the object, adjustments to the surroundings or a few seconds waiting for a process to occur.

- Adjustments to the *object* include an additional sliding or second movement of the part. The adjustment may involve guiding or aligning the object from one to three stops or marks.
- Adjustments to the *surroundings* are those brief activities required for placing the part. These adjustments include opening or closing a clamp or fixture, pushing or pulling a nearby object, alignment (up to three points) of a nearby object, an inspection of the surroundings lasting up to 4 seconds or activating a button or lever when required for or resulting from placing the object.

- Adjustments also *provide time*, up to 4 seconds, for inspecting the part. The inspection may occur before or after the placement. Waiting is also considered an adjustment when the waiting is associated with the placement of the object. The waiting may occur prior to or after the placement.

Adjustments required to *prepare for placement* include and are limited to:

- Opening a clamp or fixture by activating a lever.
- Pushing or pulling a nearby object ≤ 12 inches (30 cm).
- A process time of up to 4 seconds.
- Associated head movement and visual inspection up to 4 seconds.

Adjustments required as a *result of placement* include and are limited to:

- Pushing a button or pulling a lever to start a machine or process.
- Closing a clamp or fixture by activating a lever.
- Pushing or pulling a nearby object ≤ 12 inches (30 cm).
- A process time of up to 4 seconds.
- Associated head movement and visual inspection up to 4 seconds.
- Part alignment up to three points.
- Closing a clamp or fixture and subsequently pushing a button or pulling a lever to start a machine or process.
- Part alignment up to three points and subsequently pushing a button or activating a lever to start a machine or process.

Part Handling–Controlled Move (P)

The Part Handling–Controlled Move data card (Figure 5.5) is used for moving objects along a controlled path or over a surface. If the object is moved along an unrestricted path, the Part Handling–General Move data card should be used. The values on the Part Handling–Controlled Move data card apply to any sliding, pushing or pulling of an object over a surface. Moving objects along a surface applies to sliding objects on a solid surface, pushing or pulling a manual walking truck, wheeled table or cart across a flat surface or along rails or sliding an object along a conveyor.

The index values for a Controlled Move within Part Handling follow the same rules as stated within the General Move section. The index value for a Controlled Move will be determined by the following criteria:

1. Nature of the object
2. Difficulty of handling
3. Number of actions or objects

The activities for Controlled Move are divided into three categories: Move, Move with Adjustments and Line Handling.

Move

The index values for Move are divided into two main categories based on the distance required to move the object. Similar to General Move, the index values are then identified based on the nature of the object and the difficulty in handling.

Move ≤ 12 Inches (30 cm)

The index values apply when an object is manipulated along a controlled path ≤ 12 inches (30cm). This may apply to a single linear action, to reciprocal actions or to a series of actions along a controlled path. For a series of moves, this is cumulative up to 12 inches (30cm). The index values apply when the hand acts directly on the object. Sometimes the activity is accomplished with one hand, but heavier pushing may require the use of both hands. Count each direction as a separate action when determining the index value.

Situate

The index values for Situate include the time to get and position an object with an additional sliding movement of ≤ 12 inches (30 cm) when gaining control of and/or when positioning the object. The gain control of the object can include a more difficult gain control than a simple pickup and may include a hesitation or pause. Also included is the time for a brief visual check for location. The index value is determined by the number of objects moved. Typical part or object characteristics are heavy, large and bulky objects.

Example: Situate wheel over studs on axle: $\mathbf{P}_3$.

Manipulate

The index values for Manipulate include the time to get and position an object with an additional sliding movement of ≤ 12 inches (30 cm). The gain control of the object can include a more difficult gain control than a simple pickup and may include a hesitation or pause. Also included in the values for Manipulate is the time to align the object to two points outside the area of normal vision, inspect up to three points and a brief visual check for location. The area of normal vision is the average area covered by a single eye focus and is described by a circle 4 inches (10 cm) in diameter at a normal reading distance of 16 inches (40 cm) from the eyes. The index value is determined by the number of objects moved. Typical part or object characteristics are heavy, large and bulky objects.

Example: Manipulate bumper onto mounting brackets: $\mathbf{P}_3$.

Shove

The index values for Shove include the time to get the object and move the object along a controlled path or over a surface. The Shove values are used when little resistance is encountered in performing the actions. A slight hesitation is noted at the beginning of the action and slowing due to resistance is observed during the action. When the Shove actions apply, the resistance is light enough to be overcome with one hand. The index value is determined by the number of actions. Typical part or object characteristics are small, light parts or parts of medium size and weight.

Example: Guide piece of wood into wood chipper with three shoving actions: $\mathbf{P}_1$.

Draw

The index values for Draw include the time to get an object and move the object along a controlled path or over a surface. The Draw actions are used when there is considerable resistance. The actions are characterized by a pause in starting the action due to a building up of muscular force to overcome resistance to the action. If a reciprocating action has resistance in one direction, apply the Draw index values for all the actions. Draw actions often require the use of both hands to overcome resistance. The index value is determined by the number of actions. Typical part characteristics are medium- or heavy-weight parts.

Example: Open and close a heavy tool chest drawer: $\mathbf{P_3}$.

Move > 12 Inches (30 cm)

The index values apply when an object is manipulated along a controlled path, greater than 12 inches (30 cm). This may apply to a single linear action, to reciprocal actions or to a series of actions along a controlled path. The index value is selected by the *distance moved in feet (meters)*. The index values apply when the hand acts directly on the object. Sometimes the activity is accomplished with one hand, but heavier pushing may require the use of both hands. Index values are classified as Push or Pull or Slide.

Push or Pull

The index values for Push or Pull include the time to get the object and push or pull it along a controlled path or over a surface greater than 12 inches (30 cm). Push or Pull values are used when little difficulty or resistance is encountered in pushing or pulling the object. When Push or Pull actions apply, the resistance is light enough to be overcome with one hand. The index value is selected by the distance moved in feet (meters). Typical part characteristics are lightweight parts or small wheeled carts that are moved along a low friction surface or objects pushed on a conveyor.

Example: Push a lightweight carton distance of 5 feet (1.5 m) along a roller conveyor: $\mathbf{P_1}$.

Slide

The index values for Slide include the time to get and slide a part along a controlled path or over a surface greater than 12 inches (30 cm). Slide values are used when the part requires two hands to slide it on a surface or when the force required to start the sliding motion results in a noticeable pause or hesitation prior to the movement of the part with one hand. The index value is selected by the distance moved in feet (meters). Typical part characteristics are medium- or heavy-weight parts or objects such as loaded walking trucks or heavy parts slid across a table or on the floor. An example of a walking truck is shown in Figure 5.6. If a hand truck is used, the loading and unloading of the hand truck is analyzed as a General Move within the Part Handling Sequence Model.

Example: Slide a full, heavy hand truck 20 feet (6 m): $\mathbf{P_6}$.

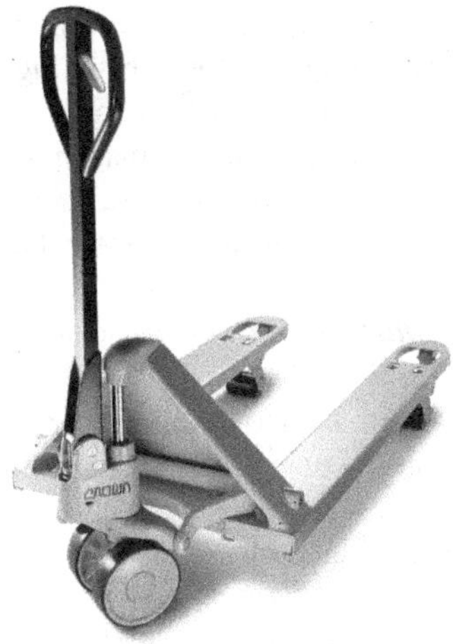

FIGURE 5.6 Example of a walking truck.

Move with Adjustments

The adjustments definitions are identical for a General Move or Controlled Move. Adjustments include and provide time for a number of brief activities, usually requiring 4 seconds or less, that are associated with the primary activity.

Adjustments apply when additional adjustments are required prior to completing the move or after the completion of the move. The adjustments referred to here are not those adjustments required for the initial placement in locating the object. This time is included in the moving of the object. These adjustments refer to the activity that prepares the surroundings for the object, turns or relocates the part after the original placement is complete or makes changes to the surroundings necessitated by moving the object, up to 4 seconds.

The definitions for the values for Move with Adjustments are the same as those for the Move values. The only difference between the two cases is the added adjustment. The index values for both categories are still selected based on the nature of the object, the difficulty of handling and the number of objects or actions as shown on the Part Handling–Controlled Move data card (Figure 5.5). As noted earlier, the definitions for adjustments for General Move and Controlled Move are the same. The definitions of adjustments are repeated below as a convenience for the reader.

Adjustments may include adjustments to the object, adjustments to the surroundings or a few seconds waiting for a process to occur.

- Adjustments to the *object* include an additional sliding or second movement of the part. The adjustment may involve guiding or aligning the object from one to three stops or marks.
- Adjustments to the *surroundings* are those brief activities required for placing the part. These adjustments include opening or closing a clamp or fixture, pushing or pulling a nearby object, alignment (up to three points) of a nearby

object, an inspection of the surroundings lasting up to 4 seconds or activating a button or lever when required for or resulting from placing the object.
- Adjustments also *provide time,* up to 4 seconds, for inspecting the part. The inspection may occur before or after the placement. Waiting is also considered an adjustment when the waiting is associated with the placement of the object. The waiting may occur prior to or after the placement.

Adjustments required to *prepare for placement* include and are limited to:

- Opening a clamp or fixture by activating a lever.
- Pushing or pulling a nearby object ≤ 12 inches (30 cm).
- A process time of up to 4 seconds.
- Associated head movement and visual inspection up to 4 seconds.

Adjustments required as a *result of placement* include and are limited to:

- Pushing a button or pulling a lever to start a machine or process.
- Closing a clamp or fixture by activating a lever.
- Pushing or pulling a nearby object ≤ 12 inches (30 cm).
- A process time of up to 4 seconds.
- Associated head movement and visual inspection up to 4 seconds.
- Part alignment up to three points.
- Closing a clamp or fixture and subsequently pushing a button or pulling a lever to start a machine or process.
- Part alignment up to three points and subsequently pushing a button or activating a lever to start a machine or process.

Line Handling

Line Handling includes the manual motions required for activities performed to handle lines, hoses, cables or any long, flexible object. The actions considered are those particular to these objects. Lines are pulled from place to place, through openings, from reels and into coils either in the operator's hand or on the floor. The index values include time to gain control of the line and manipulate the line for the purpose of relocating or coiling the line.

Types of Line Handling

There are two types of Line Handling:

1. Straight (Tugs)
2. Into Hand or On Floor (Coils)

Straight (Tugs)

Straight line handling refers to handling the line with relatively straight strokes of the hand or hands. This involves very little control of the line after the tugging action that pulls a section of the line to a new location. This action may occur to *pull* a line through an opening, to *clear* a line from an area, to provide *slack* in a line at the operator location or to *remove* a line from a reel when little or no arrangement of the line is required. The action may be performed with alternate actions of the hands, 'hand over hand' or with simultaneous use of both hands. The index value selected is based on the number of tugs.

Example: A maintenance worker moves a hose with five tugs: $\mathbf{P_3}$.

Into Hand or On Floor (Coils)

Winding an object into the hand or on the floor describes a coiling activity. These index values are used when the operator coils or arranges the lines in a coil-like arrangement in the hand, on the floor or on another flat surface (Figure 5.7). The first part of the action is much like Straight line handling in that the line is generally pulled toward the operator. The significant difference is the added arranging of the line in the operator's hand or on the surface. When the coil is formed in the hand, time is included in this parameter for asiding the coiled line to a hook or to a surface. The index value selected is based on the number of coils.

Example: A fireman makes eight coils on the floor with a hose: $\mathbf{P_{10}}$.

Parameter Frequencies

Frequency

Frequency is an entire sequence occurring more than once. If an activity occurs more than once, the frequency will be specified in a MOST analysis and the time for the activity multiplied by the frequency indicated. The time calculation for a frequency activity is as follows:

1. Add all index values for the entire sequence model.
2. Multiply this total by the number of frequency occurrences.
3. Calculate the total TMU by multiplying by 100.

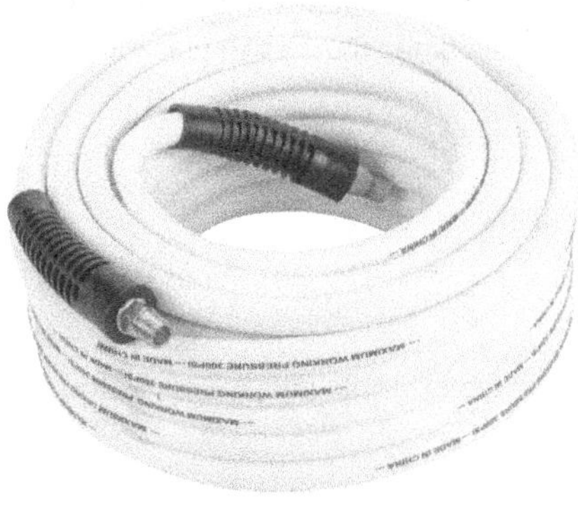

FIGURE 5.7 Example of On Floor (Coils)

Example: An associate walks a total of 15 steps to restock one part at a workstation. The process is completed for five workstations.

$\mathbf{A_3}$ $\mathbf{B_0}$ $\mathbf{P_1}$ **5**

$\mathbf{A_3}$ Walk 15 steps
$\mathbf{B_0}$ No body motion
$\mathbf{P_1}$ Put part

The time calculation for the example above is as follows:

1. $3+0+1=4$
2. $4\times5=20$
3. $20\times100=2000$ TMU

Partial Frequency

Partial frequency is one or more parameters within a sequence model occurring more than once. These types of activities do not occur often in MaxiMOST because the system design accounts for the multiple activities which often occur in applications where MaxiMOST is used. However, it is helpful to the analyst to understand this concept.

This activity is shown in the sequence model by placing parentheses around the parameters that are repeated and documenting the number of occurrences as a partial frequency in a MOST analysis, also within parentheses. The time calculation for partial frequency is shown below:

1. Add all index values for the parameters within parentheses.
2. Multiply this value by the number of occurrences (the number noted as a partial frequency at the end of the sequence model).
3. Add this total to the remaining parameter index values.
4. Convert the total to TMU by multiplying by 100.

Example: A construction site foreman walks a total of 550 feet (170 m) and goes through two mechanical doors to place a tool box on a workbench.

$\mathbf{A_{32}}$ $\mathbf{(B_3)}$ $\mathbf{P_1}$ **(2)**

$\mathbf{A_{32}}$ Walk 550 feet (170 m)
$\mathbf{B_3}$ Mechanical door
$\mathbf{P_1}$ Place tool box

The time calculation for the example above is as follows:

1. $3\times2=6$
2. $6+32+1=39$
3. $39\times100=3900$ TMU

Note: A frequency could be applied to any one or any combination of parameters. A frequency or partial frequency can be a whole number, decimal or fraction.

Writing Method Descriptions

One of the advantages of MOST is using a standard sequence model to accurately determine time values. Another advantage is that the method description that accompanies each sequence model can be written in such a manner to consistently and clearly define the activity. It is recommended that the analyst follow a prescribed sentence structure and use consistent wording when writing method descriptions. This will provide other analysts and future readers of the analysis a clear understanding of the process.

Below is the recommended method description format for a clear and concise method description for Part Handling. Additional words for Action Distance or Body Motion are optional and may be used to enhance the method description.

The recommended sentence structure for Part Handling is:

Activity Object ⟨From Location⟩ To Location

⟨If the From Location is apparent, it is not necessary to indicate it in the method description.⟩

Part Handling Examples

1. An operator walks five steps and puts a light part in a machine.

   Put part in machine

   $\mathbf{A_1}$ $\mathbf{B_0}$ $\mathbf{P_1}$

   $(1 + 1)\times 100 = 200$ TMU

2. A worker walks 220 feet (67 m) through the warehouse to pick two products from a bottom shelf and returns 220 feet (67 m) to place the products in a box.

   Place products in box

   $\mathbf{A_{32}}$ $\mathbf{B_1}$ $\mathbf{P_1}$

   $(32+1+1)\times 100 = 3400$ TMU

3. An operator places a part into machine 10 steps away and then pushes one button to start the machine.

   Place part with subsequent adjustments (push button)

   $\mathbf{A_3}$ $\mathbf{B_0}$ $\mathbf{P_3}$

   $(3 + 3)\times 100 = 600$ TMU

4. An aircraft maintenance worker coils a hose with five coils and then places the hose on a hook 16 steps away.

   Coil hose with 5 coils

   $\mathbf{A_3}$ $\mathbf{B_0}$ $\mathbf{P_6}$

   $(3 + 6) \times 100 = 900$ TMU

5. An operator walks 20 steps to a ladder, climbs five rungs and puts a light part onto a shelf.

   Put part onto shelf 20 steps away using a ladder

   $\mathbf{A_3}$ $\mathbf{B_3}$ $\mathbf{P_1}$

   $(3 + 3 + 1) \times 100 = 700$ TMU

6. An operator walks a total of 55 steps and bends to position four objects to a machine.

   Position 4 objects to machine with bend

   $\mathbf{A_{10}}$ $\mathbf{B_1}$ $\mathbf{P_6}$

   $(10 + 1 + 6) \times 100 = 1700$ TMU

7. A warehouse worker walks 10 steps to a table and pushes a box 8 feet (2.5 m).

   Push box 8 feet (2.5 m) on table

   $\mathbf{A_3}$ $\mathbf{B_0}$ $\mathbf{P_1}$

   $(3 + 1) \times 100 = 400$ TMU

8. An operator walks six steps to a bench, grasps a medium-weight object and positions it on a low four-wheeled hand truck five steps away. The operator then gets the handle of the truck from 3 feet (1 m) away and pulls the object 36 feet (11 m). The truck is parked, and the operator walks 150 feet (46 m) to another work area.

   Position object to four-wheeled hand truck

   $\mathbf{A_3}$ $\mathbf{B_1}$ $\mathbf{P_1}$

   $(3 + 1 + 1) \times 100 = 500$ TMU

   Pull part to workplace using hand truck and return to work area

   $\mathbf{A_{10}}$ $\mathbf{B_0}$ $\mathbf{P_6}$

   $(10 + 6) \times 100 = 1600$ TMU

   $$\begin{array}{r} 500 \text{ TMU} \\ 1600 \text{ TMU} \\ \hline 2100 \text{ TMU} \end{array}$$

9. A worker walks four steps to the work area to gain control and position six heavy bags of plastic pellets onto a hand truck, bending three times.

   Walk 4 steps to position 6 bags of plastic pellets onto hand truck with three bends

   $\mathbf{A_1}$ $\mathbf{B_3}$ $\mathbf{P_6}$

   $(1+3+6)\times 100 = 1000$ TMU

10. A technician grasps a network cable and forms eight coils in her hand.

    Form 8 coils of network cable in hand

    $\mathbf{A_0}$ $\mathbf{B_0}$ $\mathbf{P_{10}}$

    $10\times 100 = 1000$ TMU

THE TOOL USE SEQUENCE MODEL

The Tool Use Sequence Model is applied to the analysis of the use of common hand tools or equipment or the use of the fingers or hand as a tool such as to tighten a bolt with a wrench or drive 10 nails with a hammer. Tool Use also includes activities related to cutting, surface treating, measuring and recording. As with the Part Handling Sequence Model, the Tool Use Sequence Model accounts for total walking distance and total body motions required for the complete use of common hand tools.

The Tool Use Sequence Model is used for the analysis of such activities as:

- Get, use and aside or return a tool.
- Get, make ready, use and aside a tool.
- Get, place and fasten fasteners by hand.
- Loosen and place or aside fasteners by hand.

Examples: Fasten bolts or nuts after placing washers.
Place nut and turn nut on bolt.
Use grease gun to lubricate axle.
Fasten threaded fasteners with final tightening.
Turn wire nut on wire ends.
Apply masking tape to window ledge.
Attach fasteners by hand.
Use hand or power tool and return it.
Change socket on tool and then use tool.
Use pliers to cut medium-gauge wire.
Use spray cans to apply cleaner.
Use hand as a tool for cranking a handwheel.
Deburr rough edge of metal sheeting with file.
Place tool and tighten.

The Sequence Model

The sequence model consists of a series of letters (parameters) representing each of the various sub-activities of Tool Use. The **A** and **B** parameters shown below were introduced in an earlier section. The new parameter for Tool Use (**T**) will be presented in this section.

A B T

where: **A** = Action Distance
B = Body Motion
T = Tool Use

Parameter Definitions

The Tool Use (**T**) parameter applies when a tool is used to perform work or when hands or fingers are used as tools. The use of eyes (think or read) and process times are also classified as Tool Use. All Tool Use index values include the time to walk one to two steps to get and aside the tool or object. To reduce the number of method steps in an analysis, certain additional activities are included within the Tool Use parameter, including:

- Change socket on tool.
- Use counter or holding tool to hold part in place.
- Place fastener onto tool or onto object.

The Tool Use parameter is broken into seven data cards:

1. Assemble or Disassemble Standard Fasteners
2. Tighten or Loosen Standard Fasteners
3. Assemble or Disassemble Long Fasteners
4. Tighten or Loosen Long Fasteners
5. General Tools I
6. General Tools II
7. Measuring Tools

Parameter Indexing

Assemble or Disassemble Standard Fasteners (T)

A *Standard Threaded Fastener* may be defined as a fastener that, when secure, has been turned in a distance one to two times its diameter (see Figure 5.8). Any fastener run in or out a distance greater than one to two times its diameter is considered a *long fastener*. Long fasteners will be discussed in the next section.

The Assemble or Disassemble Standard Fasteners data card (Figure 5.9) is used when a fastener is placed and started whether or not it is fully run in or tightened. This data is also used when a fastener, either tight or loose, is removed and laid aside.

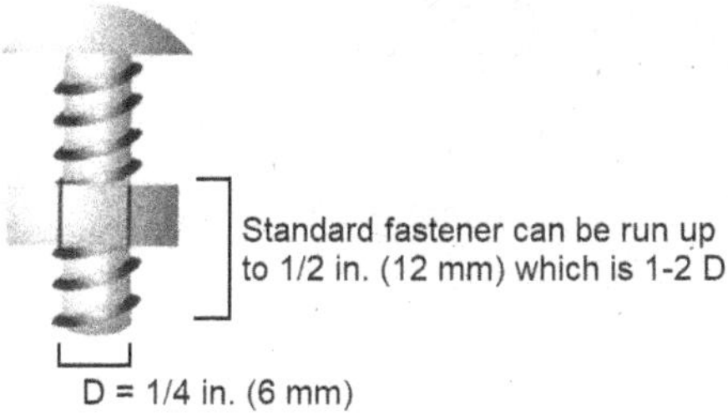

FIGURE 5.8 Example of a Standard Threaded Fastener.

MaxiMOST System	Assemble or Disassemble Standard Fasteners																	T – Tool Use
	Screwdriver		Wrench			Ratchet		Power Tool			Hand							
											Start Only	Run Down						
	Machine Screw	Sheet Metal Screw										Loose				Tight	
Thread Diameter	All	All	≤ 3/4 in. (20 mm)	≤ 1 1/2 in. (40 mm)	> 1 1/2 in. (40 mm)	≤ 3/4 in. (20 mm)	≤ 1 1/2 in. (40 mm)	≤ 1/4 in. (6 mm)	≤ 1 in. (25 mm)	> 1 in. (25 mm)	N/A	≤ 1/4 in. (6 mm)	≤ 1 in. (25 mm)	≤ 1 1/2 in. (40 mm)	> 1 1/2 in. (40 mm)	≤ 1/4 in. (6 mm)	Thread Diameter
Index x 100	Fasteners																	Index x 100
3	-	-	-	-	-	-	-	1	1	-	2	1	-	-	-	-	**3**
6	1	1	-	-	-	-	-	2	2	1	4	2	1	-	1	1	**6**
10	2	2	1	-	-	1	-	4	3	3	6	3	2	1	2	2	**10**
16	3	3	-	1	-	2	1	6	6	4	10	5	4	2	3	3	**16**
24	4	5	2	-	1	3	-	9	8	6		8	6	3	4	5	**24**
32	6	6	3	2	-	4	2	12	11	9		10	8	4	6	7	**32**
42	8	9	4	-	-	6	3			12			11	5	7	9	**42**
54	10	11	6	3	2	8	4							7	9	11	**54**
67			7	4	3	9	5							8	12		**67**
81			9	5		11	6							10			**81**

FIGURE 5.9 Tool Use data card for Assembling or Disassembling Standard Fasteners.

The values on this data card apply when the threaded fastener is moved to the assembly and fastened or loosened and removed from the assembly. The index values for Assemble or Disassemble Fasteners include time for some brief associated additional activities:

- Assembling washers and opposing fasteners.
- Changing sockets.
- Handling counter or holding tools to hold a part or bolt in place as shown in Figure 5.10.
- Adjusting wrenches.

These activities are included because they reflect typical occurrences such that no adjustment is required for the presence or absence of these activities. The subactivity index values may be applied without regard for these activities. It should be noted that the index values make no provision for connecting or disconnecting power

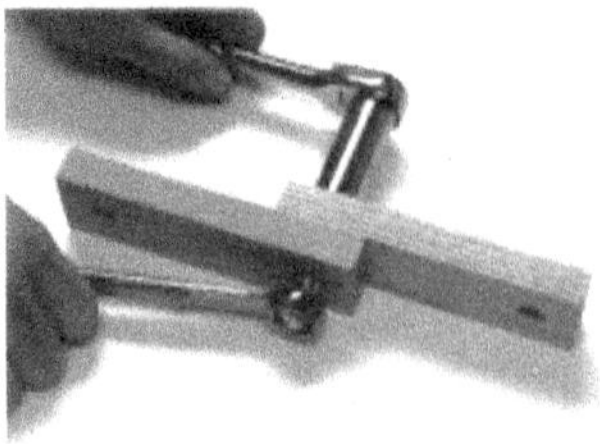

FIGURE 5.10 Holding a wrench on the back side of a bolt to keep it in place is an example of a counter tool.

tools, as well as the associated cord or hose handling, which must be analyzed separately using the Part Handling Sequence Model.

The index values for Assemble or Disassemble Fasteners are selected by the tool used, type of fastener, how it is assembled or disassembled and the number of fasteners. For simplification of the text, when one fastener is referenced it can mean one or more fasteners unless it specifically states only one in the definition. The tools and application details that follow describe the rules to assemble. The values also apply to disassembling.

Tool	Application Detail (Thread Diameter)	Explanation
Screwdriver (Figure 5.11)	**Machine screw** (Figure 5.12)	Get screwdriver, get screw, place screw, run in and completely tighten and aside screwdriver (all sizes); select value by number of fasteners. *Example:* Assemble six machine screws using screwdriver and aside: $\mathbf{T_{32}}$.
	Sheet metal screw (Figure 5.12)	Get screwdriver, get screw, place screw, run in and completely tighten and aside screwdriver (all sizes); select value by the number of fasteners. *Example:* Assemble two sheet metal screws using screwdriver and aside: $\mathbf{T_{10}}$.
Wrench (Figure 5.13)	**≤ 3/4 inch (20 mm)** **≤ 1 1/2 inch (40 mm)** **> 1 1/2 inch (40 mm)**	Get wrench, get fastener, start fastener, run in by hand, tighten with wrench and aside wrench; select value by thread diameter and number of fasteners. *Example:* Assemble four bolts 1 inch (25 mm) diameter using wrench and aside: $\mathbf{T_{67}}$.
Ratchet (Figure 5.14)	**≤ 3/4 inch (20 mm)** **≤ 1 1/2 inch (40 mm)**	Get ratchet, change socket as required, get fastener, start, run in by hand, tighten fastener with ratchet and aside ratchet; select value by thread diameter and number of fasteners. *Example:* Assemble eight spark plugs 5/8 inch (15 mm) diameter using ratchet and aside: $\mathbf{T_{54}}$.

Tool	Application Detail (Thread Diameter)	Explanation
Power Tool (Figure 5.15)	**≤ 1/4 inch (6 mm)** **≤ 1 inch (25 mm)** **> 1 inch (25 mm)**	Get power tool, change socket as required, get fastener, start, run in by hand, tighten fastener with power tool and aside power tool; select value by thread diameter and number of fasteners. *Example:* Assemble five lug nuts 5/8 inch (15 mm) diameter using power tool and aside: $\mathbf{T_{16}}$.
Hand	**Start only**	Get fastener, place, start threads up to two revolutions of the fastener; select value by the number of fasteners. *Example:* Start nut using hand: $\mathbf{T_3}$.
	Loose **≤ 1/4 inch (6 mm)** **≤ 1 inch (25 mm)** **≤ 1 1/2 inch (40 mm)** **> 1 1/2 inch (40 mm)**	Get fastener, place, start threads until resistance is encountered. Thread diameter must be considered for running down or removal of fastener by hand; select value by thread diameter and the number of fasteners. *Example:* Assemble nut 3/8 inch (10 mm) diameter using hand: $\mathbf{T_6}$.
	Tight **≤ 1/4 inch (6 mm)**	Get fastener, place, start threads, run in and tighten by hand; select value by the number of fasteners. *Example:* Assemble two wing nuts 1/4 inch (6 mm) diameter tight using hand: $\mathbf{T_{10}}$.

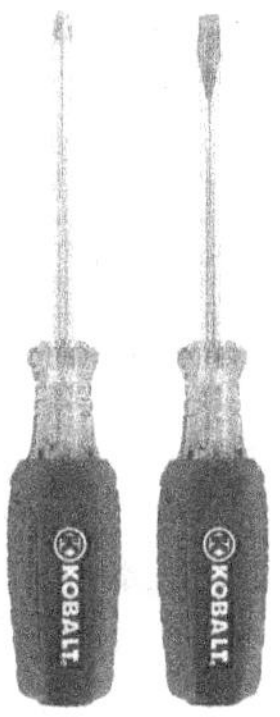

FIGURE 5.11 Example of a Screwdriver.

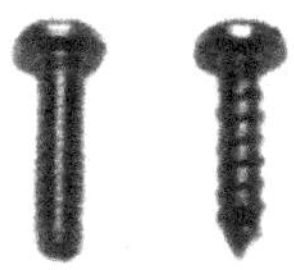

FIGURE 5.12 A machine screw is shown on the left and the sheet metal screw is shown on the right.

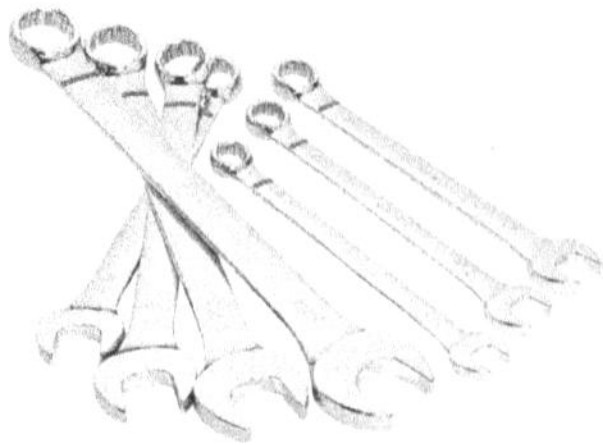

FIGURE 5.13 Example of a Combination Wrench.

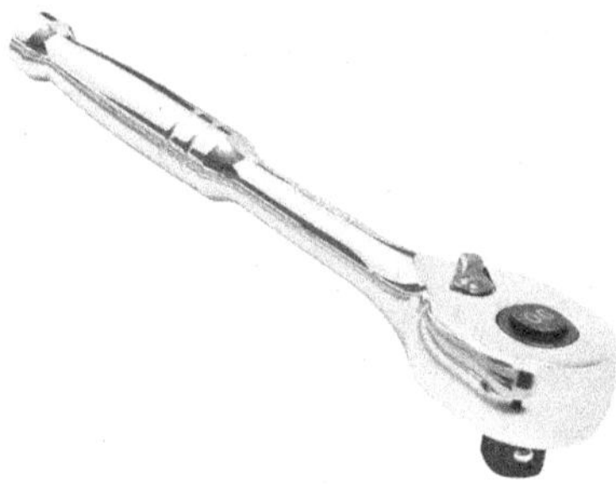

FIGURE 5.14 Example of a Ratchet.

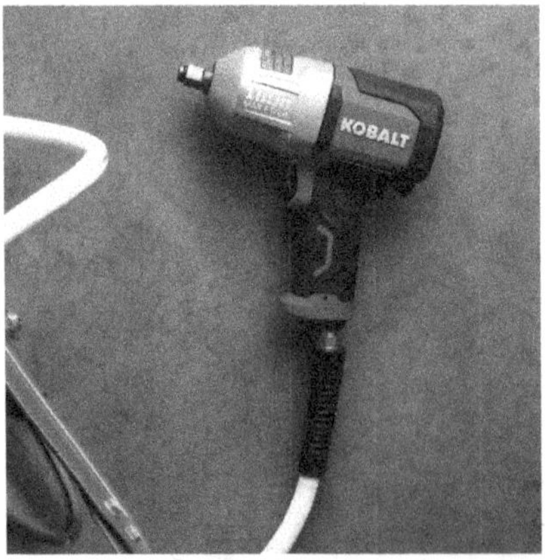

FIGURE 5.15 A Pneumatic Wrench is an example of a power tool.

Tighten or Loosen Standard Fasteners (T)

The Tighten or Loosen data card (Figure 5.16) is used when a standard fastener, already in place, is tightened or loosened with up to five revolutions but not removed from an assembly. The Tighten or Loosen index values include time for the following common related additional activities:

- Changing sockets
- Handling counter or holding tools
- Adjusting wrenches

Line handling and line connecting for power tools must be analyzed separately using the Part Handling Sequence Model.

The index values for Tighten or Loosen Fasteners are selected by the tool used, the type of fastener and the number of fasteners. The tools and application details that follow describe the rules to tighten. The values also apply to loosen:

Tool	Application Detail (Thread Diameter)	Explanation
Screwdriver	**Applies to any screw**	Get screwdriver, place on screw on assembly, run in and completely tighten screw and aside screwdriver (all sizes); select value by the number of fasteners. *Example:* Tighten screw using screwdriver and aside: $\mathbf{T_3}$.
Wrench	**≤ 3/4 inch (20 mm)** **≤ 1 1/2 inch (40 mm)** **> 1 1/2 inch (40 mm)**	Get wrench, place on fastener, run in and aside wrench; select value by thread diameter and the number of fasteners. *Example:* Tighten six bolts 1 inch (25 mm) diameter using wrench and aside: $\mathbf{T_{54}}$.
Ratchet	**≤ 3/4 inch (20 mm)** **≤ 1 1/2 inch (40 mm)**	Get ratchet, change socket as required, place on fastener, run in and aside ratchet; select value by thread diameter and the number of fasteners. *Example:* Loosen eight bolts 1/2 inch (12 mm) diameter using ratchet and aside: $\mathbf{T_{42}}$.
Power Tool	**≤ 1/4 inch (6 mm)** **≤ 1 inch (25 mm)** **> 1 inch (25 mm)**	Get power tool, change socket as required, place on fastener, run in and aside power tool; select value by thread diameter and the number of fasteners. *Example:* Tighten five nuts 1 inch (25 mm) diameter using power tool and aside: $\mathbf{T_6}$.
Hand	**≤ 1/4 inch (6 mm)**	Get fastener (already started), run in; select value by the number of fasteners. *Example:* Loosen two wing nuts 1/4 inch (6 mm) diameter using hand: $\mathbf{T_3}$.

MaxiMOST System	Tighten or Loosen Standard Fasteners									T – Tool Use	
	Screwdriver	Wrench			Ratchet		Power Tool			Hand	
Thread Diameter	All	≤ 3/4 in. (20 mm)	≤ 1 1/2 in. (40 mm)	> 1 1/2 in. (40 mm)	≤ 3/4 in. (20 mm)	≤ 1 1/2 in. (40 mm)	≤ 1/4 in. (6 mm)	≤ 1 in. (25 mm)	> 1 in. (25 mm)	≤ 1/4 in. (6 mm)	Thread Diameter
Index x 100	Fasteners										Index x 100
1	-	-	-	-	-	-	1	1	-	-	**1**
3	1	-	-	-	-	-	5	3	2	2	**3**
6	3	1	-	-	1	1	10	7	3	4	**6**
10	6	2	1	-	2	-	17	12	6	7	**10**
16	9	3	-	1	4	2			10	10	**16**
24	13	4	3	-	5	3					**24**
32		6	4	2	7	5					**32**
42		8	5	-	10	6					**42**
54		10	6	3		8					**54**
67			8	4		10					**67**
81			9								**81**

FIGURE 5.16 Tool Use data card for Tightening or Loosening Standard Fasteners.

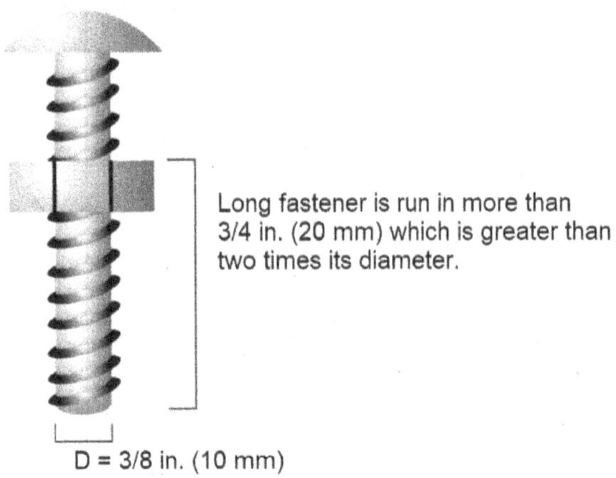

FIGURE 5.17 Example of a long fastener.

Assemble or Disassemble Long Fasteners (T)

Situations may occur in an activity in which a long fastener is used. A long fastener is one that is run in or out a considerably longer distance than a standard fastener. For example, a 3/4 inch (20 mm) fastener could be run in up to 1 1/2 inches (40 mm) to be considered a standard fastener. If the fastener is run in longer than two times its diameter (see Figure 5.17), it is considered a *long fastener.*

The data card (Figure 5.18) for Assemble or Disassemble Long Fasteners is used in a manner similar to that for assembling or disassembling standard fasteners.

The application criteria are the tool, details about the fastener and the number of fasteners. The length considered is not the total length of the fastener, but the length run in for assembling or out for disassembling. The lengths for installation and removal of long fasteners are up to and including 2 inches (5 cm) and up to and including 4 inches (10 cm). When a fastener is run in or out more than 4 inches (10 cm), allow the installation or removal for a 4 inch (10 cm) length, and in a separate sequence model, allow the additional length from the Tighten or Loosen Long Fastener data card. The tools and application details that follow describe the rules to assemble. The values also apply to disassembling:

Tool	Application Detail (Thread Diameter)	Explanation
Installation Length		
Screwdriver		
≤ 2 inches (5 cm)	All types	Get screwdriver, get screw, place screw, run in and completely tighten and aside screwdriver (all sizes); select value by the length of screw and number of fasteners.
≤ 4 inches (10 cm)	All types	*Example:* Assemble six machine screws 2 inches (5 cm) in length using screwdriver and aside: $\mathbf{T_{67}}$.
Wrench		
≤ 2 inches (5 cm)	≤ 3/4 inch (20 mm)	Get wrench, get fastener, start fastener, run in by hand, tighten fastener with wrench and aside wrench; select value by thread length, thread diameter and the number of fasteners.
≤ 4 inches (10 cm)	≤ 3/4 inch (20 mm)	*Example:* Assemble four bolts 3 inches (7.5 cm) in length and 1 inch (25 mm) diameter using wrench and aside: $\mathbf{T_{96}}$.
	> 3/4 inch (20 mm)	
Ratchet		
≤ 2 inches (5 cm)	≤ 3/4 inch (20 mm)	Get ratchet, change socket as required, get fastener, start, run in by hand and tighten fastener with ratchet and aside ratchet; select value by thread length, thread diameter and the number of fasteners.
≤ 4 inches (10 cm)	≤ 3/4 inch (20 mm)	*Example:* Assemble two bolts 1 1/2 inches (3.75 cm) in length and 5/8 inch (15 mm) diameter using ratchet and aside: $\mathbf{T_{54}}$.
	> 3/4 inch (20 mm)	
Power Tool		
≤ 2 inches (5 cm)	≤ 1/4 inch (6 mm)	Get power tool, change socket as required, get fastener, start, run in by hand, tighten fastener with power tool and aside power tool; select value by thread length, thread diameter and the number of fasteners.
	≤ 3/4 inch (20 mm)	
≤ 4 inches (10 cm)	≤ 1/4 inch (6 mm)	*Example:* Assemble five nuts 2 inches (5 cm) in length and 5/8 inch (15 mm) diameter using power tool and aside: $\mathbf{T_{16}}$.
	≤ 3/4 inch (20 mm)	

MaxiMOST System	Assemble or Disassemble Long Fasteners											T – Tool Use	
	Screwdriver		Wrench			Ratchet			Power Tool				
Length	≤ 2 in. (5 cm)	≤ 4 in. (10 cm)	≤ 2 in. (5 cm)	≤ 4 in. (10 cm)		≤ 2 in. (5 cm)	≤ 4 in. (10 cm)		≤ 2 in. (5 cm)		≤ 4 in. (10 cm)		Length
Thread Diameter	All	All	≤ 3/4 in. (20 mm)	≤ 3/4 in. (20 mm)	> 3/4 in. (20 mm)	≤ 3/4 in. (20 mm)	≤ 3/4 in. (20 mm)	> 3/4 in. (20 mm)	≤ 1/4 in. (6 mm)	≤ 3/4 in. (20 mm)	≤ 1/4 in. (6 mm)	≤ 3/4 in. (20 mm)	Thread Diameter
Index x 100	Fasteners												Index x 100
6	-	-	-	-	-	-	-	-	1	1	1	1	**6**
10	1	-	-	-	-	-	-	-	2	3	-	2	**10**
16	-	-	-	-	-	-	-	-	4	5	3	3	**16**
24	2	1	-	-	1	1	-	-	6	7	4	5	**24**
32	3	-	1	-	-	-	-	1	8	10	6	7	**32**
42	4	2	-	-	2	-	-	-	11		7	9	**42**
54	5	-	-	-	-	2	1	2			10	12	**54**
67	7	3	2	1	3	-	-	-					**67**
81	8	4	-		-	3	-	3					**81**
96	10	5	3		4	-	2	-					**96**
113		6			5	4		4					**113**

FIGURE 5.18 Tool Use data card for Assembling or Disassembling Long Fasteners.

Tighten or Loosen Long Fasteners (T)

The Tighten or Loosen Long Fasteners data card (Figure 5.19) provides for the tightening or loosening of long fasteners or long threaded devices. These index values apply when a long fastener is tightened or loosened in place. The long fastener is not assembled or removed using these values.

The index value for tightening or loosening long fasteners is chosen by the tool used, details about the fastener (length of thread adjustment and thread diameter) and the number of fasteners. The lengths shown are up to and including 2 inches (5 cm) and up to and including 4 inches (10 cm). When the adjustment length exceeds 4 inches (10 cm), divide the total adjustment by 4 inches (10 cm), round to the next highest whole number and allow the 4 inch (10 cm) value with a frequency determined by the division and rounding. Some of the tools also require that thread diameter be considered in selecting the index value. The ranges allowed for thread diameter are selected as appropriate for the tool considered.

The Tighten or Loosen Long Fasteners data card is also used for the installation or removal of fasteners when the fastener is run in or out a distance exceeding 4 inches (10 cm). When this occurs, use the Assemble or Disassemble Long Fasteners values (Figure 5.18) to account for the first 4 inches (10 cm) of run in or run out distance. In a separate sequence model, use the Tighten or Loosen Long Fastener (Figure 5.19) values to allow for the additional run in or run out distance. The tools and application details that follow describe the rules to tighten. The values also apply to loosen and are as follows:

Tool	Application Detail (Thread Diameter)	Explanation
Installation Length		
Screwdriver		
≤ 2 inches (5 cm)	All types	Get screwdriver, place on screw on assembly, run in and completely tighten screw (all sizes) and aside screwdriver; select value by thread length and the number of fasteners.
≤ 4 inches (10 cm)	All types	*Example:* Tighten 1 1/2 inch (3.75 cm) long screw using screwdriver and aside: $\mathbf{T_6}$.
Wrench		
≤ 2 inches (5 cm)	≤ 3/4 inch (20 mm)	Get wrench, place on fastener, run in and aside wrench; select value by thread length, thread diameter and the number of fasteners.
	> 3/4 inch (20 mm)	
≤ 4 inches (10 cm)	≤ 3/4 inch (20 mm)	*Example:* Loosen one bolt 3 inches (7.5 cm) in length and 1 inch (25 mm) diameter using wrench and aside: $\mathbf{T_{42}}$.
	> 3/4 inch (20 mm)	
Ratchet		
≤ 2 inches (5 cm)	≤ 3/4 inch (20 mm)	Get ratchet, change socket as required, place on fastener, run in and aside ratchet; select value by thread length, thread diameter and the number of fasteners.
	≤ 1 1/2 inch (40 mm)	
≤ 4 inches (10 cm)	≤ 3/4 inch (20 mm)	*Example:* Loosen two bolts 2 inches (5 cm) in length and 1/2 inch (12 mm) diameter using ratchet and aside: $\mathbf{T_{54}}$.
	≤ 1 1/2 inch (40 mm)	
Power Tool		
≤ 2 inches (5 cm)	≤ 1/4 inch (6 mm)	Get power tool, change socket as required, place on fastener, run in and aside power tool; select value by thread length, thread diameter and the number of fasteners.
	> 1/4 inch (6 mm)	
≤ 4 inches (10 cm)	≤ 1/4 inch (6 mm)	*Example:* Tighten five 2 inch (5 cm) long and 1/4 inch (6 mm) diameter nuts using power tool and aside: $\mathbf{T_{10}}$.
	> 1/4 inch (6 mm)	

Writing Method Descriptions

Below is the recommended format for a clear and concise method description for the Tool Use Sequence Model. Additional words may be used to enhance the method description. These could be Action Distances, Body Motions or adjectives. Examples of this structure can be found in the Tool Use examples listed below and throughout the Tool Use section.

MaxiMOST System	Tighten or Loosen Long Fasteners														T – Tool Use		
	Screwdriver		Wrench				Ratchet				Power Tool				
Length	≤ 2 in. (5 cm)	≤ 4 in. (10 cm)	≤ 2 in. (5 cm)		≤ 4 in. (10 cm)		≤ 2 in. (5 cm)		≤ 4 in. (10 cm)		≤ 2 in. (5 cm)		≤ 4 in. (10 cm)		Length
Thread Diameter	All	All	≤ 3/4 in. (20 mm)	> 3/4 in. (20 mm)	≤ 3/4 in. (20 mm)	> 3/4 in. (20 mm)	≤ 3/4 in. (20 mm)	≤ 1 1/2 in. (40 mm)	≤ 3/4 in. (20 mm)	≤ 1 1/2 in. (40 mm)	≤ 1/4 in. (6 mm)	> 1/4 in. (6 mm)	≤ 1/4 in. (6 mm)	> 1/4 in. (6 mm)	Thread Diameter
Index x 100	Fasteners														Index x 100
3	-	-	-	-	-	-	-	-	-	-	1	2	-	1	**3**
6	1	-	-	-	-	-	-	-	-	-	3	4	1	2	**6**
10	2	-	-	-	-	-	-	-	-	-	5	7	3	4	**10**
16	-	1	-	-	-	-	-	1	-	-	8	12	4	7	**16**
24	3	-	-	1	-	-	-	-	-	-	12		7	10	**24**
32	5	2	-	-	-	-	1	2	-	1			9		**32**
42	6	-	1	-	-	1	-	-	-	-			12		**42**
54	8	3	-	2	-		2	3	1	2					**54**
67	10	4	-	3	-		-	4		-					**67**
81		5	2		1		3	5		3					**81**

FIGURE 5.19 Tool Use data card for Tightening or Loosening Long Fasteners.

The recommended sentence structure for Tool Use is:

Activity	Number of fasteners (items)	Details of fastener (items)	Tool	At Location

Tool Use Examples—Fasteners

1. Remove four sheet metal screws from access cover using a screwdriver and aside all to bench.

   Disassemble 4 sheet metal screws with screwdriver and aside to bench

   $\mathbf{A_0}$ $\mathbf{B_0}$ $\mathbf{T_{24}}$

   $24 \times 100 = 2400$ TMU

2. Start six lug nuts by hand while bending at wheel.

   Start 6 lug nuts using hand

   $\mathbf{A_0}$ $\mathbf{B_1}$ $\mathbf{T_{10}}$

   $(1+10) \times 100 = 1100$ TMU

3. An operator walks six steps to get a wrench from a tool bin, returns six steps to loosen four 1/2 inch (12 mm) nuts on pipe clamp and asides the wrench back in the tool bin.

   Loosen 4 nuts using wrench on pipe

   $\mathbf{A_3}$ $\mathbf{B_0}$ $\mathbf{T_{24}}$

   $(3+24) \times 100 = 2700$ TMU

4. A worker assembles a water pump to a mounting bracket with four 3/4 inch (20 mm) bolts and lock washers using a ratchet.

   Assemble 4 bolts and washers with a ratchet

   $\mathbf{A_0 \quad B_0 \quad T_{32}}$

   $32 \times 100 = 3200$ TMU

5. A mechanic disassembles two 3/4 inch (20 mm) bolts, 5 inches (12.5 cm) long from a bracket using a pneumatic tool.

   Disassemble 2 long bolts, 5 inches (12.5 cm) in length using pneumatic tool

   $\mathbf{A_0 \quad B_0 \quad T_{10}}$

   $10 \times 100 = 1000$ TMU

   Loosen 2 long bolts—additional 1 inch (2.5 cm)

   $\mathbf{A_0 \quad B_0 \quad T_3}$

   $3 \times 100 = 300$ TMU

   1000 TMU
   300 TMU
   1300 TMU

General Tools I (T)

The index values included in the General Tools I data card (Figure 5.20) include the time to walk one to two steps to get and aside the tool. The General Tools I data card includes the following activities:

- Turn by hand
- Pry
- Strike with hand, hammer, mallet or sledge
- Apply material with tools

Turn by Hand

The index values for Turn by Hand (Figure 5.20) are selected by the type of turn and number of actions. The index values include time for up to two steps to get the object and time for manipulating the object.

Finger Spins

Index values from the Finger Spins column are selected when an object is turned or manipulated by the fingers and thumb while the position of the hand does not change significantly. These short finger movements are characterized by rolling or spinning

MaxiMOST System								General Tools I										T – Tool Use	
	Turn By Hand			Pry	Strike						Apply Material with Tool								
					Hand		Hammer		Mallet	Sledge									
Index x 100	Finger Spins	Wrist Turns	Arm Crank	Pry Bar	Wrist Taps	Arm Taps	Wrist Strikes	Arm Strikes	Arm Strikes	Arm Strikes	Seal Gun	Grease Gun	Squeeze Bottle	Tube	Brush, Stick, Hand or Finger	Squirt Can	Aerosol Can	Tape Roll	Index x 100
	Actions		Revs.	Actions	Actions						Pulls	Lever Actions	Drops	1 in. (2.5 cm) Spots	1 in. (2.5 cm) Spots	Squirts	Sq. Ft (0.1 m²)	Foot or Strip	
1	3	4	2	3	10	5	9	4	3	-	-	-	1	-	-	-	-	-	**1**
3	26	15	16	16	31	15	30	16	11	4	1	4	7	-	2	8	-	1	**3**
6	58	31	36	34		30		32	22	10	5	14	16	6	6	20	1	3	**6**
10	102	53	63	60					37	18	11	28		16	12		3	5	**10**
16										29	19	48					6	8	**16**
24										43	28	72					9	12	**24**

FIGURE 5.20 Tool Use data card for General Tools I.

an object between the thumb and index finger. The index value is selected by the number of total actions. The time to reach back and obtain a new grip is included in the index value and should not be counted when applying the value.

Example: Gain control of small cap and turn with seven finger spins: $\mathbf{T_3}$.

Wrist Turns

Index values from the Wrist Turns column are selected when the object is turned or manipulated by wrist actions. Wrist actions occur when the hand is turned by rotations about the forearm. The index value is selected by the number of total actions. The time to reach back and obtain a new grip is included in the time for each action and should not be counted as an action when selecting the index value.

Example: Tighten a light bulb using four turning actions pivoted at the wrist: $\mathbf{T_1}$.

Arm Crank

Arm Crank index values apply when an object is manipulated by circular motions of the forearm. The index value for Arm Cranks is selected by the number of revolutions.

Example: With the arm, wind a piece of wire around a spool with 15 revolutions: $\mathbf{T_3}$.

Pry

Pry actions occur when a tool, such as a Prybar (Figure 5.21), is pushed or pulled to exert a prying force on an object. Pry index values include the time to take one to two steps to get, use and aside the tool. The index value is selected by the number of active prying actions. Backstrokes and resetting the tool are included in the time for the Pry index values.

Example: Remove a floorboard by prying it loose with five active actions: $\mathbf{T_3}$.

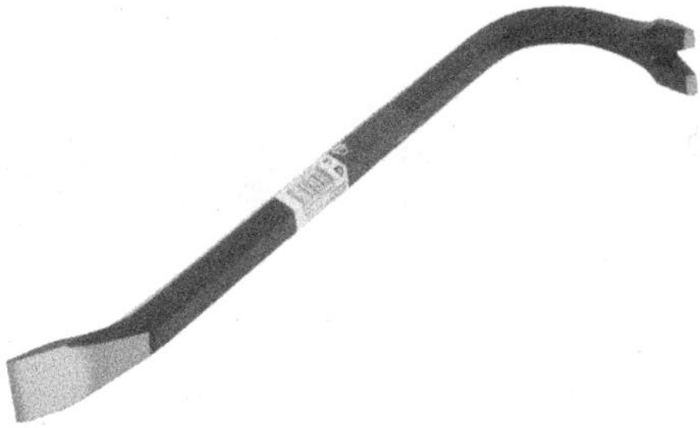

FIGURE 5.21 Example of a Prybar.

Strike

Strike index values are also located on the General Tools I data card (Figure 5.20). Strike applies to blows delivered by the hand or with a tool. The index values are determined by the number of times the hand or tool impacts against the object. The backswing that occurs during a strike activity is included in the time per strike.

Hand—Taps

Hand index values apply to striking down with the empty hand. The hand may be opened or closed into a fist. The muscles of the hand may be tensed or relaxed. The point of impact may be any part of the hand. The Hand striking index values are divided into Wrist Taps and Arm Taps.

Wrist Taps

Wrist Taps are those blows pivoted primarily at the wrist with the arm held relatively stationary. These short tapping motions are characteristic of tapping a tool or object to move it slightly. Data in this column includes the number of tapping actions made with the hand. The index value is determined by counting the number of wrist taps. Do not count the backstroke as this action is included in the values.

Example: Tap rubber stopper eight times with wrist to secure: $\mathbf{T_1}$.

Arm Taps

An Arm Tap is performed primarily by a motion of the arm pivoted at the elbow or the shoulder. The index value is determined by counting the number of arm taps. Do not count the backstroke as this action is included in the values.

Example: Tap tires with arm 20 times to fit onto wheelbase: $\mathbf{T_6}$.

Hammer Strikes

Hammer blows are delivered with the aid of a hammer (Figure 5.22) or any tool or object used in the manner of a hammer. The index values include the time to walk one to two steps to get or aside the hammer. Hammer striking, like hand striking, is divided into two groups, Wrist Strikes and Arm Strikes, based on the method of striking.

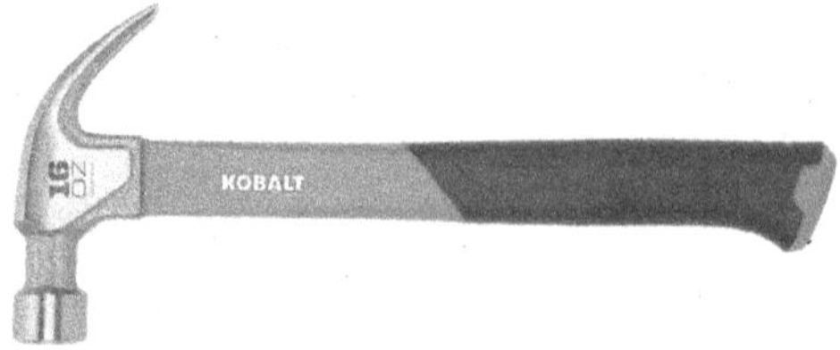

FIGURE 5.22 Example of a Hammer.

Wrist Strikes

Wrist Strikes are those blows pivoted primarily at the wrist with the arm held relatively stationary. The index value is determined by counting the number of wrist strikes. Do not count the backstroke as this action is included in the values.

Example: Strike fastener with hammer using 10 wrist actions: $\mathbf{T_3}$.

Arm Strikes

An Arm Strike is performed primarily by motion of the arm pivoted at the elbow or the shoulder. The index value is determined by counting the number of arm strikes. Do not count the backstroke as this action is included in the values.

Example: Strike pegs with hammer using 18 arm actions: $\mathbf{T_6}$.

Mallet Strikes

Mallet index values apply when the object is struck with a large hammer or mallet (Figure 5.23). Because of the tool size, an arm strike is normally used. The arm strike is performed primarily by a motion of the arm pivoted at the elbow or the shoulder. The index value is determined by counting the number of strikes with a mallet. Do not count the backstroke as this action is included in the values.

Example: Using a mallet, strike several lids to secure with a total of 15 actions: $\mathbf{T_6}$.

Sledge Strikes

Sledge index values apply to blows delivered with a sledge (Figure 5.24) of up to 10 pounds (5 kg) weight. Because of the tool size, an arm strike is normally used. The arm strike is performed primarily by motion of the arm pivoted at the elbow or the shoulder. The index value is determined by counting the number of strikes with a sledge. Do not count the backstroke as this action is included in the values.

Example: Repair train track using a sledgehammer with 30 strikes: $\mathbf{T_{24}}$.

Apply Material with Tool

The index values for Apply Material with Tool are located on the General Tools I data card (Figure 5.20). These index values provide for the analysis of material application with a variety of commonly used tools. The data is categorized by the tool used and the application which is defined as follows:

Tool	Application Detail	Explanation
Seal Gun	Pull	Get gun, place onto surface and apply sealing compound along a seam; the process time can be up to 2 seconds; place to next spot and aside gun; index values also include refilling gun with new canister when required; select value by the number of pulls. *Example:* Apply sealant around headlight three pulls using seal gun and aside: $\mathbf{T_6}$.
Grease Gun	Lever action	Get gun, place onto fitting and apply grease by pushing lever (per power stroke); place to next fitting and aside gun; also includes refilling gun with new canister when required; select value by the number of lever actions. *Example:* Apply grease to one fitting with six lever actions using grease gun and aside: $\mathbf{T_6}$.
Squeeze Bottle	Drop	Pick up squeeze bottle, open and shut cap, apply drop on spot and aside bottle; select value by the number of drops or squeezes applied. *Example:* Apply thread sealant on stud four drops using squeeze bottle and aside: $\mathbf{T_3}$.
Tube	1 inch (2.5 cm) spot	Pick up grease or glue tube, open and shut cap, apply grease or glue on spot and aside tube; select value by the number of spots (up to 1 inch, 2.5 cm, across) to which the material is applied. This is not appropriate for laying a bead of material. *Example:* Apply glue on gasket four 1 inch (2.5 cm) spots using tube and aside: $\mathbf{T_6}$.
Brush, Stick, Hand or Finger	1 inch (2.5 cm) spot	Pick up brush or stick, clean it against the can, get grease or glue from can and apply on surface or spot; aside tool and wipe hand if necessary; select value by the number of spots (up to 1 inch, 2.5 cm) to which the material is applied. *Example:* Apply grease on shaft with three 1 inch (2.5 cm) spots using stick and aside: $\mathbf{T_6}$.
Squirt Can	Squirt	Pick up can or bottle and squirt liquid on one spot by pulling trigger, pushing pump or squeezing the bottle; select value by the number of squirts required to apply the proper amount of material. *Example:* Apply window cleaner with two squirts to rearview mirror with squirt can and aside: $\mathbf{T_3}$.
Aerosol Can	Square foot (m^2)	Get aerosol can, remove and replace cap, shake can initially and during spraying, aim, spray and aside can; select values by per square foot ($0.1 m^2$) of application area. *Example:* Spray 1 square foot ($0.1 m^2$) layout ink to plate: $\mathbf{T_6}$.
Tape Roll	Foot or strip	Get tape roll, such as masking tape, open end, pull tape and apply on surface (up to 12 inches, 30 cm) or wrap around object (three to six revolutions); tear off and aside; select value by per foot (30 cm) or per strip, whichever is greater. *Example:* Apply 10 inch (25 cm) strip of masking tape to name plate prior to painting operation: $\mathbf{T_3}$.

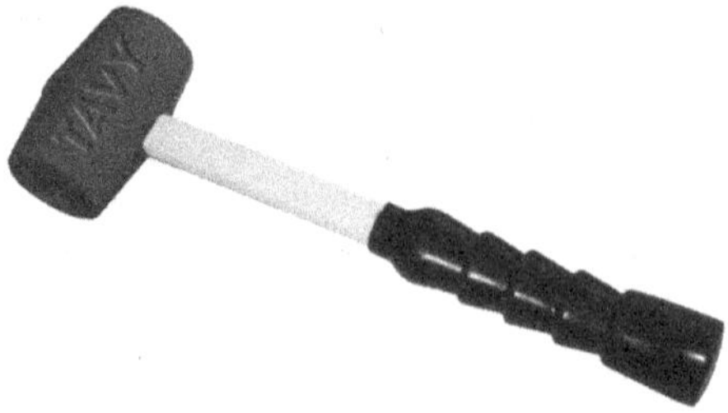

FIGURE 5.23 Example of a Mallet.

FIGURE 5.24 Example of a Sledge.

Tool Use—General Tools I Examples

The Tool Use examples below follow the recommended sentence structure for Tool Use which is:

Activity Object Tool Action Tool At Location

1. An operator cranks a wheel with five arm cranks using hands.

   Turn wheel 5 arm cranks using hands

   $\mathbf{A_0}$ $\mathbf{B_0}$ $\mathbf{T_3}$

   $3 \times 100 = 300$ TMU

2. A quality assurance associate walks eight steps to get a squeeze bottle from a workbench, bends and uses it to apply six drops of oil to lubricate a part. He then asides the bottle back at the workbench.

   Use squeeze bottle to apply 6 drops of oil

   $\mathbf{A_3}$ $\mathbf{B_1}$ $\mathbf{T_3}$

   $(3+1+3) \times 100 = 700$ TMU

3. An operator grasps a cap and turns it with two finger spins using fingers.

   Turn cap 2 finger spins using fingers

A_0 B_0 T_1

$1 \times 100 = 100$ TMU

4. An operator turns an eyebolt with six wrist turns using hand.

Turn eyebolt 6 wrist turns using hand

A_0 B_0 T_3

$3 \times 100 = 300$ TMU

5. Mechanic lies down on creeper and slides under truck where he strikes the differential with four arms strikes using a mallet.

Strike differential 4 arm strikes using mallet

A_0 B_3 T_3

$(3+3) \times 100 = 600$ TMU

6. A worker uses an aerosol can to spray adhesive on the back of a piece of canvas measuring 3 square feet (0.3 m^2).

Use aerosol can to spray adhesive on canvas

A_0 B_0 T_{10}

$10 \times 100 = 1000$ TMU

7. An operator strikes an object with two arm taps using hand.

Strike object with 2 arm taps using hand

A_0 B_0 T_1

$1 \times 100 = 100$ TMU

8. A worker lifts up the plate edges using four prying actions with a prybar and asides the tool six steps away.

Pry plate edges 4 actions using prybar and aside

A_1 B_0 T_3

$(1+3) \times 100 = 400$ TMU

9. An operator walks seven steps to get a sledge, strikes a brace and retainer with 10 arm strikes and asides the tool 12 steps away.

Strike brace and retainer 10 arm strikes using sledge and aside

A_3 B_0 T_6

$(3+6) \times 100 = 900$ TMU

10. A worker uses painter's tape to mask off a 12 foot (36 m) section of a wall to be painted.

Use tape to mask off 12 foot (36 m) wall

$A_0\ B_0\ T_{24}$

$24 \times 100 = 2400$ TMU

General Tools II (T)

The index values in the General Tools II data card (Figure 5.25) include the time to walk one to two steps to get or aside the tool. The General Tools II data card includes the following activities:

- Clean Surface
- Cut or Slice
- Twist or Bend with Pliers
- Record
- Stamp
- Think
- Deburr with File
- Free with Drift Pin
- Tap or Thread by Hand
- Process Time

Clean Surface

Clean Surface may be done with an air hose, brush, cloth or similar tools used in the same way as these tools. Clean Surface data includes getting the tool, using the tool for cleaning and asiding the tool with up to two steps. The index values for Clean Surface are determined by the method or tool employed and the size of the area being cleaned in square feet (0.1 m²). The values for Clean Surface may also be applied as an application or treatment to an object or area in addition to cleaning.

MaxiMOST System									General Tools II											T – Tool Use	
Index x 100	Clean Surface			Cut or Slice					Twist or Bend with Pliers	Record		Stamp	Think		Deburr with File	Free with Drift Pin	Tap or Thread by Hand		Process Time		Index x 100
	Air-Clean	Brush-Clean	Wipe	Pliers Soft	Pliers Medium	Pliers Hard	Scissors	Knife		Writing Tool Write	Writing Tool Mark	Hammer and Die	Inspect	Read			Diameter				
	Sq. Ft. (0.1 m²)			Cuts or Slices					Loops or Bends	≤ 1 in. (2.5 cm) Characters	≤ 3 in. (7.5 cm) Characters	Letter or Figure	Points	Words or Values	Feet (30 cm) or Edge	Tools	≤ 5/16 in. (8 mm) Occurrences	≤ 3/4 in. (20 mm) Occurrences	Sec.	Min.	
1	-	-	-	2	1	-	3	1	1	-	-	-	2	4	-	-	-	-	3.5	-	**1**
3	4	3	2	6	4	2	16	7	4	9	5	1	7	13	1	2	-	-	11	-	**3**
6	11	9	8	12	8	5	35	16	8	23	13	4	13	25	2	6	-	-	22	-	**6**
10	23	16	15		14	9		27	15	41	24	7	22	43	3	10	-	-	36	-	**10**
16	39	27	25			15				86	40	12	34	68	5	16	1	-	58	1	**16**
24	57	40	37									17			8		-	1		-	**24**
32		53	51									23			11		-	-		2	**32**
42																	2	-		-	**42**
54																	3	2		3	**54**

FIGURE 5.25 Tool Use data card for General Tools II.

Air-Clean

Using an air hose or nozzle to blow small particles or chips out of a hole, cavity or from a surface. The Air-Clean index values include time to get an air hose (within two steps), activate the air hose, direct the air over the surface for cleaning and aside the air hose. The index value is selected based on the area cleaned in square feet (0.1 m^2). Line handling activities are not included in the Air-Clean index values. Tugging or coiling of the air hose would be analyzed with a separate Part Handling Sequence Model.

Example: Use an air hose to remove excess wood shavings from a 16 square foot (1.6 m^2) area: $\mathbf{T_{10}}$.

Brush-Clean

Using a brush to remove particles, chips or other debris from an object or surface. Brush-Clean index values include time to gain control of a brush (within two steps), move the brush to a surface, clean the surface by brushing and aside the brush. The index value is selected based on the area cleaned in square feet (0.1 m^2).

Example: Use a brush to remove metal shavings from a counter measuring 8 square feet (0.8 m^2): $\mathbf{T_6}$.

Wipe

The Wipe index values apply to cleaning by rubbing the surface with a cloth, sponge or other suitable material. Time is included for getting the tool or material (within two steps), cleaning the surface by wiping and asiding the tool or material. The index value is selected based on the area cleaned in square feet (0.1 m^2).

Example: Clean a finished 2×2 foot (0.2×0.2 m^2) part with a rag: $\mathbf{T_6}$.

Cut or Slice

Cut describes the manual actions employed to separate, divide or remove part of an object using a sharp-edged hand tool. As Figure 5.25 indicates, the index values cover the use of pliers, scissors or a knife for general cutting activities. In addition, pliers are used for gripping and bending activities. These cutting tools and their use are described as follows.

Pliers

Three different methods may be used to cut through wire using pliers (Figure 5.26). The method used largely depends on the firmness of the wire material and the diameter or gauge of the wire. The data for cutting with pliers includes three categories for cutting wire. The index value is determined by number of cuts made with the pliers.

Soft

This parameter applies to cutting a soft steel, copper or other small-gauge wire and is recognized by using the pliers with one hand and making one cut.

Example: Cut excess wire with one cut from assembly: $\mathbf{T_1}$.

FIGURE 5.26 Example of Pliers.

Medium

This parameter applies to cutting a steel wire or cable and can be recognized by using the pliers with one hand and making two cuts. Following the initial cut, the pliers are rotated around the wire and repositioned over the cut before completely cutting through the wire.

Example: Cut exposed medium-gauge wire with three cuts before connecting to electrical service: $\mathbf{T_3}$.

Hard

This parameter applies to cutting a heavier wire (approximately 10 gauge) and can be recognized where both hands are needed to apply sufficient force to cut through the wire to make two cuts. Similar to medium wire, following the initial cut, the pliers are rotated around the wire and repositioned over the cut before completely cutting through the wire with two hands.

Example: Electrician makes four cuts on a heavy gauge wire at electric meter: $\mathbf{T_6}$.

Scissors

Using scissors (Figure 5.27) applies to cutting paper, fabric, light cardboard or other similar material using scissors. The index values include the time to get the scissors (within two steps), place for cutting, cut, relocate scissors and aside the scissors. Opening the scissors and relocating the scissors when required are included in the time per cut and are not counted. Index values are selected according to the number of cuts or scissor actions made during the cutting activity.

Example: An associate uses scissors to make 15 cuts on a piece of fabric: $\mathbf{T_3}$.

FIGURE 5.27 Example of Scissors.

Note: If the scissors are being held open following an initial cut to make one long cut (e.g., cutting through a piece of plastic), the Part Handling—Controlled Move values should be used to analyze the long cut.

Knife

Cut with a Knife (Figure 5.28) can be used for cutting string, material and light cord or to cut through corrugated material or cardboard. The index values include time to get a knife (within two steps), place the knife to cut, make the cuts, place the knife for additional cuts when required and aside the knife. The index value is determined by counting the number of cutting slices. Count only the actual cutting slices, as the move back and additional placement are included in the time for the cut.

Example: A maintenance worker uses a utility knife to cut through three separate pieces of plastic banding: $\mathbf{T_3}$.

Twist or Bend with Pliers

Twist or Bend with Pliers includes time to get the pliers (within two steps), place the pliers to the wire, form a loop or bend, or twist two ends of wire together, place the pliers for additional forming when required and aside the pliers. The index value is selected by the number of loops or bends.

Example: Using pliers, secure a wire around a piece of pipe with eight loops: $\mathbf{T_6}$.

Record

Record covers the manual actions performed with a writing or marking tool for the purpose of recording information. Two categories of data are found in Figure 5.25 for Record. The index values for Write apply to the normal-size [up to 1 inch (2.5 cm) in height] handwriting operations (script or print) performed with a pen, pencil or other writing instrument such as a stylus (Figure 5.29). The Mark values cover the use of such marking tools as a scribe or marker for the purpose of identifying or making a larger mark (1–3 inches, 2.5–7.5 cm) on an object. The index value is selected by the instrument used, the size of the character and the number of characters marked on the surface.

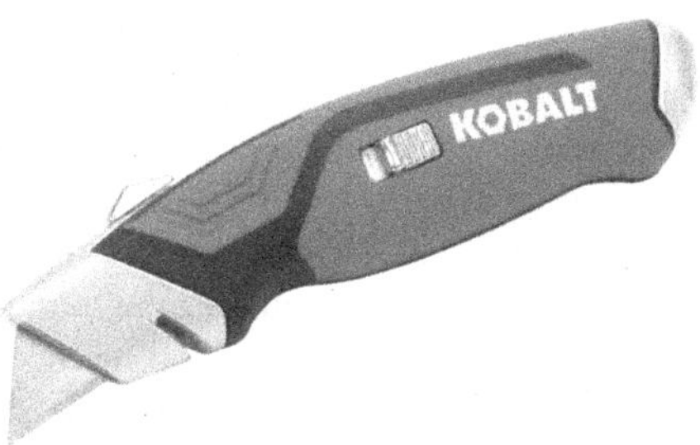

FIGURE 5.28 Example of a Utility Knife.

FIGURE 5.29 Example of a Stylus.

Write—≤ 1 Inch (2.5 cm)

The Write data is provided to cover the routine administrative activities encountered in virtually every industry. These activities may include filling out forms or writing a part number. The index values for Write include the time to get the pen, remove and replace the cap, write characters, check for correctness and aside pen. The index value is chosen by the number of characters written.

Example: Write four numerals (7502) to show the quantity on job ticket using pen and aside: $\mathbf{T_3}$.

Mark—≤ 3 Inches (7.5 cm)

The Mark data applies to marking or identifying an object or container using a marking tool, such as a scribe or marker. Each mark is counted as a character. The index values for Mark apply to printed characters (letters and numerals) 1–3 inches (2.5–7.5 cm) high. The values include the time to get the marker, remove and replace the cap, mark characters, check for accuracy and aside marker. The index value is chosen by the number of characters marked.

Example: Mark 10 large characters (K314B08Z73) for part number on steel plate using chalk and aside: $\mathbf{T_6}$.

Stamp (Hammer and Die)

The values for Stamp include the time to get the hammer, select and position die, strike with hammer, return die to die set, aside hammer when finished and inspect completed job. Index values are chosen by the number of letters or figures stamped.

Example: Stamp six digits on plate using die and aside: $\mathbf{T_{10}}$.

Think

Think refers to the use of sensory mental processes, particularly those involving visual perception. The Think data in Figure 5.25 is designed to cover only those types of reading and inspection activities that occur as a necessary part of a worker's job. Although these activities usually occur internally to the manual work and therefore have no effect on the duration of the work cycle, on some occasions these activities must be considered in the overall work content of the job. The analyst should exercise care in determining the extent to which these activities affect the total analysis time.

Inspect

The values for Inspect include the time to walk up to two steps to position the body, focus and check the object by looking at the surface. The index value is determined by the number of points inspected.

The data in this column applies to inspection work designed for making simple decisions regarding certain characteristics of the object under inspection. The activity involves first locating the inspection points and then making a quick yes-or-no decision concerning the existence of a defect. These mental processes presume that the inspector possesses a clear understanding of the characteristic being judged and that the presence of any defect, such as a scratch, stain, scar or color variance, is readily apparent to the inspector.

The index values for Inspect refer to the number of inspection points examined on the object. For each point, a yes-or-no decision is made concerning the presence or absence of readily distinguishable characteristics.

Caution should be exercised in using these or any inspection values. In practical work situations, inspection time is rarely external but usually occurs during the manual handling of elements. Whenever possible, work should be designed to make inspections internal to other activities.

Example: Inspect three points on a part: $\mathbf{T_3}$.

Read

To read is to locate and interpret characters or groups of characters. The data for Read is based on reading single words or values such as reading a scale or gauge value. This data is also to be used for reading data such as item numbers, codes, quantities or dimensions from a blueprint. To index the T parameter, simply count the number of words or values read and choose the appropriate index value from the data card (Figure 5.25).

Note: These index values do not include time to pick up and/or aside the item, which must be analyzed separately with the Part Handling Sequence Model.

Example: Read a part number (one value) on work order: $\mathbf{T_1}$.

Deburr with File

The Deburr with File index values are used to analyze filing with a hand file (Figure 5.30) to remove burrs from a part. Time is allowed for getting the file (within two steps), placing the file to the part, filing, front strokes and backstrokes and laying the file aside. The count is based on the number of edges or feet deburred, whichever is greater. For each edge 1 foot (30cm) or less in length, count the edge and allow 1 foot (30cm). For each edge greater than 1 foot (30cm), count each foot (30cm) of length plus one for any remaining partial foot (30cm).

Example: A part requires deburring on three edges that are:

- 3 inches (7.5cm) long
- 22 inches (55cm) long
- 37 inches (92.5cm) long

FIGURE 5.30 Example of a hand file.

Allow the following for each edge:

- 1 foot (30 cm) for the 3 inch (7.5 cm) edge
- 2 feet (60 cm) for the 22 inch (55 cm) edge
- 4 feet (120 cm) for the 37 inch (92.5 cm) edge

Find the index value for 7 feet (210 cm) on the data card and allow a $\mathbf{T_{24}}$ index value.

Free with Drift Pin

Free with Drift Pin includes the time to get a drift pin and a hammer, place the drift pin onto the base of the tool, loosen the tool by tapping the drift pin with the hammer up to three arm strikes and asiding the drift pin and the hammer. No direct handling of the tool or manipulation of the machine is included in the index value. Select the index value by the number of tools or objects freed. If more than three arm strikes are required, use the Hammer data under the General Tools I data card (Figure 5.20).

Example: An operator loosens three broken bolts by using arm strikes on a drift pin with a hammer: $\mathbf{T_6}$.

Tap or Thread by Hand

The values for Tap or Thread by Hand apply to hand tapping using a solid tap affixed to a handle or to hand threading using a solid threading die affixed to a handle. These index values include time to walk up to two steps to get the tool, tap into or thread onto the part, back the tool clear of the part and aside the tool. The length tapped or threaded is great enough to accept a standard fastener plus an added distance for clearance. No time is provided in these index values for handling the part or for attaching the handle to the cutting tool. Use Apply Material with Tool index values for the necessary application of oil while tapping.

The value for Tap or Thread by Hand is determined by the thread diameter and the number of places tapped or threaded. Values are read up to and including.

For example, the values for the ≤ 3/4 inch (20 mm) diameter column of data applies to the thread diameter greater than 5/16 inch (8 mm) but not exceeding 3/4 inch (20 mm).

≤ 5/16 Inch (8 mm) diameter
≤ 3/4 Inch (20 mm) diameter

Example: Operator uses a tap to form a thread on the inside of a hole for two 1/2 inch (12 mm) fasteners: $\mathbf{T_{54}}$.
Worker uses a die on one 1/4 inch (6 mm) rod: $\mathbf{T_{16}}$.

Process Time

Process Time is defined as the portion of work that is controlled by electronic or mechanical devices or machines, not by manual actions. The T parameter in the Tool Use Sequence Model is intended to cover process times of relatively short duration. These process times will normally have minor variations and are often difficult to time. Power fluctuations and minor differences in parts can affect the process time. Process Time is indexed by selecting the appropriate index value that corresponds to the observed or calculated 'actual time.'

Longer process times, such as machining times based on feeds and speeds, are normally calculated and entered separately as a process time in the analysis. The actual clock time is *never* placed on the T parameter of the sequence model. Only the index value that statistically represents the actual time should be placed in the sequence model.

Example: A punch press cycles for 4 seconds after the palm buttons are pressed: $\mathbf{T_3}$.

Tool Use—General Tools II Examples

The Tool Use examples below follow the recommended sentence structure for Tool Use which is:

Activity Object Tool Action Tool At Location

1. Air-clean the surface of a part which is 2 square feet (0.2 m²) at milling machine.

   Air-clean 2 square feet (0.2 m²) of part

   $\mathbf{A_0 \quad B_0 \quad T_3}$

   $3 \times 100 = 300$ TMU

2. Brush metal chips from 4 square feet (0.4 m²) surface of lathe bed.

   Brush chips 4 square feet (0.4 m²)

   $\mathbf{A_0 \quad B_0 \quad T_6}$

   $6 \times 100 = 600$ TMU

3. Wipe storefront window with two bends that is 8 square feet (0.8 m^2).

   Wipe window 8 square feet (0.8 m^2)

   A_0 B_1 T_6

   $(1+6)\times 100=700$ TMU

4. A truck assembler walks four steps to get pliers and then returns four steps to make two cuts to medium-gauge wire.

   Cut wire with 2 cuts

   A_1 B_0 T_3

   $(1+3)\times 100=400$ TMU

5. A receiving clerk gets a marker from the receiving desk 10 steps away, returns 10 steps and bends to mark six large numbers on the side of the box.

   Mark 6 large numbers on box

   A_3 B_1 T_6

   $(3+1+6)\times 100=1000$ TMU

6. Operator walks three steps to a tool box and gets a drift pin and hammer and returns. He then frees the lift arm from the frame by driving the connecting pin out with three arm strikes of the hammer onto the drift pin.

   Free lift arm with drift pin

   A_1 B_0 T_3

   $(1+3)\times 100=400$ TMU

7. A maintenance worker grasps a file from a workbench eight steps away, returns and deburrs three rough edges [14 inches (35 cm); 8 inches (20 cm); 20 inches (50 cm)] from a piece of wood cut with a circular saw.

   Deburr 3 edges with file

   A_3 B_0 T_{16}

   $(3+16)\times 100=1900$ TMU

8. A plastic extruding machine operates for approximately 32 seconds.

   Process time

   A_0 B_0 T_{10}

   $10\times 100=1000$ TMU

9. An operator bends to grasp a pair of scissors from an open tool box on the floor. She then cuts a piece of material with eight cuts.

   Cut material with scissors

   A_0 B_1 T_3

   $(1+3)\times 100=400$ TMU

10. A worker walks 13 steps and goes through one door to read instructions consisting of 65 words.

    Read instructions

    $\mathbf{A_3 \quad B_1 \quad T_{16}}$

    $(3+1+16)\times 100 = 2000$ TMU

MEASURING TOOLS (T)

The index values to Measure are located on the Measuring Tools data card (Figure 5.31), which also includes values for Prepare to Measure. Measure index values have been developed for the most common standard measuring devices found in manufacturing. Measure includes the actions employed to determine a certain physical characteristic of an object using a standard measuring tool. Index values for the Measure elements cover all actions necessary to align, adjust and examine both the measuring tool and the object during the measuring activity. The index value is determined by the tool used, the number of measurements taken and the time to walk one to two steps to get and aside the tool. The values include time to properly align the part and the measuring device, adjust the tool when required and determine the fit to the tool or reading a scale value. Time is allowed for handling either the tool or the part, but not both.

Note: Measuring tools continue to evolve and may include easier to read scale values or digital readouts. In those cases, validation of the applied index value is required.

The data from the Measuring Tools data card (Figure 5.31) covers the measuring tools listed below.

Flat Rule or Scale

The Flat Rule or Scale column provides time to use a machine graduated scale up to 18 inches (45 cm) in length or to use a printed or etched flat rule up to 4 feet (1.2 m) in length. This column also applies to the use of a protractor (Figure 5.32).

MaxiMOST System										Measuring Tools										T – Tool Use	
Index x 100	Flat Rule or Scale	Tape Rule	Wood Rule	Profile Gauge	Caliper ≤ 12 in. (30 cm)	Caliper ≤ 36 in. (90 cm)	Feeler Gauge	Micrometer ≤ 4 in. (10 cm)	Micrometer ≤ 36 in. (90 cm)	Ring Gauge	Plug Gauge ≤ 1 in. (25 mm)	Plug Gauge ≤ 8 in. (200 mm)	Thread Gauge ≤ 4 in. (100 mm)	Thread Gauge ≤ 8 in. (200 mm)	Set to Measure Telescope Gauge	Set to Measure Caliper Spring Joint	Set to Measure Caliper Firm Joint	Snap Gauge ≤ 16 in. (400 mm)	Dial Indicator	Taper Gauge	Prepare to Measure	Index x 100
	Measurements																				Device	
3	2	1	-	3	2	-	1	-	-	1	2	-	-	-	1	-	-	2	-	-	-	**3**
6	3	2	3	5	3	1	2	1	1	2	3	1	1	-	2	1	1	5	-	1	Telescope Gauge Firm Joint Caliper Vernier Caliper	**6**
10	6	3	7	10	6	2	4	2	-	3	6	-	2	-	3	2		9	-	2	Combination Square 4 in. (10 cm) Bevel Protractor	**10**
16	10	6	14	15	10	3	7	4	3	5	10	3	4	1	5	4	2	14	1	4	4 in. (10 cm) Micrometer Spring Joint Caliper	**16**
24		8	21		15	5	10	5	4	8	15	4	6	2	8	6	3		-	6	36 in. (90 cm) Micrometer	**24**
32						7		7	5	11		5	8		10	8	5		2	9		**32**
42						9			7			7	10	3		11	6		-	12		**42**
54						12			9			9		4			8		3	16		**54**
67									11					5			10		4	19		**67**

FIGURE 5.31 Tool Use data card for Measuring Tools.

FIGURE 5.32 A protractor is an example of a Flat Rule.

Example: A setup worker uses a flat rule to center a die on a printing press by taking four measurements: $\mathbf{T}_{10}$.

Tape Rule (Tape Measure)

Tape Rule index values apply to the use of concave tapes housed in a case suitable for carrying in the pocket or clipped to the belt (Figure 5.33). These devices vary from 6 feet (1.8 m) to 25 feet (7.6 m) in length. These values may also be applied to flat pocket tapes no more than 6 feet (1.8 m) in length. The index values should not be applied to the use of flat, hand-wound, engineers' tapes regardless of length. The index values include time to get or aside the tape rule (within two steps).

Example: An assembly worker uses a tape measure to determine the location to place hinges on a cabinet before drilling holes: $\mathbf{T}_3$.

Wood Rule

The Wood Rule data applies to measurements taken with a 6 or 8 foot (1.8 or 2.5 m) folding rule. The index values may also be applied to aluminum or steel folding rules provided the measurement is to the same accuracy as a wood rule. Index values include the time to unfold up to eight sections to use the rule and fold up to eight sections to store the rule.

Example: A cabinet maker uses a wood rule to take six different measurements (top, middle, bottom) of the length and width of a cabinet opening: $\mathbf{T}_{10}$.

FIGURE 5.33 Example of a Tape Rule (Tape Measure)

Profile Gauge

The Profile Gauge index values apply to the use of gauges, such as a square, level, angle, radius and screw-pitch gauges that are used to compare the shape of the part to that of the gauge. This value includes adjusting the gauge to the object as well as the visual actions to compare the configuration of the object with that of the gauge. A level and a square are shown as examples of profile gauges in Figures 5.34 and 5.35.

Example: A carpenter uses a level to determine if a sliding door frame is level: $\mathbf{T_3}$.

Caliper

The index values to measure with a Caliper apply to outside and inside measurements and include setting the caliper legs to the object dimension, operating one locking device and reading the Vernier scale to determine the measurement. The index values are based on the size of the Caliper, the number of measurements taken and the use of a caliper with a Vernier scale (Figure 5.36).

Note: If the caliper includes a dial (Figure 5.37) or digital readout (Figure 5.38) of the measurement, the index value selected *must* be validated.

≤ 12 Inch (30 cm) Caliper

This column applies to measurements of no more than 12 inches (30 cm) using a caliper. The index value is based on the use of the lighter calipers normally used for smaller dimensions and includes the operation of one locking device.

Example: An operator confirms the thickness of a plate is 0.375 inch (9.5 mm) using a caliper: $\mathbf{T_3}$.

FIGURE 5.34 A level is an example of a Profile Gauge.

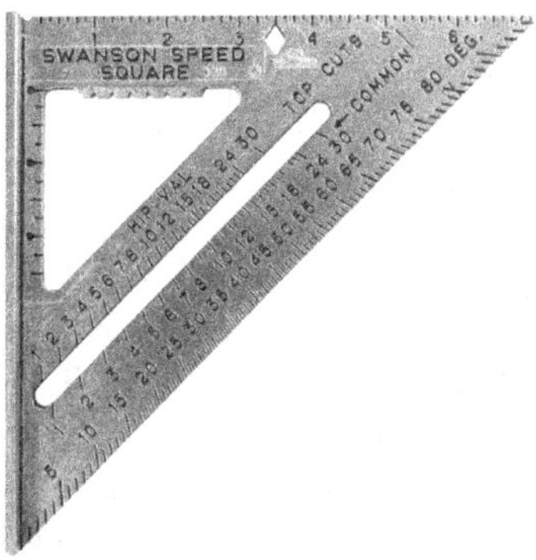

FIGURE 5.35 A square can be used as a Profile Gauge.

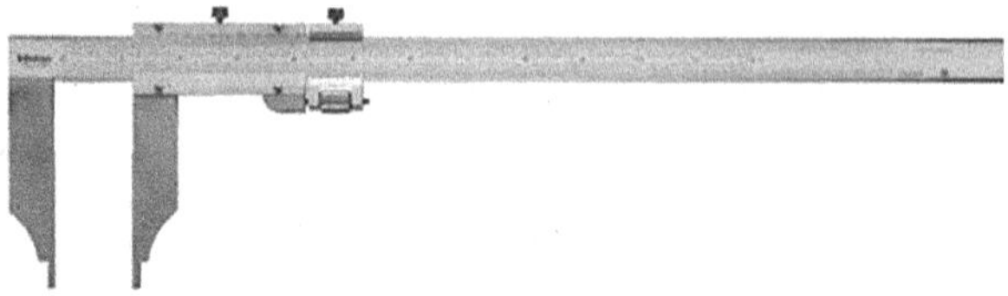

FIGURE 5.36 Caliper with Vernier scale.

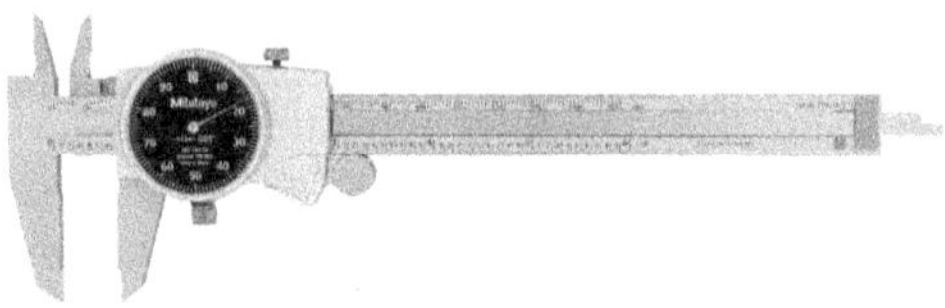

FIGURE 5.37 Caliper with dial indicator.

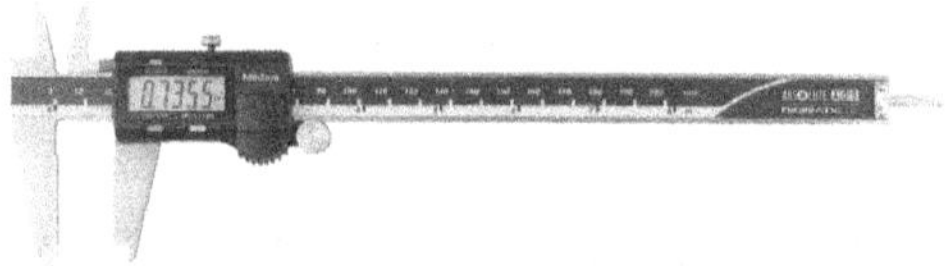

FIGURE 5.38 Caliper with digital readout.

≤ 36 Inch (90 cm) Caliper

This column applies to measurements of more than 12 inches (30 cm) but not exceeding 36 inches (90 cm). The index value is based on the use of the larger, heavier calipers associated with measuring greater dimensions. Time is included for separately identifying and correcting the location of each end of the caliper. Time is allowed for the operation of two locking devices.

Example: A lathe operator checks the finished diameter of a cylindrical object confirming it is 14.000 inches (35 cm) using a caliper: $\mathbf{T_6}$.

Feeler Gauge

These index values apply to the use of a Feeler Gauge (Figure 5.39) to measure the gap between two points or surfaces. Time is included to get the gauge, fan out and select the appropriate blade, insert the blade, make visual and tactile checks of the fit and aside the gauge. The index value is selected by the number of gaps checked.

Example: A mechanic takes one measurement to check the gap between the center and side electrodes of a spark plug: $\mathbf{T_3}$.

Micrometer

Micrometer index values apply to the use of outside (Figure 5.40), inside (Figure 5.41) and depth (Figure 5.42) micrometers. The index values include time to get the micrometer, set it on the part, adjust the micrometer to the part dimension, lock the

FIGURE 5.39 Example of a Feeler Gauge.

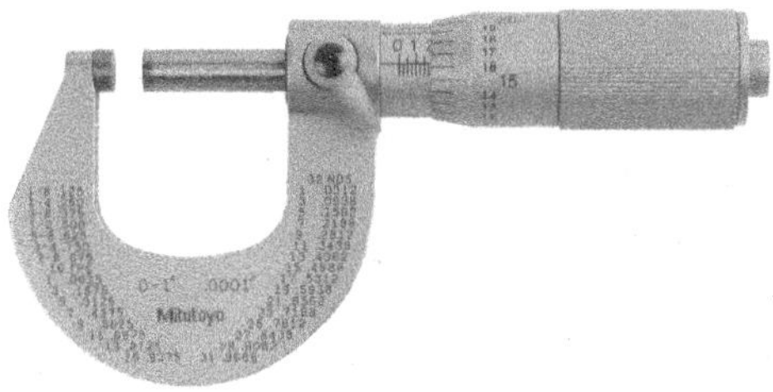

FIGURE 5.40 Outside Micrometer.

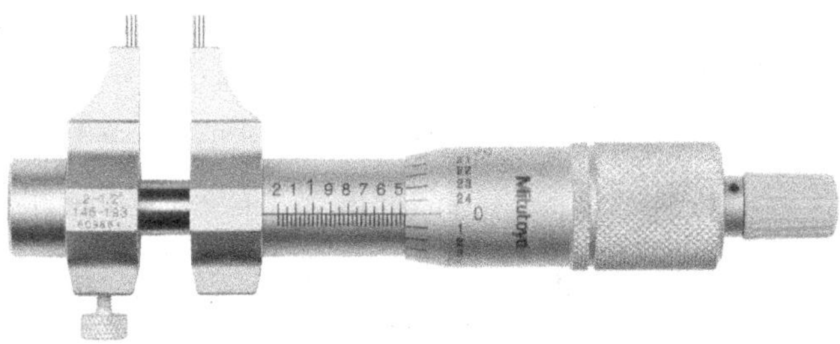

FIGURE 5.41 Inside Micrometer.

setting, pick up and read the micrometer, unlock the micrometer and set it aside. These index values do not include preparation activities such as getting the micrometer from the case, changing the anvil or the initial coarse adjustments to approximate the size of the part. The index value is selected by the size of the dimension and the number of measurements taken.

≤ 4 Inch (10 cm) Micrometer

This column applies to the measurement of dimensions no greater than 4 inches (10 cm).

Example: A quality assurance worker uses an outside micrometer to measure five parts, confirming each measurement is 0.500 inches (12 mm): $\mathbf{T}_{24}$.

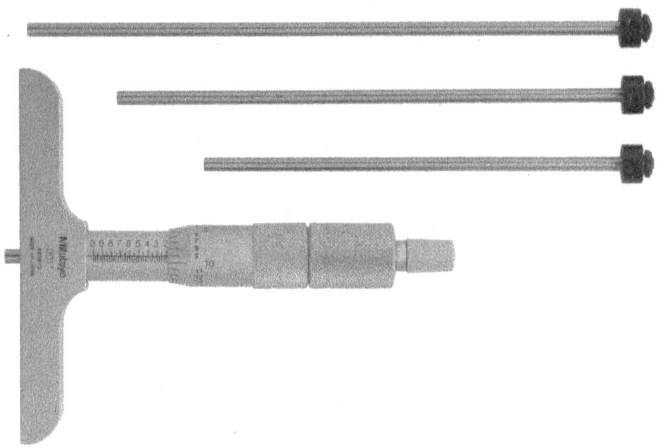

FIGURE 5.42 Depth Micrometer.

≤ 36 Inch (90 cm) Micrometer

This column applies to the measurement of dimensions exceeding 4 inches (10 cm) but not exceeding 36 inches (90 cm).

Example: A machinist measures the depth of a slot that is milled into a steel beam with a depth micrometer to confirm it is 10 inches (25 cm) deep: $\mathbf{T_6}$.

Ring Gauge

The index values for Ring Gauge apply to the comparison of an outside diameter of a part to a standard ring designed to match the required part diameter. The index values include time to get the ring gauge, fit the ring to the part, check the fit of the ring to the part and remove and aside the part. Select the index value by the number of places checked.

Example: A worker examines one part of an extruded plastic tube with a ring gauge: $\mathbf{T_3}$.

Plug Gauge

The index values for Plug Gauge include time to get the gauge, fit the go-end of the gauge into the opening in the part, turn the gauge, fit the no-go-end of the gauge to the part, determine that the no-go-end will not fit into the opening and aside the gauge. The index value is determined by the size of the gauge and the number of holes checked.

≤ 1 Inch (25 mm) Plug Gauge

This column is used to analyze the use of Plug Gauges that have a diameter or thickness of 1 inch (25 mm) or less.

Example: A machine operator inserts a plug gauge into four different holes in a metal mold to determine if the diameters are within specifications before continuing with the tapping operation to add threads: $\mathbf{T_{10}}$.

≤ 8 Inch (200 mm) Plug Gauge

This column is for analyzing the use of Plug Gauges with a diameter or thickness greater than 1 inch (25 mm) but less than 8 inches (200 mm).

Example: A machine operator inserts a plug gauge into a reamed hole 6 inches (150 mm) in diameter in an auto part to check the tolerance to accept the pin that will be inserted later: $\mathbf{T_6}$.

Thread Gauge

The Thread Gauge index values apply to the use of threaded plug gauges (Figure 5.43) and threaded ring gauges for the purpose of inspecting a threaded section (male or female) of a part. These index values should not be used for screw-pitch gauges as these are analyzed as profile gauges. The index values apply whether the gauge is moved to the part or the part to the gauge. The index value is selected by the diameter of the gauge and the number of threaded openings or threaded protrusions.

≤ 4 Inch (100 mm) Thread Gauge

This column is used to determine the index value for inspection with a threaded ring or threaded plug gauge having a diameter no larger than 4 inches (100 mm).

Example: A mechanic uses a threaded ring gauge on eight small bolts to ensure the correct bolts are used in the assembly: $\mathbf{T_{32}}$.

≤ 8 Inch (200 mm) Thread Gauge

The index values in this column apply to the use of threaded gauges greater than 4 inches (100 mm) but less than 8 inches (200 mm) in thread diameter.

Example: A machine operator uses a threaded plug gauge to confirm that the threads tapped in four different holes will accept the large bolts that will be inserted during assembly $\mathbf{T_{54}}$.

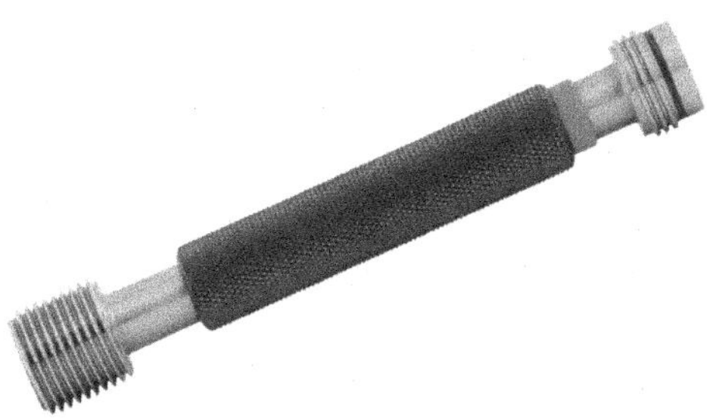

FIGURE 5.43 Threaded Plug Gauge.

Set to Measure

The Set to Measure values are for analyzing the tools used for indirect measurement. These tools are set to a dimension on the part, the tool removed from the part and then a second tool used to measure the setting.

The index values for Set to Measure include the time to get the tool, to locate the tool on the part, to set the tool to the dimension being measured, to remove the set tool and to aside the tool when the measuring is complete. There is no time allowed for measuring the setting with a second tool. The measurement of the setting with a second tool must be analyzed separately. Allow a sequence model for the tool set on the part and an additional sequence model for measuring the setting with a second tool. The index value is selected by the type of tool used and by the number of measurements required.

Telescope Gauge

A Telescope Gauge (Figure 5.44) is a gauge inserted into an opening, expanded to fill the opening, the setting locked and then the gauge removed from the opening for subsequent measurement. The index value is determined by the number of openings measured.

Example: A mechanic sets a telescope gauge to check the bore size of a crankcase in an engine: $\mathbf{T_3}$.

Caliper (Spring Joint or Firm Joint)

The index values for Spring Joint (Figure 5.45) or Firm Joint (Figure 5.46) Calipers include the time to get the caliper, place the caliper on the part, adjust the caliper for size and remove the caliper for measurement. The time for measuring the caliper setting is not included in the index value and must be allowed in a separate sequence.

Example: A quality assurance technician uses a set of inside spring joint calipers to check three keyways that have been machined on the inside of a cylinder: $\mathbf{T_{16}}$.

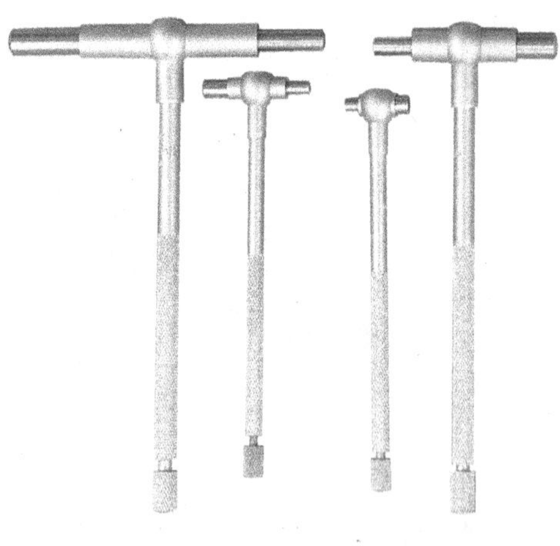

FIGURE 5.44 Telescope Gauge.

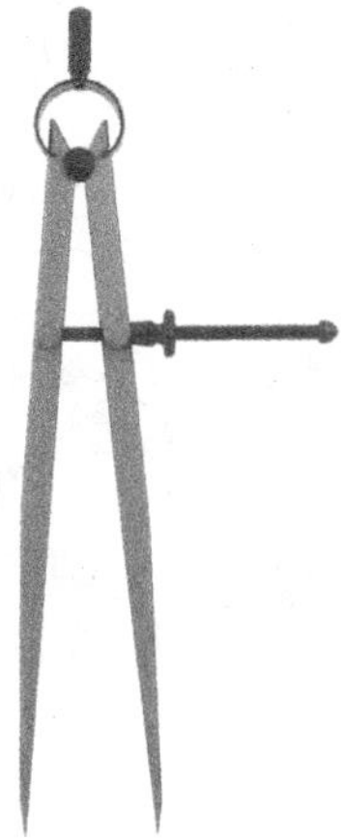

FIGURE 5.45 Example of an inside Spring Joint Caliper.

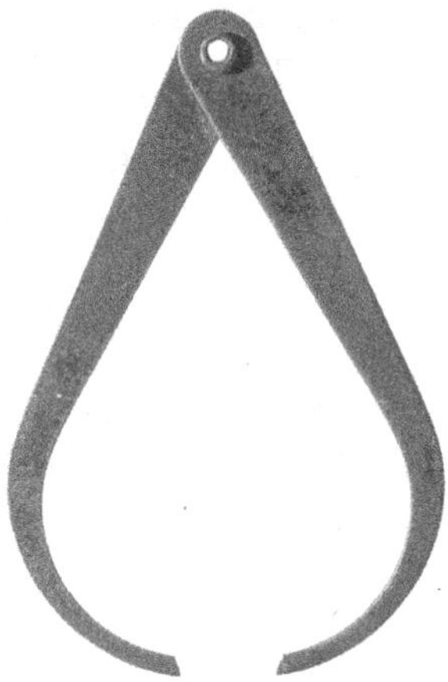

FIGURE 5.46 Example of an outside Firm Joint Caliper.

Snap Gauge

Snap Gauges (Figure 5.47) are fit on the part or the part fit into the gauge to compare the actual dimension of the part to the standard gauge dimension. The index value is determined by the number of times the gauge is fit over a part (or a part into a gauge). Step-type, go-no-go snap gauges can determine maximum and minimum requirements in one fitting between the gauge and the part. However, when separate gauges are set to minimum and maximum tolerances, each gauge must be counted separately. The size of the gauge can be up to 16 inches (400 mm).

Example: An operator gets a go-no-go snap gauge and fits it over a part twice to determine if it is within the specified tolerances: $\mathbf{T_3}$.

FIGURE 5.47 Snap Gauge.

Dial Indicator

The index values for Dial Indicator (Figure 5.48) include the time to stop the machine, wipe the workpiece, position the dial indicator in place, set the dial to zero, turn the part, check the dial as required, clear the dial indicator from the part and restart the machine.

Example: A machine operator periodically assures that the drill press does not have runout to make sure that hole diameters do not end up oversize. The operator uses a dial indicator to note the maximum variation from zero on the dial before removing the gauge from the machine: $\mathbf{T}_{16}$.

Taper Gauge

The time for measuring with a Taper Gauge includes opening the bluing containers, getting the Taper Gauge, brushing bluing on the gauge, positioning the Taper Gauge to the part and removing the gauge. Time is also included for inspecting the

FIGURE 5.48 Example of a Dial Indicator.

displacement of the bluing to determine the contact area between the gauge and the part, wiping the bluing from the gauge and setting aside the gauge.

Example: A machinist uses a taper plug gauge to check the cylinder bore taper that has been machined on the inside diameter of a hub to confirm the interference fit will achieve at least 85% contact with the shaft that will be inserted and assembled later: $\mathbf{T_6}$.

Prepare to Measure

Some measuring devices require preparation before use. This preparation usually includes removal from a case and coarse adjustments to the approximate size of the dimension of the part. The preparation of some measuring tools will require changing fittings or parts of the tool for the general size range of the dimensions to be measured. The time required to adjust the tool subsequent to use and return it to its case is also included in the preparation time value when required. The Prepare to Measure index values do not include any time for use of the measuring device. The index value for Prepare to Measure is determined by the tool required.

In volume production work, the measuring device need only be prepared during the setup for each order. However, in low-volume work, especially when greatly varied dimensions must be made with the same tool, Preparation for Measurement may be required each time the tool is used. All the routine activities necessary to prepare these tools for use and return them to their storage locations are covered in the index values.

The tools described in Prepare to Measure are:

$\mathbf{T_6}$	Telescope Gauge
	Firm Joint Caliper
	Caliper with a Vernier Scale
$\mathbf{T_{10}}$	Combination Square (Figure 5.49)
	4 inch (10 cm) Bevel Protractor
$\mathbf{T_{16}}$	4 inch (10 cm) Micrometer
	Spring Joint Caliper
$\mathbf{T_{24}}$	36 inch (90 cm) Micrometer

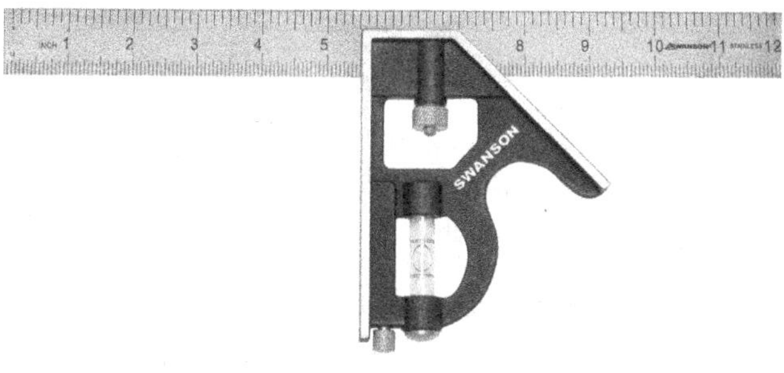

FIGURE 5.49 Example of a Combination Square.

Tool Use—Measuring Tools Examples

The Tool Use examples below follow the recommended sentence structure for Tool Use which is

Activity Object Tool Action Tool At Location

1. A mechanic walks 10 steps, bends and checks the gap in eight spark plugs with a feeler gauge prior to installation. He then walks five steps to put the gauge back in the toolbox.

   Measure gap in 8 spark plugs with feeler gauge

   $\mathbf{A_3}$ $\mathbf{B_1}$ $\mathbf{T_{24}}$

   $(3+1+24)\times 100=2800$ TMU

2. A furniture maker using firm joint calipers sets the measurement on the diameter of a table leg.

   Set measurement to diameter with firm joint calipers

   $\mathbf{A_0}$ $\mathbf{B_0}$ $\mathbf{T_6}$

   $6\times 100=600$ TMU

3. Prior to making a cut, a carpenter unfolds four sections of a wood rule and measures two places on the saw guide.

   Unfold 4 sections of wood rule and measure 2 places on saw guide

   $\mathbf{A_0}$ $\mathbf{B_0}$ $\mathbf{T_6}$

   $6\times 100=600$ TMU

4. Prior to tack welding, a ship fitter uses a profile gauge to check if two pieces (one measurement) are square.

   Ship fitter takes 1 measurement with profile gauge

   $\mathbf{A_0}$ $\mathbf{B_0}$ $\mathbf{T_3}$

   $3\times 100=300$ TMU

5. Before beginning a new job, the milling machine operator walks four steps and gets a dial indicator from an open tool box and returns to the mill to set the head perpendicular to the table.

   Operator walks 8 steps to get and use dial indicator

A_1 B_0 T_{16}

$(1+16)\times 100=1700$ TMU

6. A worker walks 10 steps to a counter, grasps a telescope gauge and walks back to prepare the gauge for measurement on a part located on an empty pallet on the floor.

Prepare to measure with telescope gauge

A_3 B_1 T_6

$(3+1+6)\times 100=1000$ TMU

7. An operator walks 40 steps to grasp a snap gauge and uses it to take four separate measurements, asiding the gauge two steps away when finished.

Take 4 measurements with snap gauge

A_6 B_0 T_6

$(6+6)\times 100=1200$ TMU

8. Using a thread gauge, a job shop operator determines the correct diameter of a bolt at 6 inches (15 cm). He walks 10 steps to put the thread gauge aside.

Take 1 measurement with a thread gauge

A_3 B_0 T_{16}

$(3+16)\times 100=1900$ TMU

9. A technician uses a flat rule to measure the size of several components with a total of 10 measurements.

Take 10 measurements with a flat rule

A_0 B_0 T_{16}

$16\times 100=1600$ TMU

10. A technician walks eight steps to grasp spring joint calipers and walks an additional 12 steps to set the calipers to measure four extruded parts.

Set to measure 4 measurements with spring joint calipers

A_3 B_0 T_{16}

$(3+16)\times 100=1900$ TMU

THE MACHINE HANDLING SEQUENCE MODEL

Machine Handling is used for analysis of the manual operations associated with manipulating the controls of machines. As with Part Handling and Tool Use, the Machine Handling Sequence Model accounts for total walking distance and total body motions required for the completion of the use of machine controls. The Machine Handling Sequence Model is used for the analysis of situations such as:

- Set controls on a machine or equipment.
- Secure the workpiece for machining.
- Set feed and/or speed.
- Activate controls.

The Sequence Model

The sequence model consists of a series of letters (parameters) representing each of the various sub-activities of Machine Handling. The **A** and **B** parameters shown below were introduced in an earlier section. The new parameter for Machine Handling (**M**) will be presented in this section.

A B M

where: **A** = Action Distance
B = Body Motion
M = Machine Handling

Parameter Definitions

The Machine Handling (**M**) parameter accounts for activities associated with the manipulation of machine controls, the changing of cutting tools and the securing or releasing of a workpiece. This parameter covers grasping and operating machine controls or fixed machine equipment, such as:

- Buttons and switches
- Cranks and handwheels
- Slides

Parameter Indexing

The **M** value in Machine Handling includes the time to grasp and operate machine controls and is broken down into two data cards:

1. *Operate Machine Controls*—For the manipulation of machine controls and the activities associated with the changing of the cutting tools (Figure 5.50).
2. *Secure or Release Parts*—For the activities associated with securing or releasing a workpiece (Figure 5.51).

MaxiMOST System	Operate Machine Controls					M – Machine Handling			
Index x 100	Button or Switch	Lever		Crank	Knob	Handwheel		Change Tool	Index x 100
		1 or 2 Stages	Difficult 3 or 4 Stages			Normal	Heavy		
	Controls			Revs.	Actions			Device	
1	4	2	1	2	4	3	1	Quick Change Post	**1**
3	12	7	3	16	17	11	4	-	**3**
6		13	6	36	34	23	9	Jacobs Chuck	**6**
10			10	63			16	Carbide Insert	**10**
16									**16**

FIGURE 5.50 Machine Handling data card—Operate Machine Controls. Values are read up to and including.

MaxiMOST System			Secure or Release Parts					M – Machine Handling			
Index x 100	Open or Close	Install or Remove Lathe Dog	Engage or Disengage Tail Stock Center	Install or Remove Jack Screw	Install or Remove C-Clamp	Tighten or Loosen Part in Fixture			Clamp or Unclamp Part on Bed		Index x 100
						With Wrench	By Hand	With Cam or Eccentric Clamp	Tighten or Loosen Clamp or Relieve Strain	Assemble or Disassemble Clamp	
	Device			Devices					Clamps		
1	Collet Hand Vise Air Vise	Cam Type	Lever Operated	-	-	-	1	2	-	-	**1**
3	Mallet Vise 3-Jaw Chuck	Standard	Crank Operated	1	1	1	3	6	1	-	**3**
6	4-Jaw Chuck			2	2	3	6	12	2	1	**6**
10	6-Jaw Chuck			4	3	5	10		4	2	**10**
16				6	5	8			6	4	**16**
24				9	7	12			9	6	**24**
32				16	10				12	8	**32**
42										10	**42**

FIGURE 5.51 Machine Handling data card—Secure or Release Parts. Values are read up to and including.

Operate Machine Controls (M)

The data for Operate Machine Controls covers the manipulation of buttons, switches, levers, cranks, knobs and handwheels. The data card values include time for walking to the control with one to two steps, getting the control, manipulating the control and relinquishing the control.

Button or Switch

Index values for operating a button or switch are typically based on the number of buttons or switches that are activated. The notable exceptions are palm buttons. When two palm buttons are simultaneously activated, these two buttons should be counted as only one for the purpose of establishing the index value.

The values for button or switch cover any stationary (panel-mounted) or pendant button. Activation of the button or switch will be characterized by a single action of the finger or hand.

Example: Operator flips 10 switches to activate a machine: $\mathbf{M_3}$.

Lever

Operate Lever includes the displacement of a lever in either one or two stages or three to four stages. Operate Lever index values are based on the number of levers manipulated.

Example: A worker manipulates a single four-stage lever on a machine to engage the feed: $\mathbf{M_1}$.

Crank

The data for Crank is identical to the turn by hand Arm Crank data found on the General Tools I data card. Duplication on the Machine Handling data card is for convenience only. Arm Crank index values apply when an object is manipulated by circular motions of the forearm. The index value is selected by the number of revolutions.

Example: An operator unwinds a hose by turning the hose reel four revolutions: $\mathbf{M_3}$.

Knob

Knob is used to analyze the rotation of a device using the fingers or the hand. The Knob index values are based on the total number of positive actions involved in turning knobs and include an alignment of one point or an alignment to a workpiece.

Example: A technician operates a total of six knobs to adjust the pressure on a stamping machine: $\mathbf{M_3}$.

Handwheel

Operate Handwheel includes moving the circumference of a circular device with either wrist or arm actions, where following each positive action the handwheel is regrasped. Operate Handwheel index values include the time to align to a scale mark. The index values for Handwheel are based on the total number of positive actions.

Normal. The movement for Handwheel—Normal is characterized by little or no resistance.

Heavy. The movement for Handwheel—Heavy includes moving the circumference of a circular device with either wrist or arm actions. The movement is characterized by resistance that requires the application of muscular force.

Example: With resistance, a worker rotates a handwheel six times, regrasping the handwheel with each revolution: $\mathbf{M_6}$.

Change Tool

The Change Tool data covers a single tool change involving a Quick Change Post, a Jacobs Chuck or a Carbide Insert. The data card index values cover walking to the tool holder with one or two steps, removing the existing tool and installing the next tool.

M_1 Quick Change Post

The Quick Change Post index value covers removing the existing tool and installing the next tool.

M_6 Jacobs Chuck

The Jacobs Chuck index value covers obtaining the chuck key, loosening the chuck with the key, removing the key and loosening the chuck by hand, removing the tool, installing the next tool, tightening the chuck by hand, tightening the chuck with the key and asiding the key.

M_{10} Carbide Insert

The Carbide Insert index value covers obtaining the screwdriver, loosening the retainer with the screwdriver, removing the retainer, removing the old insert, installing the new insert in the retainer, installing the retainer and asiding the screwdriver.

Secure or Release Parts (M)

The second set of data for the Machine Control (**M**) parameter is for Secure or Release Parts. The values on this data card cover opening or closing a holding device and relieving or increasing clamp pressure (strain) on a workpiece. All values include the time for walking one or two steps to the holding device.

Open or Close

M_1 Collet

The index value for Collet covers opening or closing a collet by utilizing a lever, handwheel or hydraulic switch.

M_1 Hand Vise

The index value for Hand Vise covers obtaining the vise handle, placing the vise handle in the vise, opening or closing the vise with the handle and asiding the handle.

M_1 Air Vise

The index value for Air Vise covers opening or closing a vise by activating a pneumatic device by hand or foot.

M_3 Mallet Vise

The index value for Mallet Vise covers opening or closing a vise by utilizing a mallet with three arm strikes to initially loosen or final tighten the vise. The vise handle may be in place or placed in the vise. The index value includes time for placing the handle in the vise, obtaining the mallet, striking the handle to loosen or tighten the vise up to three arm strikes, asiding the mallet and opening or closing the vise by hand.

M_3 3-Jaw Chuck

The index value for 3-Jaw Chuck covers obtaining the wrench, placing the wrench on the chuck, tightening or loosening the chuck with the wrench and asiding the wrench.

M_6 4-Jaw Chuck

The index value for 4-Jaw Chuck covers obtaining the wrench, placing the wrench to the first jaw, tightening or loosening the jaw, revolving the chuck to the second jaw, placing the wrench to the jaw, tightening or loosening the jaw and asiding the wrench.

M_{10} 6-Jaw Chuck

The index value for 6-Jaw Chuck covers obtaining the wrench, placing the wrench to the first jaw, tightening or loosening the jaw, revolving the chuck to the second jaw and two additional jaws, placing the wrench to each jaw, tightening or loosening the jaw and asiding the wrench.

Install or Remove Device

The index values for Install or Remove Device cover the installation or removal of lathe dogs, jack screws or C-clamps used for the purpose of securing or stabilizing a workpiece. These index values include the time for walking to the device with one or two steps and installing or removing the device.

Install or Remove Lathe Dog

M_1 Cam Type Lathe Dog

The index value for the Cam Type Lathe Dog covers obtaining the dog, opening the cam and placing the dog to the workpiece or reversing this to remove the dog from the workpiece.

M_3 Standard Lathe Dog

The index value for Standard Lathe Dog covers obtaining the wrench, placing the wrench on the lathe dog bolt, loosening or tightening the bolt, asiding the wrench and removing the dog from the workpiece or placing it on the workpiece.

Engage or Disengage Tail Stock Center

The index values for Engage or Disengage Tail Stock Center cover engaging or disengaging tail stock centers that are either spring-loaded, lever-activated or crank-activated. The data card index values include the time for walking one or two steps to the activation device and engaging or disengaging the device in place. Index values do not include sliding the tail stock into place. This would be handled in a separate Part Handling Sequence Model.

M_1 Lever Operated

The index value for Lever Operated covers either engaging or disengaging the activation arm.

M_3 Crank Operated

The index value for Crank Operated covers cranking the center in or out up to six revolutions, engaging and disengaging the center, adjusting the crank pressure, setting the lock lever and tightening or releasing the lock lever.

Install or Remove Jack Screw

The index values for Install or Remove Jack Screw cover running the jack screw in or out by hand, setting the jack screw to the workpiece, locating the jack screw head, obtaining a wrench, placing the wrench on the jack screw, locking or unlocking the jack screw with the wrench, asiding the wrench and asiding the jack screw. Install or Remove Jack Screw index values are based on the number of jack screws manipulated.

Install or Remove C-Clamp

The index values for C-clamp cover placing the C-clamp on the workpiece or removing it from the workpiece, tightening or loosening the C-clamp by hand with up to three wrist turns, obtaining a wrench, placing the wrench on the C-clamp and tightening or loosening the C-clamp with the wrench with up to five wrist strokes and asiding the wrench. Install or Remove C-clamp index values are based on the number of C-clamps manipulated.

Tighten or Loosen Part in Fixture

With Wrench

The index values for Tighten or Loosen with Wrench include using a wrench to tighten or loosen bolts on a fixture for the purpose of securing or releasing a workpiece. The data includes the time for obtaining the wrench, placing the wrench on one or more bolts, tightening or loosening the bolt with up to three wrist strokes each, asiding the wrench and running the bolt in or out by hand. Tighten or Loosen with Wrench index values are based on the number of bolts that are run in or out.

By Hand

The index values for Tighten or Loosen by Hand cover using the hand to tighten or loosen thumb screws, handwheels or star wheels on a fixture. The purpose is to secure or release a workpiece. Tighten or Loosen by Hand index values are based on the number of thumb screws, handwheels or star wheels that are run in or out.

With Cam or Eccentric Clamp

The index values for Tighten or Loosen with Cam or Eccentric Clamp cover placing cams or clamps on a fixture to tighten it or removing cams or clamps from a fixture to loosen it. Tighten or Loosen part with Cam or Eccentric Clamp index values are based on the number of cams or clamps manipulated.

Clamp or Unclamp Part on Bed

The index values for Clamp or Unclamp Part on Bed cover loosening and sliding clamps or sliding and tightening clamps for the purpose of releasing or securing a workpiece on a bed. The index values include time for obtaining the wrench (within two steps), placing the wrench on the nut, tightening or loosening or assembling or disassembling the nut on the stud, asiding the wrench, running the nut out by hand to obtain clearance or in to tighten and pushing the clamp clear of the workpiece or locating the clamp on the workpiece. The index values are based on the number of clamps manipulated.

Figure 5.52 illustrates the relationship of the workpiece, clamp, stud and nut.

Tighten or Loosen Clamp or Relieve Strain

The index values for Tighten or Loosen Clamp or Relieve Strain cover tightening or loosening a nut for the purpose of relieving or increasing the clamp pressure (strain) on the workpiece, which rests on a bed. With these index values, the nut and the clamp remain on the stud. The index values are based on the number of clamps manipulated.

Assemble or Disassemble Clamp

The index values for Assemble or Disassemble Clamp cover loosening and removing clamps and nuts or placing and tightening clamps and nuts for the purpose of releasing or securing a workpiece on a bed. These index values include the time for obtaining the wrench, placing the wrench on the nut, assembling or disassembling the nut on the stud, asiding the wrench and removing the nut and clamp from the stud or placing the clamp and nut on the stud and running down the nut. The index values are based on the number of clamps manipulated.

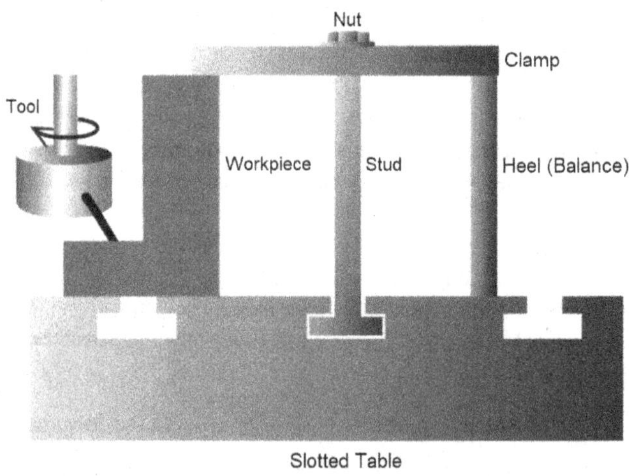

FIGURE 5.52 Relationship of the workpiece, clamp, stud and nut.

Writing Method Descriptions

Below is the recommended format for a clear and concise Machine Handling method description. Additional words for Action Distance or Body Motion are optional and may be used to enhance the method description.

The recommended sentence structure for Machine Handling is:

Activity	Object	At Location

Machine Handling Examples

1. An operator pushes three buttons to set and start machine.

   Push 3 buttons to set and start machine

   $\mathbf{A_0}$ $\mathbf{B_0}$ $\mathbf{M_1}$

   $1 \times 100 = 100$ TMU

2. A machine operator moves a lever two stages to retract tool.

   Move lever 2 stages to retract tool

   $\mathbf{A_0}$ $\mathbf{B_0}$ $\mathbf{M_1}$

   $1 \times 100 = 100$ TMU

3. A worker walks 24 steps and cranks 10 revolutions on a normal handwheel bringing the tool to the workpiece.

   Crank 10 revolutions and bring tool to workpiece

   $\mathbf{A_6}$ $\mathbf{B_0}$ $\mathbf{M_3}$

   $(6+3) \times 100 = 900$ TMU

4. An operator changes the carbide insert on a milling cutter.

   Change carbide insert on milling cutter

   $\mathbf{A_0}$ $\mathbf{B_0}$ $\mathbf{M_{10}}$

   $10 \times 100 = 1000$ TMU

5. An operator secures a part in a 3-jaw chuck.

   Secure part in 3-jaw chuck

   $\mathbf{A_0}$ $\mathbf{B_0}$ $\mathbf{M_3}$

   $3 \times 100 = 300$ TMU

6. Operator tightens part with four clamps to machine bed.

Tighten part to machine bed with 4 clamps

$\mathbf{A}_0$ $\mathbf{B}_0$ $\mathbf{M}_{10}$

$10 \times 100 = 1000$ TMU

7. The machine operator installs four C-clamps.

Install 4 C-clamps

$\mathbf{A}_0$ $\mathbf{B}_0$ $\mathbf{M}_{16}$

$16 \times 100 = 1600$ TMU

8. A worker loosens five bolts in a fixture with a wrench.

Loosen 5 bolts in fixture with wrench

$\mathbf{A}_0$ $\mathbf{B}_0$ $\mathbf{M}_{10}$

$10 \times 100 = 1000$ TMU

9. A technician walks 12 steps and engages the tail stock center with a lever.

Engage tail stock center with lever

$\mathbf{A}_3$ $\mathbf{B}_0$ $\mathbf{M}_1$

$(3+1) \times 100 = 400$ TMU

10. An operator uses her hand to tighten three thumb screws to secure a part in a fixture.

With hand, tighten 3 thumb screws to secure part

$\mathbf{A}_0$ $\mathbf{B}_0$ $\mathbf{M}_3$

$3 \times 100 = 300$ TMU

THE POWERED CRANE SEQUENCE MODEL

Transport with Crane is used for the analysis of the movement of objects with the aid of a powered bridge crane. The Powered Crane Sequence Model describes such activities as:

- Walk to the crane.
- Start and transport empty crane to the location to hook-up the object.
- Hook-up object to crane hook.
- Transport object to the location for placement.
- Place object with necessary manipulations of the crane.
- Unhook and transport crane to another location.
- Return to workplace by walking.

The Powered Crane Sequence Model is appropriate for cranes that move the load laterally and longitudinally under power and may resemble an overhead, pendant-operated bridge crane (Figure 5.53).

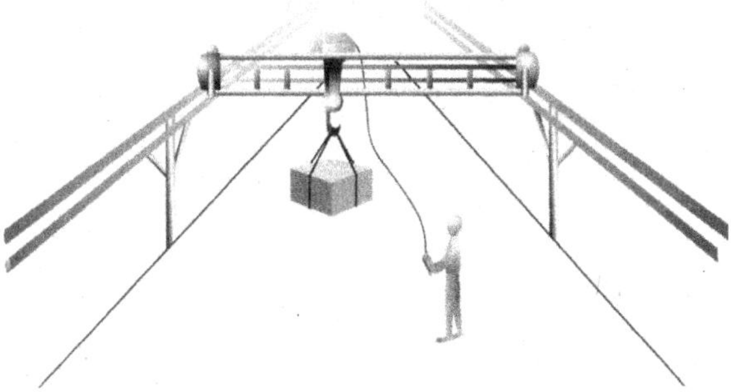

FIGURE 5.53 Overhead bridge crane.

The Sequence Model

The movement of an object with a powered crane is described by the following sequence model:

A T K T P T A

where: **A** = Action Distance
T = Transport
K = Hook-up and Unhook
P = Placement

The Powered Crane Sequence Model consists of the following activities:

1. The operator walks to the control panel (Action Distance).
2. The operator grasps the controls, elevates the crane hook, moves the crane so the hook reaches the position for coupling and then releases the controls (Transport).
3. The object is fastened either directly with the crane hook or with a sling or chain. The operator grasps the controls and elevates the crane hook to the correct position for hooking and then adjusts the controls so that the chain or other holding device is tight and secure (Hook-up, Unhook). The holding device is subsequently removed from the object following the placement.
4. The crane hook with the object is freed from its surroundings and elevated so the object can be moved. The object is then moved horizontally to the desired location (Transport).
5. The object is lowered and placed in the desired location (Placement).
6. The empty crane is moved aside (Transport).
7. The operator returns to the starting point after moving the crane aside (Action Distance).

Figure 5.54 illustrates the sequence of events that occurs when an object is moved with a powered crane following the Powered Crane Sequence Model.

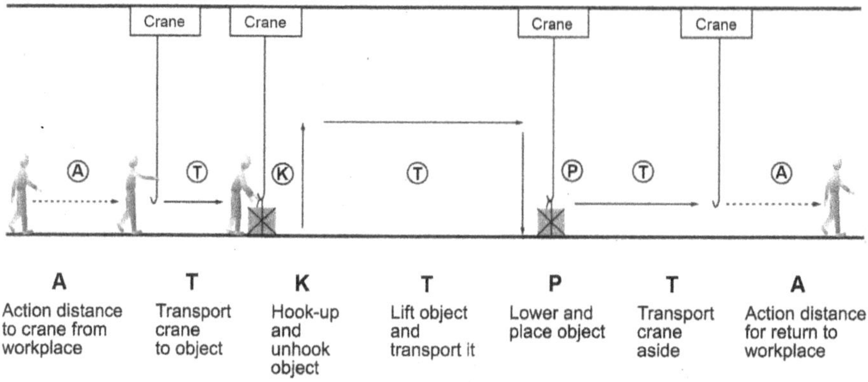

FIGURE 5.54 Illustration of the Powered Crane Sequence Model.

Parameter Definitions

A Action Distance

The Action Distance is the horizontal distance the operator walks to or from the bridge crane control panel.

T Transport

This parameter covers the movement of the crane with or without a load. The time values are based on a sample of typical cranes operated under average conditions. Note that *all time values must be validated* for the cranes actually being used prior to any analysis work involving the Powered Crane Sequence Model.

K Hook-Up and Unhook

This parameter includes the activities involved in both connecting and disconnecting the object from the crane. The parameter begins when the hook has been transported close to the hooking position and is completed when the holding device has been disconnected from the object.

P Placement

This parameter involves all actions necessary to lower the object with a combination of high speed and creep speed and to place the object in the desired location.

If the local conditions call for a designated bridge crane operator (riding with the crane), only the **T**, **K** and **P** parameters are needed for analysis purposes. The index value for the **A** parameter will be equal to zero.

Powered Crane Data Card Backup Information

It is recognized that there are many manufacturers, models, capacities, etc., of cranes available. As a result, the information presented on the Powered Crane data card (Figure 5.55) should be treated as sample information. The *method must be verified,* and the *process times must be validated* to fit a company's particular equipment.

MaxiMOST System	Powered Crane		ATKTPTA	
Index x 100	T Transport	K Hook-Up and Unhook	P Placement	Index x 100
	Feet (m)	Holding Device	Difficulty	
3	-	-	With or Without Single Change of Direction	**3**
6	-	Single Hook or Electromagnet	-	**6**
10	2 (0.5)	-	-	**10**
16	25 (8)	-	With Double Change of Direction	**16**
24	50 (16)	1 Hook Plus Slings or Chains	With Several Changes of Direction	**24**
32	80 (25)	2 Hooks Plus Slings or Chains		**32**

FIGURE 5.55 Powered Crane data card.

The Transport parameter in Powered Crane is best developed using time study and regression analysis based on specific working conditions. The data for Loaded Transport and Unloaded Transport needs to be analyzed separately and a regression analysis developed for each. Guidelines for validating process times for powered cranes can be found in the Application of the MaxiMOST Work Measurement System section.

Parameter Indexing

A Action Distance

Action Distance is defined in The Sequence Models section and the values found on the Action Distance and Body Motions data card. Choose the index value by the distance the operator walks to get to or move away from the crane.

T Transport

Choose the index value by the distance the crane is moved horizontally, either loaded or unloaded. All vertical distances are included in the (**T**) index values; separate vertical analyses are not necessary. The values for the **T** parameter provided on the data card (Figure 5.55) are sample data and must be *verified* and *validated* to fit particular situations and/or cranes.

K Hook-up and Unhook

Hook-up and Unhook covers the time to connect and disconnect one or more objects to a crane. Adjustments of the crane for hooking and for tightening and securing the holding device are included. Hooking and Unhooking includes fastening chains, slings or other holding devices to the crane hook or to the object.

Getting and moving aside chains or slings, for example, and fastening them to the object (or to the crane hook) initially are not covered by the **K** parameter. Such activities are analyzed using the Part Handling Sequence Model.

Hook-up and Unhook index values are chosen by the type of holding device used.

K_6 Single Hook or Electromagnet
K_{24} 1 Hook Plus Slings or Chains
K_{32} 2 Hooks Plus Slings or Chains

P Placement

Choose the proper index value by the difficulty involved in lowering the object the last 2–3 inches (5–7.5 cm) and placing it in the desired location.

P_3 With or Without Single Change of Direction

The load is placed with or without a single change of direction. This index value includes lowering the load up to 20 inches (50cm) at high speed, then placing the load with either a longitudinal or lateral traversing action of up to 4 inches (10cm) followed by lowering the load the last 2–3 inches (5–7.5cm) at low speed.

P_{16} With Double Change of Direction

The load is placed with a double change of direction (up to four total direction changes). This index value includes lowering the load up to 5 feet (1.5m) at high speed, then placing the load with a double change of longitudinal and lateral traversing actions (up to 4 inches or 10cm each direction and 20 inches or 50cm total) followed by lowering the load the last 20 inches (50cm) at low speed.

P_{24} With Several Direction Changes

The load is placed with several changes of direction (up to eight total direction changes). This index value includes lowering the load up to 6.5 feet (2m) at high speed, then placing the load with several changes of longitudinal and lateral traversing actions (up to 4 inches or 10cm each direction and 3.3 feet or 1m total) followed by lowering the load the last 20 inches (50cm) at low speed.

Writing Method Descriptions

Below is the recommended sentence structure for a clear and concise method description for the Powered Crane Sequence Model. Additional words may be used to enhance the method description. These could be Action Distances or adjectives. This structure can be found in the example listed below.

The recommended sentence structure for Powered Crane is:

Transport	Object	Holding Device	To Location	Placement

Powered Crane Example

1. An operator walks 90 feet (27m) to a powered crane control panel and transports the crane to a heavy part 25 feet (7.5m) away. The part is connected

to the crane with one hook and a sling and transported 2 feet (0.6 m), where it is placed with a double change of direction. The operator then moves the crane 9 feet (3 m) out of the way and walks back to the part.

Transport part with one hook and sling 2 feet (0.6 m) and place with a double change of direction

$\mathbf{A_6 \quad T_{16} \quad K_{24} \quad T_{10} \quad P_{16} \quad T_{16} \quad A_1}$

$(6+16+24+10+16+16+1)\times 100=8900$ TMU

THE POWERED TRUCK SEQUENCE MODEL

Powered Truck is used for the analysis of the movement of objects with the aid of a powered truck. The Powered Truck Sequence Model describes the following activities:

- Walk to the truck.
- All activities necessary to start and park truck.
- Move the empty truck to a location to load the truck.
- Load one or more objects mechanically.
- Move the loaded truck to a location to unload the truck.
- Unload the truck mechanically.
- Move the empty truck to another location.
- Return by walking.

The Powered Truck Sequence Model is primarily used for determining a time for the horizontal transportation of material from one location to another using a Powered Truck. Equipment covered by this sequence falls within two general categories:

1. Riding Trucks—Trucks operated from a riding position
   Forklift (Figure 5.56)
   High Stacker (Figure 5.57)

2. Walking Trucks—Mechanized trucks operated from a walking position
   Stacker (Figure 5.58)
   Low Lift Pallet Truck (Figure 5.59)

The Sequence Model

The movement of an object with a powered truck is described by the following sequence model:

A S T L T L T A

where **A** = Action Distance
S = Start *and* Park
T = Transport
L = Load *or* Unload

FIGURE 5.56 Example of a Forklift.

FIGURE 5.57 Example of a High Stacker

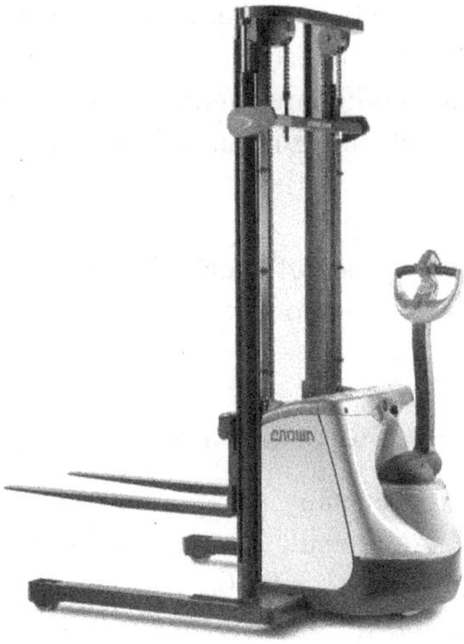

FIGURE 5.58 Example of a Stacker.

FIGURE 5.59 Example of a Low Lift Pallet Truck.

Transportation of material with trucks consists of the following activities:

1. The operator walks to the truck (Action Distance).
2. The operator takes a seat (if riding) and starts the truck (Start).
3. The truck is driven or transported to the material (Transport).
4. The material is loaded by the fork or lifting attachment (Load).

5. The material is transported to another location (Transport).
6. The material is unloaded (Unload).
7. The truck is driven to another area and parked (Transport).
8. The operator switches off the ignition and/or parks the truck.
9. The operator returns to the original (or another) location (Action Distance).

Figure 5.60 illustrates the sequence of events that occurs when an object is moved with a powered truck following the Powered Truck Sequence Model.

Powered Truck Data Card Backup Information

It is recognized that there are many manufacturers, models, capacities, etc., of powered trucks available. As a result, the information presented on the Powered Truck data card (Figure 5.61) should be treated as sample information. The *method must be verified*, and the *process times must be validated* to fit a company's particular equipment. The Transport and Load parameters in Powered

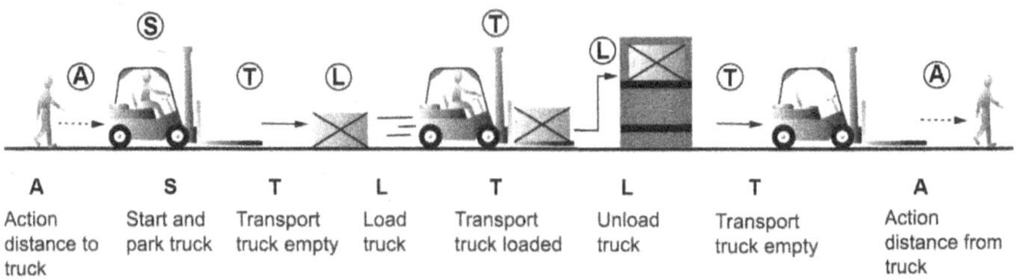

FIGURE 5.60 Illustration of Powered Truck Sequence Model.

MaxiMOST System	Powered Truck					ASTLTLTA	
Index x 100	S Start and Park	T Transport With or Without Load				L Load or Unload	Index x 100
		Riding		Walking			
		Forklift	High Stacker	Stacker	Low Lift Pallet Truck		
		Feet (m)					
1	-	27 (8)	21 (6)	11 (3)	14 (4)	-	**1**
3	Walking Truck	67 (20)	50 (15)	27 (8)	34 (10)	Floor-Simple	**3**
6	Riding Truck	132 (40)	100 (30)	50 (15)	67 (20)	Floor	**6**
10		198 (60)	165 (50)	83 (25)	100 (30)	Pallet Rack	**10**
16		329 (100)	247 (75)	116 (35)	165 (50)		**16**
24		460 (140)	362 (110)	182 (55)			**24**

FIGURE 5.61 Powered Truck data card.

Truck are best developed using time study and regression analysis based on specific working conditions. The data for Loaded Transport and Unloaded Transport needs to be analyzed separately with a regression analysis developed for each. These values would then have a frequency applied based on workplace conditions to arrive at one transport constant for determining index values. Guidelines for validating process times for powered trucks can be found in the Application of the MaxiMOST Work Measurement System section.

Parameter Definitions

A Action Distance

The Action Distance is the horizontal distance the operator walks to or from the truck.

S Start and Park

This parameter includes the actions to prepare the truck for moving plus the parking activity following the final transport.

T Transport

This parameter applies to the movement of the truck with or without a load. The time values are based on a sample of typical trucks (riding and walking) operated under average conditions. Note that *all time values must be validated* for the trucks actually being used prior to any analysis work involving the Powered Truck Sequence Model.

L Load *or* Unload

This involves either picking up the material at the original location or placing the material at the destination using the forks or other lifting attachments. Load or Unload Truck covers the activity to mechanically load or unload an object using a truck.

Parameter Indexing

A Action Distance

Action Distance is defined in The Sequence Models section and the values found on the Action Distance and Body Motions data card. Choose the index value by the distance the operator walks to get to or move away from the truck.

S Start *and* Park

Choose the index value by the type of truck used.

S_3 Walking Truck

For the *walking truck,* the activities include grasping the handle, starting and stopping the power and tilting the body or handle.

S_6 Riding Truck

For the *riding truck,* the activities include climbing in and out of the seat, starting and stopping the engine and releasing and engaging the hand brake.

T Transport

This parameter applies to the movement of the truck with or without a load. The time values are based on a sample of typical trucks (riding and walking) operated under average conditions. Note that *all time values must be validated* for the trucks actually being used prior to any analysis work involving the Powered Truck Sequence Model. To determine the index values, do the following:

1. Choose the correct column by the general truck type (riding or walking).
2. Select the specific kind of truck (forklift, high stacker, stacker or low-lift pallet truck) being used.
3. Select the index value based on the distance (in feet or meters) that the truck is transported.

L Load or Unload

Choose the correct index value by the location of the object when mechanically loading or unloading. Loading or unloading parts to or from a pallet already on the truck should be analyzed with the Part Handling Sequence Model.

L_3 Floor-Simple

The Floor-Simple value is used when loading or unloading an object either from or to the floor when no adjustments are required.

L_6 Floor

The Floor value is used when loading or unloading an object either from or to the floor with adjustments.

L_{10} Pallet Rack

The value for Pallet Rack is used when loading or unloading an object either from or to a pallet rack (above the floor).

Again, due to the large number of manufacturers and seemingly infinite configuration of trucks, the data provided in Figure 5.61 should be treated as *sample information*. Before its use in establishing a labor standard, the *method must be verified*, and the *process time must be validated* to a company's particular trucks and conditions. Once index values are selected from the data card, they are placed on the Powered Truck Sequence Model, added and multiplied by 100 to convert to TMU.

Writing Method Descriptions

Below is the recommended sentence structure for a clear and concise method description for the Powered Truck Sequence Model. Additional words may be used to enhance the method description. These could be Action Distances or adjectives. This structure can be found in the example listed below.

The recommended sentence structure for Powered Truck is:

Transport	Object	From Location (method of load)	To Location (method of unload)

Powered Truck Examples

1. An operator walks 120 feet (36 m) to a forklift, climbs into the seat and starts the engine. It is driven 12 feet (4 m), where a pallet is picked up from the floor requiring some adjustments and then transported 75 feet (23 m) and placed in a pallet rack. The truck is then parked 30 feet (9 m) away and the operator returns 60 feet (18 m) to the workplace.

   Transport part from workplace floor to raised pallet-rack using forklift and return to workplace

   $A_6 \quad S_6 \quad T_1 \quad L_6 \quad T_6 \quad L_{10} \quad T_3 \quad A_3$

   $(6+6+1+6+6+10+3+3)\times 100=4100$ TMU

2. An operator walks 15 feet (4.5 m) to a low-lift pallet truck, starts it and transports it 21 feet (6 m) to a pallet located on the floor (simple). The pallet is loaded on the truck and then transported 100 feet (30 m) to a warehouse where it is then simply placed on the floor. The operator then transports the truck 89 feet (27 m), parks it and walks 30 feet (9 m) to a workbench.

   Transport part from floor using truck to warehouse floor and return to workbench

   $A_1 \quad S_3 \quad T_3 \quad L_3 \quad T_{10} \quad L_3 \quad T_{10} \quad A_3$

   $(1+3+3+3+10+3+10+3)\times 100=3600$ TMU

DEVELOPING NEW ELEMENTS

Because of the wide variations in the tools, conditions and methods described with MaxiMOST, developing new elements is a common and necessary procedure. Knowledge of the element development technique is important. The following should be noted:

- The element backup may be in the form of a time formula.
- The backup analysis should preferably be made with BasicMOST.
- Activity words should be assigned to new values for (1) type of activity or object and (2) application details, such as the nature of the part or tool, the nature of the surroundings and a measure of magnitude, such as number of items or occurrences, size, weight or distances.
- A data card showing the activity words and a suitable number of variables should be designed.
- Document and preserve the backup information supporting the new elements.

The majority of the MaxiMOST elements were calculated using the time formula

$$y = mx + c$$

where: y = total time in TMU
m = TMU per unit
x = number of activities
c = constant time

Constants and TMU per unit were developed for the activities shown on the data cards and index values. The formula shows a straight-line application. Data can also be developed for curved-line applications by following the same format but using different formulas.

Selecting and evaluating the unit of measure is sometimes the most difficult aspect of establishing a formula basis. Suppose a company needs to develop index values to 'fasten a 5/32 inch (4 mm) plate screw using a spiral screwdriver.' The BasicMOST analysis of the activities provides the following times as shown in Figure 5.62.

Assuming at least two strokes, the time for the first and last strokes (60 TMU) can be added together and then added to the time for placing the screwdriver to the screw (40 TMU) to get a 'per screw' value.

The number of screwdriver strokes is studied and a reliable average of four strokes per screw is developed. It can then be determined that the strokes needed beyond the first and last are two.

With this data, the times for fastening any number of 5/32 inch (4 mm) plate screws with a spiral screwdriver can be developed using the following formula:

$$y = mx + c$$

where: y = total time to fasten x screws
m = per screw average time (190 TMU) (40 TMU per screwdriver placement + 60 TMU for first and last stroke combined + 30 TMU for final tighten or initial loosen + 2 strokes per screw × 30 TMU per stroke)
x = number of screws to be fastened or loosened
c = constant for getting and asiding screwdriver (60 TMU) (to get or aside screwdriver up to two steps)

$$y = 190x + 60$$

Step	Method Description	Seq. Model	Freq.	Simo	TMU	Total
1	GRASP AND ASIDE SCREWDRIVER	A3 B0 G1 A1 B0 P1 A0	1.000	N	60	60
2	PLACE SCREWDRIVER TO SCREW	A0 B0 G0 A1 B0 P3 A0	1.000	N	40	40
3	FIRST AND LAST STROKES COMBINED	A0 B0 G0 M3 X0 I0 A0	2.000	N	30	60
4	FINAL TIGHTEN OR INITIAL LOOSEN	A0 B0 G0 M3 X0 I0 A0	1.000	N	30	30
5	TWO ADDITIONAL STROKES	A0 B0 G0 M3 X0 I0 A0	2.000	N	30	60

FIGURE 5.62 BasicMOST Analysis to fasten plate with spiral screwdriver.

The maximum number or screws for each index value can then be determined by solving this formula for x, assigning the maximum interval limits to y and truncating the results. The formula to solve for x is:

$$x = (y - 60) / 190$$

Taking the maximum interval limit values from the Index Value Table from Appendix A (Figure A.3) and multiplying by 10 to get the upper interval limit for MaxiMOST, the data produced is shown in Figure 5.63.

Data tables should only be extended to the practical limits of their application. Theoretically, this spiral screwdriver data table could be extended to cover 181 screws (index value 330), but the upper ranges of the table would rarely be used. A table with the supplementary index values for the spiral screwdriver is found in Figure 5.64. The values are read up to and including.

Index Value	Upper (y value) interval limit	Number of screws (x value) x = (y-60)/190	Number of screws x (truncated)
1	170	0.58	0
3	420	1.89	1
6	770	3.74	3
10	1260	6.32	6
16	1960	10.00	10
24	2770	14.26	14
32	3660	18.95	18
42	4760	24.74	24
54	6010	31.32	31

FIGURE 5.63 Data table for spiral screwdriver.

Spiral Screwdriver	
Index Value	*Number of Screws*
T_3	1
T_6	3
T_{10}	6
T_{16}	10

FIGURE 5.64 Supplementary index values for spiral screwdriver. Values are read up to and including.

APPLICATION OF THE MAXIMOST WORK MEASUREMENT SYSTEM

THE MAXIMOST ANALYSIS

The application of MOST is consistent among all three systems. There are several key factors to a MOST analysis which are listed below and detailed in Chapter 2. Analysts need to consider the following factors and the content included in this section when developing a MaxiMOST analysis:

- Identifying the best method
- Practical analysis procedures
- Analyst consistency
- General application rules for all MOST Systems

MaxiMOST is used for the analysis of longer-cycle, non-repetitive activities and less detail is needed than a typical BasicMOST analysis. The sequence models represented in Figures 5.65 and 5.66 provide a comparison of the same activity using BasicMOST and MaxiMOST as shown from the StandardsPro software. The MaxiMOST analysis requires less detail than the BasicMOST analysis. Note that seven method steps are needed to describe the activities in the BasicMOST analysis and only two method steps are needed for the MaxiMOST analysis. The inclusion of the round trip Action Distance and the use of multiple objects or tools in MaxiMOST may require only one sequence model in MaxiMOST where BasicMOST will need several sequence models to show more detail for each object movement.

MAXIMOST ANALYSIS DOCUMENTATION

The MaxiMOST Analysis form is shown in Figure 5.67 to provide the user with a historical perspective for a MaxiMOST analysis. Today, most companies use software applications, such as StandardsPro, to develop and maintain engineered standards. MOST has progressed from a manual system to a fully integrated digital solution. Figure 5.68 shows a Method and TMU report for the activity to *Install rear cab latch assembly to cab* using the StandardsPro software application. Whether the manual form or a software application is used to document

Step	Method Description	Seq. Model	Freq.	Simo	TMU	Total
1	GRASP PRYBAR AND AND PLACE TO FIRST CAGE NUT 4 STEPS AWAY	A1 B0 G1 A6 B0 P3 A0	1.000	N	110	110
2	REGRASP PRYBAR AND PLACE TO 4 REMAINING CAGE NUTS WITHIN REACH	A1 B0 G1 A1 B0 P3 A0	4.000	N	60	240
3	REGRASP PRYBAR AND AND PULL 5 CAGE NUTS WITH RESISTANCE	A1 B0 G1 M3 X0 I0 A0	5.000	N	50	250
4	ASIDE PRYBAR TO WORKBENCH 2 STEPS AWAY	A0 B0 G0 A3 B0 P1 A0	1.000	N	40	40
5	WALK 3 STEPS TO GRASP GRINDER AND AND PLACE TO REAR CAB 4 STEPS AWAY	A6 B0 G1 A6 B6 P3 A0	1.000	N	220	220
6	CONTACT BUTTON ON GRINDER TO ACTUATE 7 SECOND PT	A1 B0 G1 M1 X16 I0 A0	1.000	N	190	190
7	ASIDE GRINDER TO COUNTER 3 STEPS AWAY	A0 B0 G0 A6 B0 P1 A0	1.000	N	70	70

FIGURE 5.65 BasicMOST sequence models to use prybar and grinder.

Step	Method Description	Seq. Model	Freq.	Simo	TMU	Total
1	LOOSEN 5 CAGE NUTS USING PRYBAR AT CAB - 6 STEPS	A1 B0 T3	1.000	N	400	400
2	GRIND OUT 1 HOLE PT - 7 SECONDS USING GRINDER 10 STEPS WITH A BEND	A3 B1 T3	1.000	N	700	700

FIGURE 5.66 MaxiMOST sequence models to use prybar and grinder.

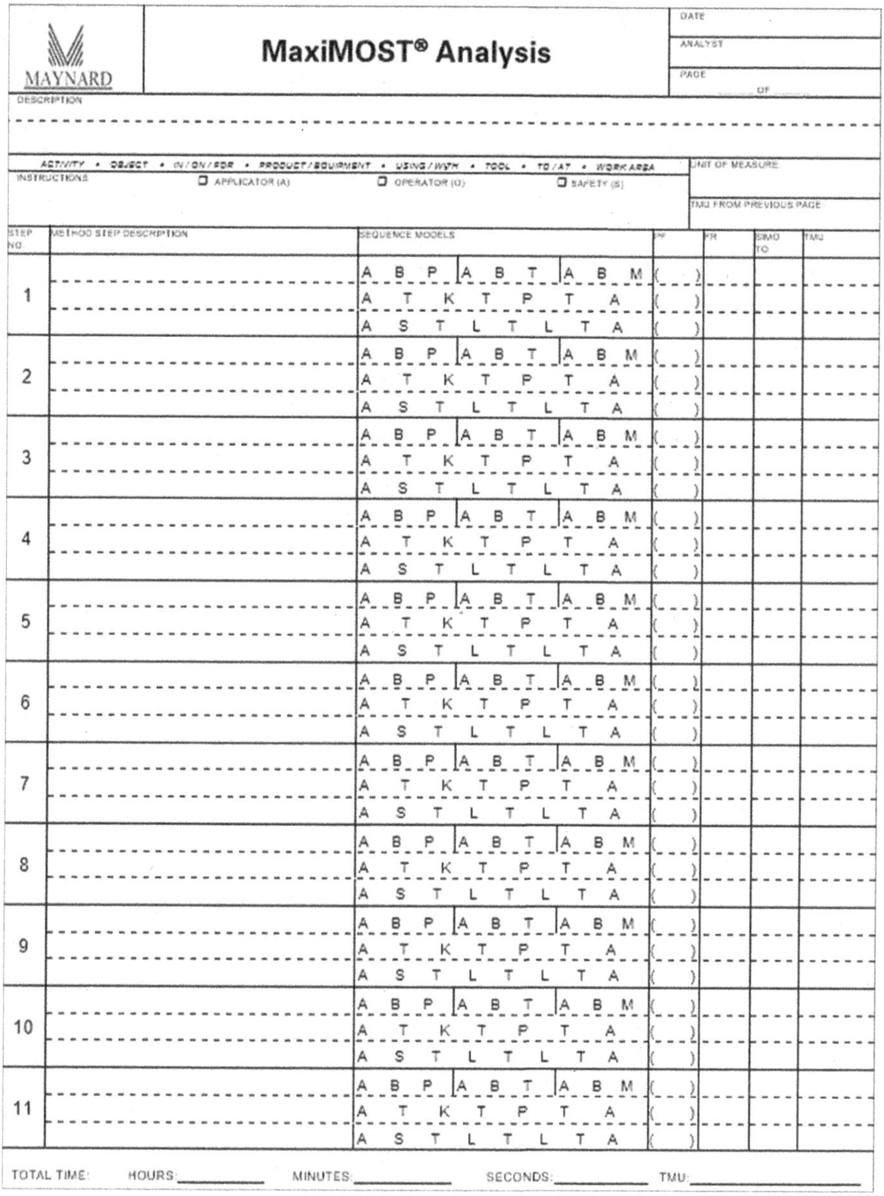

MAYNARD

MaxiMOST® Analysis

DATE

ANALYST

PAGE OF

DESCRIPTION

ACTIVITY • OBJECT • IN / ON / FOR • PRODUCT / EQUIPMENT • USING / WITH • TOOL • TO / AT • WORK AREA

INSTRUCTIONS ☐ APPLICATOR (A) ☐ OPERATOR (O) ☐ SAFETY (S)

UNIT OF MEASURE

TMU FROM PREVIOUS PAGE

STEP NO.	METHOD STEP DESCRIPTION	SEQUENCE MODELS	PF	FR	SIMO TO	TMU
1		A B P | A B T | A B M A T K T P T A A S T L T L T A	( ) ( ) ( )			
2		A B P | A B T | A B M A T K T P T A A S T L T L T A	( ) ( ) ( )			
3		A B P | A B T | A B M A T K T P T A A S T L T L T A	( ) ( ) ( )			
4		A B P | A B T | A B M A T K T P T A A S T L T L T A	( ) ( ) ( )			
5		A B P | A B T | A B M A T K T P T A A S T L T L T A	( ) ( ) ( )			
6		A B P | A B T | A B M A T K T P T A A S T L T L T A	( ) ( ) ( )			
7		A B P | A B T | A B M A T K T P T A A S T L T L T A	( ) ( ) ( )			
8		A B P | A B T | A B M A T K T P T A A S T L T L T A	( ) ( ) ( )			
9		A B P | A B T | A B M A T K T P T A A S T L T L T A	( ) ( ) ( )			
10		A B P | A B T | A B M A T K T P T A A S T L T L T A	( ) ( ) ( )			
11		A B P | A B T | A B M A T K T P T A A S T L T L T A	( ) ( ) ( )			

TOTAL TIME: HOURS: ________ MINUTES: ________ SECONDS: ________ TMU: ________

FIGURE 5.67 MaxiMOST Analysis form.

a MaxiMOST analysis, it is important to include essential information. One of the key features of the MOST Work Measurement Technique is the predefined sequence models. Using a manual system or a software application, the use of the predefined sequence models is critical for the analyst to maintain applicator accuracy and consistency.

Sub-Operation - Method / TMU Report
H. B. Maynard and Co.

Sub-Op ID:	70875	**Status:**	Private
Title:	INSTALL REAR CAB LATCH ASSEMBLY TO CAB		
Activity:	INSTALL	**Sub-Activity:**	
Object:	LATCH	**Prod/Equip:**	
Size/Capacity:		**Tool:**	
WA Number:		**WA Origin:**	
5-S Level:	SOME EFFORT	**Other:**	
Unit of Measure:	RCL CAB	**OFG:**	2
WorkArea ID:		**Starting Oper:**	
Starting Location:		**Total TMU:**	6,700
Operator Instr:			
Applicator Instr:			
Safety Instr:			
Applicator:	MCS		
Create Date:	3/31/2020	**Effect Date:**	3/24/2020

Method Steps

Step	Method Description	Freq.	Simo	TMU	Total
1	LOOSEN 5 CAGE NUTS USING PRYBAR AT CAB - 6 STEPS A1 B0 T3	1.000	N	400	400
2	GRIND OUT 1 HOLE PT - 7 SEC. USING GRINDER - 10 STEPS WITH A BEND A3 B1 T3	1.000	N	700	700
3	PLACE AIR HOSE TO IMPACT AND MOVE IMPACT TO CAB-FIXTURE A0 B0 P1	1.000	N	100	100
4	COLLECT AND MOVE 7 BOLTS TO CAB-FIXTURE - 4 STEPS WITH A BEND A1 B1 P3	1.000	N	500	500
5	PUT 7 WASHERS ONTO BOLTS A0 B0 P3	1.000	N	300	300
6	APPLY LOCTITE ON 7 BOLTS WITH SQUEEZE BOTTLE A0 B0 T3	1.000	N	300	300
7	PLACE CAB LATCH ASSEMBLY TO CAB UNDERSIDE - 12 STEPS WITH A BEND A3 B1 P1	1.000	N	500	500
8	HAND START AND RUN DOWN 5 BOLTS TO CAB LATCH ASSEMBLY - 15 STEPS A3 B0 T24	1.000	N	2700	2700
9	HAND START 2 BOLTS AND NUTS TO CAB LATCH ASSEMBLY A0 B0 T3	1.000	N	300	300
10	TIGHTEN 7 BOLTS USING IMPACT WRENCH RETURN IMPACT WRENCH TO STAND - 12 STEPS A3 B0 T6	1.000	N	900	900

FIGURE 5.68 Method and TMU report.

Regardless of the documentation method, the information listed below should be included in the analysis.

1. *Identification and Header*

   This section of the analysis should include all key information for ease of searching and updating the analysis.

   - Title or description

The title often begins with a verb of the activity and may include the object, product/equipment, tools and a work area description. An example title could be:
Install rear cab latch assembly to cab.

- Activity
  The activity should be a verb that indicates the overall context and/or main goal of the actions which are included in the analysis. The activity for the example to *Install rear cab latch assembly to cab* would be 'Install.'
- Unit of measure
  Used to designate what the activity is based on. Examples of units of measure are per unit, per part, per box, per customer, per occurrence, per pallet, etc.
- Operator, applicator or safety instructions
  Instructions can be added to clarify key points in the analysis. Examples of instructions are listed below:

  Operator Instructions—Confirm part is fully connected before adding sealant.
  Applicator Instructions—Process times calculated by machine speeds and are noted on a separate method step.
  Safety Instructions—Adjust cart to correct height before loading product.
- Date of analysis
- Identification of analyst
- Total TMU
  The total time for the activity is shown by simply adding all the numbers in the TMU column for each method step. That number is then noted in the Total Time section of the form or even better, calculated automatically in a software solution. Most electronic systems automatically convert the TMU to the organization's choice of time in seconds, minutes or hours.

2. *Method Steps and Sequence Models*
   This section of the analysis should include the step number, method step description, sequence model, including documentation of any frequencies or simultaneous motions and the TMU for each method step.
   - Method steps
     The amount of information placed in the method description section is usually a function of its eventual use; that is, the description can be used for detailed operator instructions or for an outline of the manual work for time computation only. Each method step has *only one* corresponding sequence model so the method description should be phrased in terms of moving an object or using a tool as to easily identify the sequence model used for that activity.
   - Sequence model analysis

> Regardless of the system used to document a MaxiMOST analysis, it should include the predefined sequence models for Part Handling, Tool Use and Machine Handling. This helps to ensure applicator accuracy and consistency. After the index values to the selected sequence model are applied, any frequencies if they occur in the method step or if the method step is performed simultaneously to another activity are documented.

As noted earlier in the chapter, PF represents a partial frequency. Partial frequencies were discussed earlier in the chapter and are used when one or more parameters of a sequence model occurs more or less than once. FR, or frequency, is used to note that an entire sequence model occurs more or less than once. The default for any system is a frequency of one (1).

The use of Simo is to document that a method step or a portion of the method step occurs at the same time as another step. If an entire sequence model is performed simultaneous to another, Simo is to indicate the method step number to which a certain step is simultaneous. The time for a simultaneous activity is documented in some manner to designate that time is not included in the total time for the activity.

The time for each method step is then calculated by adding the index values, applying the frequencies as needed and then multiplying by 100 to get the time value for the sequence model in TMU.

With the use of software applications, updating a MaxiMOST analysis is very simple with the functionality to update one sub-operation, determine the impact of a method change on the overall time and even complete a mass update of the new time to automatically update all sub-operations that include a particular method step.

The ease with which MaxiMOST analyses can be updated and/or new methods determined is one of the greatest features of the MOST Work Measurement Technique, making simulation and comparison easy.

Workplace Layout

It is not a requirement to define the workplace layout, but it is helpful to the analyst and other readers when trying to understand the current situation. Prior to applying sequence models for analyzing manual work using MaxiMOST, a work area layout may be documented and would include different work area information, such as:

- Workplace names
- Tools and their locations
- Objects and their locations
- Equipment and its location
- The operators and their starting location
- Body motions always associated with particular workplaces
- The distance in steps between workplaces

Multiple Operator Activities

It is not uncommon to find that many of the long-cycle assembly or machining activities studied will be multi-operator operations. Analysis of such situations can be comfortably handled in a MaxiMOST analysis.

First, it is imperative that for multi-operator operations, the analysis must include the number of operators for the operation under consideration. For example, in the MaxiMOST analysis under Description, specify '1/4 inch (6 mm) steel plate, 4×6 feet (1.2×1.8 m) operators-2.' Also, MaxiMOST analysts should locate each operator at the start of the analysis.

Example: Op-1 (Operator 1) begins at workbench. Op-2 begins at outstock. At the beginning of each method description step, identify the operator performing the activity.

Example: Op-1 push plate on conveyor.

The analysis of simultaneous motions with the right and left hand, as discussed in the Application of the BasicMOST Work Measurement System section of Chapter 3, can be applied to the analysis of simultaneous actions between multiple operators. Therefore, the techniques of 'limiting out' certain parameters or entire sequence models by documenting that the work performed is internal to an equal or longer activity are appropriate for the analysis of multi-operator operations. In some more complex multi-operator tasks, it may be advantageous to prepare an operator analysis chart based on a separate analysis for each operator.

Care must be taken when creating such analyses to keep the final application format in mind. Choose countable production units and provide the final time in the desired format (total labor hours or total elapsed time). Indicate the unit of measure if the calculated total is elapsed time, and then extend the elapsed time on a worksheet by multiplying it by the number of operators for the operation to get total labor hours.

Summary of the MaxiMOST Analysis

A MaxiMOST analysis is completed by confirming the following is included:

1. Identification and Header Information
    - Title or description
        - Document multiple operators if applicable
        - Include a work area layout if needed
    - Activity
    - Unit of measure
    - Operator, applicator or safety instructions
    - Date of analysis
    - Identification of analyst
    - Total TMU

2. Method Steps and Sequence Models
    - Method steps
    - Sequence model analysis

- Select *one* sequence model for each method step
  - Apply the correct index value for each parameter within each sequence model.
  - Note the documentation for PF, FR or Simo activities as needed.
  - Add parameter index values together and apply frequencies as needed and multiply the MaxiMOST factor of 100 to get the time for the sequence model in TMU.
- Add all method step totals to get a total TMU

Note: A MOST analysis will provide the normal time (manual time+allowance time). Allowances need to be added to generate an engineered standard.

Validation of Process Times

It will be necessary to validate elements that are based on process times such as powered cranes and powered trucks. Also, if new elements involving process times are being developed, these elements have to be validated for different types of equipment. In all cases, the validation should be carried out to ensure that the desired level of accuracy will be achieved. The analyst compares the index values on the data card with its allowed deviation range to the process time for the selected equipment determined by stopwatch time study. The steps required to perform the validation are as follows:

1. Review the specification and method used for the existing equipment.
2. Establish criteria for the time study based on the characteristics and method for the selected equipment.
3. Conduct and compile time study.
4. Compare time study results to existing index values.
5. Determine if the current data card can be applied.
6. If necessary, develop required elements and a supplementary data card for the selected equipment according to the principles described earlier in this section.
7. Document the validation process for future use.

Because it is impractical to cover the wide variety of available and potential future equipment on data cards, it will be necessary to validate all process times in order to achieve the desired level of accuracy and consistency when using MaxiMOST.

MaxiMOST Summary

The MOST Work Measurement Technique uses a structured approach to measure work based on the movement of objects. There is a consistent approach the analyst should always use prior to analyzing an activity with MOST. It begins with determining the starting and stopping points of the activity to be analyzed and ends with a total time for the activity. The MaxiMOST Analysis Process (Figure 5.69) depicts this procedure and can be used to lead the analyst through

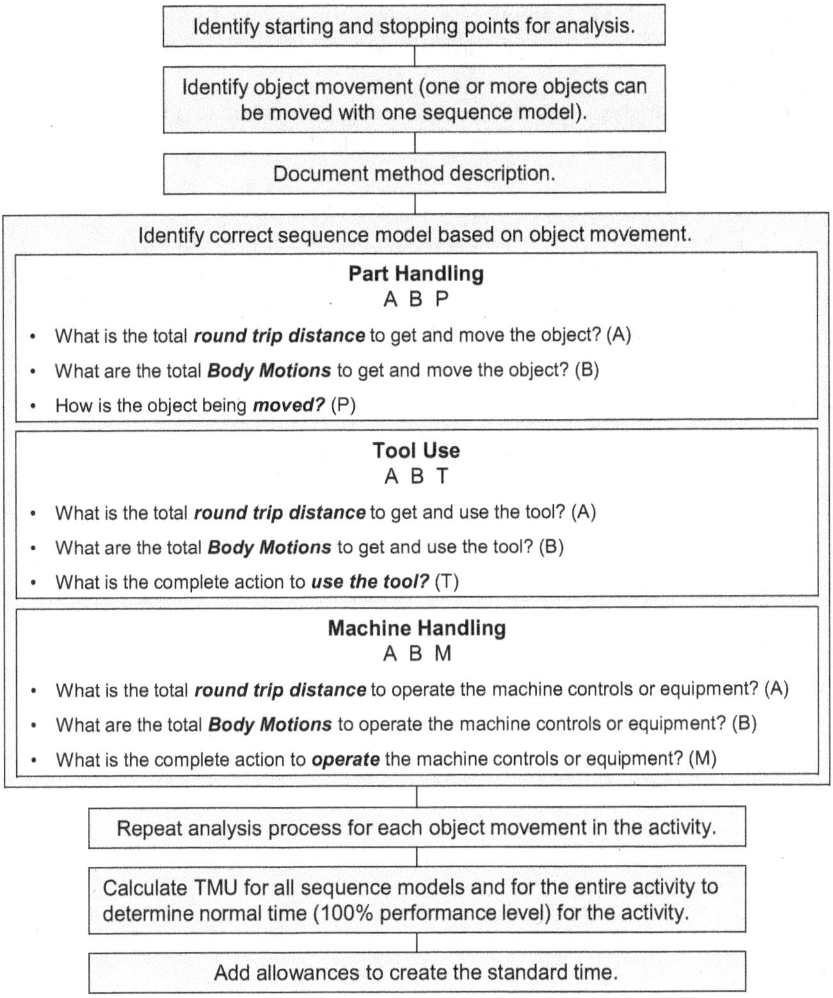

FIGURE 5.69 MaxiMOST Analysis Process.

the basic thought processes and decisions that need to be considered in order to arrive at a thorough and consistently applied MaxiMOST analysis. Follow the process to make the proper decisions to complete an analysis of an activity. The MaxiMOST Analysis Process includes the Part Handling, Tool Use and Machine Handling sequence models.

It must be remembered that measured times produced with any MOST System represent a performance level of 100%. That is, the performance of an average skilled worker following a prescribed method, working under average work conditions at a normal pace and with capable supervision. The total time value for an activity analyzed with MaxiMOST produces a normal time without allowances. Typically, the allowances as a percentage of normal time are applied as a final step to establish a standard time.

Following the analysis process and answering the questions is key to the effective application of MaxiMOST. The answers will help the analyst to:

- Determine the correct sequence model to be used.
- Determine the index value for each parameter (sub-activity).
- Avoid overlooking any other objects being moved or analyzing any unnecessary activity.
- Apply MaxiMOST consistently.

6 Using MOST Standards to Manage Performance

ACCENTURE WORKFORCE PERFORMANCE MODEL

Managing performance is critical for supervisors and team leaders in the workplace. Once an organization makes the commitment to implement a standards program, managing performance is needed to help organizations achieve objectives and realize gains from the implementation. Effectively managing performance ensures that:

- An associate is able and encouraged to perform.
- The use of available resources is optimized.
- The desired results are achieved consistently and able to be maintained.

To drive workforce performance, supervisors and team leaders need to have a comprehensive approach and practical tools.

Managing Performance is one of the disciplines that comprise the Accenture Workforce Performance Model (Figure 6.1). Originally developed by H. B. Maynard and Company (now part of Accenture), the model depicts a set of nine disciplines that enable a company to influence the productivity, quality and culture of a workforce in order to improve workforce performance and eliminate waste, therefore fostering an environment of continuous improvement.

The Workforce Performance Model should be used as a blueprint for success to improve and sustain the results in any operation. The nine disciplines are organized into three series: The Lean Workplace series, the Standard Work series and the Workforce Performance series.

Managing Performance is at the pinnacle of the Workforce Performance Model to help companies support and manage workforce performance. This discipline enables a company to fully realize the benefits achieved from the other disciplines by incorporating each into the culture of the workplace. The result is a culture of "this is the way we do things" that features sound processes and clear expectations.

Within the Workforce Performance Model, standards developed with MOST support the Engineered Standards discipline to set the expectations for associates. The Managing Performance discipline provides the framework for a successful approach using standards as the foundation, the Managing Performance Cycle and tools that enable a supervisor to:

- Use standards to establish and communicate expectations.
- Ensure associates have the conditions for success.
- Use feedback to make reality visible and recognize non-conformance.
- Effectively respond to challenges and opportunities to enable success.

FIGURE 6.1 Accenture Workforce Performance Model.

ACCENTURE MANAGING PERFORMANCE CYCLE

The approach to effectively manage performance using MOST-developed standards is shown in the Accenture Managing Performance Cycle, also originally developed by Maynard, in Figure 6.2. In this cycle, engineered standards serve as the foundation for supervisors to support and manage associate performance. Work standards provide the method for how the work is to be performed and the expectation on the time it should take to perform the task. Standards are essential to managing performance. When standards are developed with a scientific approach such as MOST and are accurate and accepted by both supervisors and workers, they can be used to confidently plan resources, guide and motivate associates' action and evaluate performance. Organizations following this process will benefit from the increased productivity and quality achieved.

USE STANDARDS TO ESTABLISH AND COMMUNICATE EXPECTATIONS

The Accenture Managing Performance Cycle begins with engineered standards, which are specifications for performing work. They are developed by a systematic process of methods analysis and a proven work measurement technique such as MOST. An engineered standard is based on the prescribed best method for

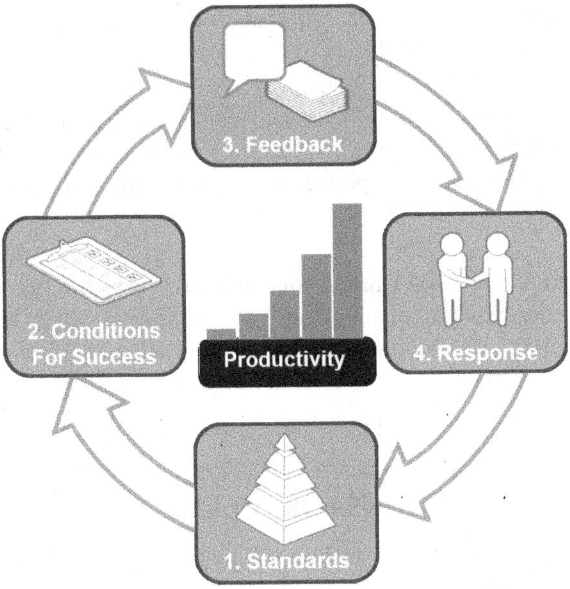

FIGURE 6.2 Accenture Managing Performance Cycle.

performing a task and also defines the time that the work should take. Using MOST, the methods are documented by the method description and the time is derived from the completed sequence model.

How do you close the gap between management, the work measurement analyst, the supervisor, the union and the worker to be sure that everyone understands and agrees on the method and time expectations?

- Use the method sensitivity of MOST to eliminate non-productive or non value-added activities.
- Use the method descriptions in the MOST analysis to instruct and train the operators on the task.
- Use the MOST standard sequence models, predetermined index values and rules to create a common language when analyzing and discussing a MOST analysis.

All of these actions allow MOST to be a unified and common tool for management, unions and workers. When a desired level of understanding is achieved between those directly involved, the organization can now function smoothly, allowing productivity and profit goals to be realized.

Engineered standards are a powerful tool in the workplace. When the process for setting standards includes a fair assessment of the best way to perform the work and there is an understanding and acceptance of the standard setting process, both the associates and supervisors can use standards to set goals that are attainable.

The acceptance of standards as accurate creates an atmosphere of trust that allows a supervisor to use the standards to:

- Manage performance to communicate method and time expectations.
- Monitor that the conditions for success are maintained.
- Use feedback to recognize problems and take corrective action to motivate performance.

An environment where MOST-based standards are viewed by all as a useful tool helps organizations *and* individuals achieve ongoing success. In this type of environment, everyone is naturally more conscious of things that interfere with performing to the standard and they are motivated to act to stay on target to achieve desired outcomes.

BUILD IN THE CONDITIONS FOR SUCCESS

In order for workers to accomplish the work that the standard requires, the appropriate conditions must be in place. The work flow, layout, storage locations, tools needed, work practices and training all need to be consistent to enable the standard method to be performed and the expected outcomes to be achieved.

When developing MOST-based standards, the four disciplines that comprise the Lean Workplace Series prepare the workplace for the standards. Times for activities developed with MOST should be based on the agreed-upon methods, which are ideally the Best Methods for the task, and the workplace should be set up following principles from the 5-S, Work Flow Design and Setup Reduction disciplines.

Establishing stable conditions despite inevitable changes requires a supervisor to rely on a system of disciplines that are built into the culture. The disciplines that drive the *Conditions for Success* as shown in the Accenture Workforce Performance Model (Figure 6.1) should be established and become part of the culture of the organization.

5-S

A fundamental condition for success is an organized, clean and safe workplace. In this environment, there is a place for everything, and everything is in its place. It is free of distractions and clutter that cause wasted effort. It allows associates to focus on what is important. The 5-S tactics describe how to create and maintain this type of workplace.

Best Methods

Associates should have prescribed methods that, when performed, consistently accomplish the desired results. These methods must be supported by a workplace layout that is efficient and safe. The principles of the Best Methods discipline help define the most effective method and workplace layout.

Work Flow Design

Associates also need a work process that is free of impediments that disrupt the flow of work. They need a steady flow of incoming materials, a fair and reasonable workload and well-maintained equipment and tools. The strategies in the Work Flow Design discipline enable supervisors to successfully establish these conditions.

Setup Reduction

To maximize productivity, associates should use minimal non-productive activities to prepare for productive work. The Setup Reduction discipline consists of strategies to minimize necessary, yet non-productive, setup or changeover activities.

Engineered Standards

For goals and feedback to be meaningful to associates, they must be based on reasonable and objective standards. The Engineered Standards discipline includes proven and straightforward work measurement techniques such as MOST to analyze the work and arrive at standard times.

Workforce Training

In order to meet performance expectations, associates should have the opportunity to develop skills to complete tasks. It is the Workforce Training discipline that provides a supervisor with strategies to help accomplish this goal. The method descriptions in a MOST analysis can be used to support this discipline to document the steps in an activity.

Staffing and Reporting

The Staffing and Reporting discipline is the bridge between Engineered Standards and Managing Performance. Integrating up-to-date, accurate and credible engineered standards that were developed with a sound work measurement technique such as MOST with staffing models and reporting capabilities enable supervisors to manage the performance of the workforce. An effective staffing model assures that the right quantity of employees is at the right place and right time to meet customer requirements.

Wage System Alignment

The correct wage structure helps to motivate behaviors that align with company objectives. The Wage System Alignment discipline contributes strategies to achieve this condition of a purposeful and equitable compensation structure.

Managing Performance

Associates are empowered to achieve success when their managers have the tools, strategies and time to focus on enabling their success.

USE FEEDBACK TO MAKE REALITY VISIBLE

Measurement provides objective information about how work is progressing. A standard generated with MOST provides the expected time for a task and then actual results can be compared to this expectation. Based on the standard, clear feedback is given to those involved so that they have an objective way to see reality. Understanding what is happening is the key to managing and improving any process.

To manage performance effectively, supervisors need to know the workforce performance and results and be able to easily compare it to the expected values.

They need to be able to quickly evaluate the performance and results against what should be happening. Defining this gap on a real-time basis is critical. Based on this information, a supervisor can provide recognition for achievement or facilitate problem solving to minimize delays and improve performance.

Real-Time Feedback

In the workplace, supervisors and associates can use visual strategies to determine whether the work is proceeding as expected, such as:

- Tally boards that show the number of items produced or delivered compared to a goal.
- Marks on storage locations showing the production goals for various time increments.
- The length of a check-out line in a retail store.

All of these indicators provide information at-a-glance to indicate if things are going as expected. Associates can use this information to track and control their own work. This form of visual management motivates and reinforces achievement of desired outcomes. It also promotes self-sufficiency, which is the ultimate form of Managing Performance.

Management Metrics & Reports

Because it is impossible for supervisors to be everywhere at once, there is a need to collect relevant data and centralize it for review and analysis. Having the ability to view the progress of some or all of an entire group over a period of time helps to identify trends and problems that might otherwise be overlooked. Supervisors can also share this data with associates to give feedback and recognition.

This type of feedback has two components: A data collection system and a reporting system. The data can either be collected as an inherent part of the process, like scanning picks in a distribution center, or can be entered by the associates as they log their completed work or delay time.

The data that is collected and reported should be carefully selected to give the supervisor useful information to manage toward achieving organizational goals. Reports, ideally delivered in a timely manner such as through a digital format, that provide feedback to the expected output from the MOST standards and the actual output can be used to identify areas for improvement. The metrics should provide a view of what is important to the organization and point to variations the supervisor and the associates can control.

RESPOND TO CHALLENGES AND OPPORTUNITIES

Knowing how to respond quickly to non-conformance can prevent delays, wasted time and errors. Supervisors and associates need to work together to investigate the cause of an issue and know how to take action to achieve desired outcomes. Whenever feedback indicates that the actual performance or results vary from what is expected, supervisors respond using a two-pronged approach. First, they investigate the situation and second, take action to provide recognition, solve a problem or improve a condition.

Investigate

The goal of the Investigate phase is to determine the fundamental cause of a variation so that it can be addressed. Each workplace has a set of variables that can affect the success of the work. They can include any of the conditions for success, such as an organized workstation, uninterrupted work flow, defect-free materials, proper methods or the availability of tools. It is also possible that the standard has not been updated to reflect method or workplace changes. For example, a supervisor notices that an associate is consistently not meeting their expected daily output based on the MOST standard. Observation of the situation and a brief discussion with the associate reveals that there is a problem with the machine that connects two components together that causes frequent delays. Further investigation reveals that the scheduled preventive maintenance has not been performed on the machine. By identifying the root cause of the problem, it can be solved and prevented from occurring in the future.

Take Action

The process of investigating and taking action can be focused on using tools for investigating productivity issues and tools for evaluating and coaching performance. Using investigative tools such as the Investigate & Take Action Chart shown in Figure 6.3 ensures that the most common problems are identified and investigation approaches, tools, and alternative solutions are recommended. Not all productivity issues are associate-driven. There may be an equipment breakdown, a supply delay or an outdated standard. Using the Investigate & Take Action Chart provides a tool for managers to review possible problem indicators and solutions and then share that information with others throughout the organization.

USING MOST-DEVELOPED STANDARDS TO IMPROVE PRODUCTIVITY

The four components identified in the Accenture Managing Performance Cycle are best applied when supervisors and associates have confidence in the standards. The MOST Work Measurement Technique is a proven system accepted by management, associates and unions. MOST-based standards are viewed as fair and attainable. When standards are accepted by all parties involved, they can be effectively used

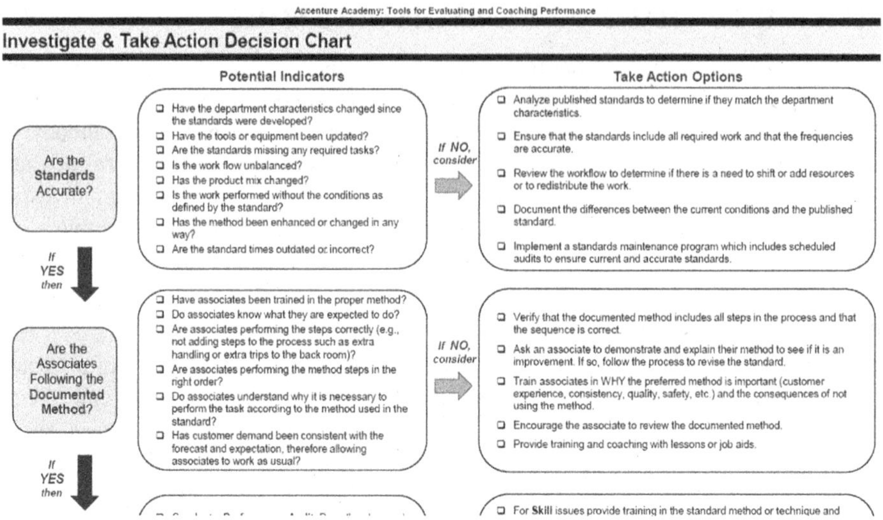

FIGURE 6.3 Investigate & Take Action Decision Chart.

to set expectations. In addition to clearly communicating expectations, the supervisors take a proactive role in maintaining the conditions for success in the workplace which allows associates to focus on the work. While the work proceeds, both the supervisors and associates need to know what is happening and how it compares to the expectations. There should be established means to measure and get feedback in order to provide encouragement and support and to quickly troubleshoot and correct non-conformances.

This management approach uses many practical tools to incorporate the discipline into the workplace and to enable an organization to effectively plan, manage and improve the operations. Accurate MOST-generated standards provide the target for performance and stable conditions provide assurance that the targets can be achieved. Measurement and accountability motivate effort for achieving expectations. Timely intervention then helps to correct non-conformances, limiting their impact on the organization.

The MOST Work Measurement Technique allows industrial engineers, work measurement analysts and others to use an efficient technique to establish a time. As stated in Chapter 1, establishing this time allows a company to do many things; mainly accomplish planning, determine performance and establish costs. It is obvious that knowing how much time it takes to perform certain tasks enables a manager to achieve and maintain a high utilization of personnel, material and equipment with high productivity and overall efficiency, which can produce sustainable organizational growth.

The concept of productivity, the ratio of output to input, is and will always be an important issue. It may be a simple concept to understand, but more difficult to measure and interpret. Companies should focus on continuous improvement to increase their productivity as technology is changing daily and new products, processes and equipment are being introduced regularly in many companies. Competition will continue to expand worldwide forcing companies to continue to improve their operations no matter how often they change. To maintain business and stay profitable, it will become necessary for companies to measure and track their productivity. Using the disciplines of the Accenture Workforce Performance Model and the key components of the Accenture Managing Performance Cycle, MOST-based standards can be used as the fundamental tool assisting in that effort.

Appendix A
Theory

With detailed predetermined motion time systems (PMTS) such as Methods Time Measurement (MTM), an 'exact' method description must be recorded; that is, *basic motions* must be expressed in terms of distinct types, distances, weights and other such variables. The result is a very detailed description of a method, but probably not the exact method actually performed in the majority of cases. Variations are inherent in any operation. The operator may follow a different motion pattern because of a lack of instructions or the variable nature of the operation; for example, reach and move distances may not be the same throughout an operation as the detailed description implies. The analyst must predict what methods will most likely be used or describe what is perceived as an average or representative method for the operation. When distances are averaged, for example, in reaching into a bin with parts at different distances from the operator, the expected accuracy of the detailed system is not achieved. Why describe a detailed method from a table of exact values, when in reality the operator will follow a method that varies from one occurrence to the next? There are situations in which this detailed approach is indeed appropriate as for highly-repetitive, short-cycle operations performed at workplaces designed to minimize any such variations. In terms of cost, however, this exact method analysis required by traditional predetermined motion time systems often seems unnecessary and impractical. In fact, there is even some question about the 'exactness' of these systems given the presence of the inherent variability in work methods.

In the design of MOST, it was recognized that these variations or deviations could be easily compensated for by using basic statistical principles. More importantly, by using these same procedures, it was also found that it was possible to greatly simplify the work measurement itself while retaining a high level of accuracy. In other words, this inherent variation in work methods has actually been used as an advantage in developing a more simplified work measurement technique with resulting accuracy surprisingly close to such systems as MTM.

ACCURACY OF A PREDETERMINED MOTION TIME SYSTEM

Can a work measurement technique produce a time that is exact? The answer to this question is *no*, because:

- Work is performed by humans and is, therefore, *variable* from one human to another. No two individuals accomplish exactly the same amount of work over a specified amount of time.

- Standard time (the result of work measurement) represents average time. It reflects the time of an *average* worker, of *average* skill, working at an *average* or normal pace, under *average* conditions with capable supervision.

Therefore, since no work measurement technique is exact, all have as either part of their original system design or as a result of their original design a *balancing time.* The calculation of a balancing time is based on the statistical principle that the variance of a sum of independent variables equals the sum of the individual variances. Simply put, balancing time is the time needed for the system's desired level of accuracy to be attained. In other words, a certain minimum amount of work must be analyzed with the system before the accuracy of the analysis can be guaranteed to a specific level of confidence.

The statistical phenomenon that occurs during the balancing time is called the *balancing effect.* The balancing effect is what causes the desired level of accuracy of a system to be attained. In other words, the balancing time is *when* the system's accuracy is attained, and the balancing effect is *how* it is attained. The balancing effect results from the combination of individual deviations for a smaller total deviation. Deviation can be defined as the difference between the 'true' time it takes for a task and the time the work measurement technique 'allows.' As independent, non-repetitive elements are combined, their total percentage deviation, due to the balancing effect, becomes less than the individual percentage deviations. This is because some of these deviations are higher than the true time and others lower. In the final result, the total relative accuracy is better than the accuracy for the individual elements.

Since the true time to perform a job is indeterminate, the widely accepted accuracy of time values determined by MTM analyses served as a point of reference in the design of MOST. Later, the same accuracy of MOST time values was calculated by applying conventional statistical formulas for standard deviation.

However, unlike MTM, whose balancing time is the result of its system design, the balancing time of BasicMOST was determined prior to its system design. To better understand how MOST produces accurate results, a look at the system design is necessary.

BASICMOST SYSTEM DESIGN

When BasicMOST was originally conceived, the decision was made that a balancing time of approximately 2 minutes would be desirable. It was reasoned that substantial simplicity in system design and application could be achieved with only a moderate reduction in accuracy. The system was originally constructed to have a consistent balancing time of approximately 2 minutes (3300 TMU).

The precise *balancing time* for BasicMOST was determined by Dr. William D. Brinckloe of the University of Pittsburgh to be 3235 TMU.* The next section of this appendix covers the actual system construction in which all calculations and index ranges

* The theoretical system accuracy of MTM-1, MTM-2 and MOST is discussed in *Comparative Precision of MTM-1, MTM-2 and MOST,* University Research Institute, June 1975.

are based on a balancing time of 3235 TMU or approximately 2 minutes. Therefore, the accuracy of BasicMOST is based on a balancing time of 3235 TMU. This means that measurements totaling 3235 TMU or more are accurate to within ±5% of a true time value with a confidence level of 95%. However, this does not mean that BasicMOST cannot be used to measure shorter activities. As will be shown in this appendix, the accuracy in the *final* result (standard time) is the deciding factor. The minimum condition of 3235 TMU applies only to the standard and not to the individual element.

MOST INTERVAL GROUPINGS

Predetermined motion time systems, such as MTM, are constructed by determining the time duration of conveniently selected basic motions. In contrast to this approach, BasicMOST starts with the construction of time intervals based on a *stated balancing time* (3235 TMU) and thereafter determines which motion patterns fall within each time interval. In contrast to MTM, the BasicMOST System provides a consistent balancing time for any combination of elements.

The construction of the BasicMOST System time intervals was influenced by the following established objectives:

1. The mean value for each time interval will be a whole number and also a multiple of 10.
2. The time intervals will cover a continuous time scale with neither gaps nor excessive overlaps.

The MOST time intervals were then calculated from the statistical formula for allowed deviation:

$$a = \pm r_{T_B}\sqrt{T_B \times t}$$

where: a = allowed deviation from interval mean in ±TMU
$\pm_{rT_B}$ = accuracy of ±5% for balancing time (± 0.05)
T_B = established balancing time of 3235 TMU
t = interval mean in TMU (a whole number and also a multiple of 10)

Figure A.1 is the result of using this formula with appropriate values.

The formula assumes a conventional normal distribution. If a uniform distribution is assumed, the formula then becomes:

$$a = \pm\ r_{T_B}\sqrt{T_B \times t \times 0.878}$$

The differences in Figure A.1 from using this second formula are quite minor and do not affect the construction of the MOST index value table (Figure A.3) or the location of the boundaries between index values.

By placing the values from Figure A.1 on a linear scale and drawing half-circles, which represent the calculated allowed deviation range of each time interval, the first

BasicMOST Time Values (TMU)	Deviation (TMU)	Interval (TMU)	
		Lower Limit	Upper Limit
10	± 9	1	19
30	± 16	14	46
60	± 22	38	82
100	± 28	72	128
160	± 36	124	196

FIGURE A.1 Allowed deviations for BasicMOST time values.

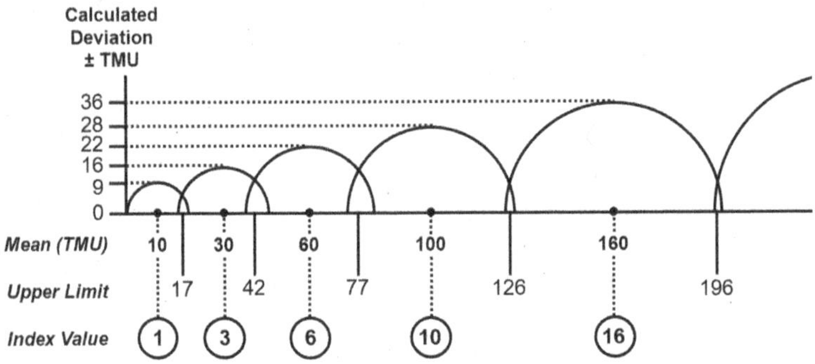

FIGURE A.2 Development of MOST time values.

five intervals or index ranges can be determined (as shown in Figure A.2). Note that since the interval means were adjusted to be divisible by 10, the application of BasicMOST is simplified by eliminating zeros, thus creating a series of index values (circled) statistically representing each time interval.

The diagram shows that adjacent half-circles overlap slightly. The overlaps have median values of 17, 42, 77, 126 and 196 TMU. These are the upper limits of the first five BasicMOST index ranges. The calculation procedure was continued to determine the index values up to 330 as shown in Figure A.3.

It is within these statistically calculated index ranges that the variation inherent in the majority of working situations is absorbed.

BACKUP DATA

The BasicMOST time intervals described in the previous section serve as the basis for all parameter index values. Motion patterns are analyzed with MTM-1 or MTM-2 and index values are assigned according to the time interval into which the detailed analysis falls. The most frequently occurring of these motion patterns are listed on data cards under appropriate sequence model parameters and comprise the variants for the sequence model sub-activities defined in earlier chapters.

Each of these motion patterns (variants), with its corresponding index value, is referred to as a 'parameter index value.' For example, MTM analyses for 'gaining

Index Value	Interval Mean (TMU)	Interval Limits (TMU)
0	0	0
1	10	1 - 17
3	30	18 - 42
6	60	43 - 77
10	100	78 - 126
16	160	127 - 196
24	240	197 - 277
32	320	278 - 366
42	420	367 - 476
54	540	477 - 601
67	670	602 - 736
81	810	737 - 881
96	960	882 - 1041
113	1130	1042 - 1216
131	1310	1217 - 1411
152	1520	1412 - 1621
173	1730	1622 - 1841
196	1960	1842 - 2076
220	2200	2077 - 2321
245	2450	2322 - 2571
270	2700	2572 - 2846
300	3000	2847 - 3146
330	3300	3147 - 3446

FIGURE A.3 MOST index values with interval limits.

control of an object requiring disengage' fall within the time interval 18–42 TMU. From Figure A.3, this translates to an index value of 3. Therefore, Gain Control with Disengage is represented in the sequence model by the parameter index value $\mathbf{G_3}$. For every value on the BasicMOST data cards, corresponding MTM-1 or MTM-2 analyses are cataloged in a backup data manual.

APPLICATOR DEVIATIONS

The total accuracy of any work measurement technique is dependent on both the system deviation and the applicator deviation. Although system deviation can be determined statistically, the deviations present, because of applicator error, must be determined empirically. Applicator deviations vary with individuals depending largely on the amount of training and experience possessed by each analyst.

One of the basic assumptions concerning the accuracy of a work measurement system is that a fine breakdown of motion variables contributes to the reduction of system deviation. This is no doubt true, but there is at the same time a greater tendency for the applicator, through inexperience, misjudgment or carelessness, to make errors in the selection of the correct time value. As a result, the systems taken to be the most precise (i.e., those with the finest sub-divisions of motions) are the most susceptible to applicator error.

Applicator deviations can be influenced by the way in which motions are classified within the system. With detailed predetermined motion time systems, as many as four variables must be considered when selecting time values. For example, MTM-1 time values for Reach are classified by distance, case and type; and time values for Move by distance, case, type and weight. Obviously, the more variables that the analyst must consider, the more likely the possibility for applicator error. Index values in BasicMOST were designed to contain one variable. To select the proper index value for Action Distance, only the distance is considered.

Perhaps the most frequent type of applicator error is that of carelessly omitting a motion from a motion pattern or erroneously including a motion that does not occur. This problem is virtually eliminated in MOST with the aid of the predefined sequence models. During the analysis procedure, the applicator's attention is focused on the sequence model as index values are applied to each parameter in the sequence model.

Surprisingly, very little research has been done on evaluating applicator deviations that are likely to occur in various work measurement systems. This is unfortunate, since applicator error probably influences the total accuracy as much as or more than the system error. However, there is analytic evidence that indicates that the total accuracy of MOST is influenced to a lesser degree by applicator deviations than other existing predetermined motion time systems. It is believed that the loss in system accuracy (the larger balancing time of BasicMOST) is compensated for by a reduction in applicator error, thus pulling the more detailed system and the more economical system together into an area of comparable total accuracy.

ACCURACY OF WORK MEASUREMENT AND STANDARD TIME

Having reviewed the theory and construction of the MOST Work Measurement Technique, let us now look at this same theory extended to a higher level; that of analyzing operations or sub-operations to establish a standard time.

A condition is said to be 'accurate' when it conforms exactly to an accepted standard; that is, the condition falls within acceptable tolerance limits. Accuracy, then, is a relative concept, relative to an accepted standard. To a carpenter, accuracy is usually expressed in inches or eighths of an inch, but to the machinist it may be expressed in thousandths of an inch. The physicist deals with even smaller tolerances. What about the work analyst? What is the 'accepted standard' for work measurement accuracy? Should operation times be accurate within thousandths of a second, or would plus or minus one day be acceptable?

Obviously, both conditions are inappropriate. In one case, the measurement would be extremely difficult to obtain, and in the other, the results would probably be meaningless. The fact is that accuracy requirements have very little to do with work measurement. The main consideration is of an economic nature. If it costs thousands of dollars to develop a standard time for an operation that seldom occurs, a rough estimate or even a guess will be sufficient. On the other hand, if substantial economic benefits can be realized from a detailed analysis providing more 'exact' times, these studies may be well worth the cost. So, the question of work measurement accuracy must first be answered in terms of the cost involved to achieve a certain level of accuracy.

The accuracy of a work measurement system is influenced by four factors, which, when assembled as a formula, explain the total relative deviation theory:

1. The *level of accuracy desired* in the final result depends on the planned use of the standard time such as incentive payment calculations (individual or group), machine loading and product costing.
2. The *time period* over which the desired level of accuracy must be attained. Do the standard times need to achieve the desired level of accuracy on a per-day basis, or will accuracy based on the 40-hour week be sufficient? This period is referred to as the calculation period, leveling period or balancing time.

   Note: The discussion now is focused on the balancing time for the standard time calculation, *not* the balancing time of the work measurement technique used to determine the standard times.
3. The *degree of repetitiveness* of the sub-operation being measured; that is, how many times the sub-operation occurs during the calculation period.
4. The *duration of the sub-operation* being measured.

These four factors are mathematically represented by the following statistical formula used to calculate 'allowed deviation.' The following formula is a derivation of the expression for standard deviation as discussed in detail below:

$$r^2 t = \text{constant}$$

Each formula variable definition is followed by a number referencing it to one of the four factors mentioned above.

$$r_t = \pm r_T \sqrt{\frac{T}{nt}}$$

where:
r_t = measured sub-operation's allowed deviation percent
r_T = total allowed deviation percent (1)
T = total time (i.e., the calculation period or balancing time) (2)
n = sub-operation's occurrence frequency over the calculation period (3)
t = the sub-operation's measured time (4)

Using the formula above, the allowed deviation of a sub-operation is calculated under two different conditions in the following example.

Example: Part of a vehicle assembly required 0.25 hours to perform according to a work measurement analysis. How accurate must this analysis be (i.e., what is the allowed deviation) if the time will be used for setting incentive rates where standards are expected to be within ±5% for a 40-hour pay period?

Case 1: The assembly is performed only twice a day.

$$r_t = \pm r_T \sqrt{\frac{T}{nt}} \qquad r_t = \pm 5\%$$

$$r_t = \pm 5 \sqrt{\frac{40}{10 \times 0.25}} \qquad T = 40 \text{ hours}$$

$$n = 2/\text{day} \times 5 \text{ days} = 10$$

$$r_t = \pm 20\% \qquad t = 0.25 \text{ hours}$$

Case 2: The assembly is performed continuously during the 40-hour calculation period.

$$r_t = \pm r_T \sqrt{\frac{T}{nt}} \qquad r_t = \pm 5\%$$

$$T = 40 \text{ hours}$$

$$r_t = \pm 5 \sqrt{\frac{40}{160 \times 0.25}} \qquad n = \frac{40 \text{ hours}}{0.25 \text{ hours}} = 160$$

$$r_t = \pm 5\% \qquad t = 0.25 \text{ hours}$$

In Case 1, a deviation of ± 20% (0.25±0.05 hours, or between 12 and 18 minutes) can be allowed for the standard time established since the job occurs infrequently and makes up only a small part of the worker's total productive time (Figure A.4). In Case 2, the worker completes the assembly throughout the entire week, causing the relative deviation to remain constant. The allowed deviation in this case is ±5%, the same as the requirement for the 40-hour calculation period. This relates to an allowed time of 0.25 ± 0.0125 hours or a 'standard time' for the task anywhere from 14.25 to 15.75 minutes (Figure A.4). These statistically calculated allowed deviations

FIGURE A.4 Allowed deviation for Case 1 and Case 2.

guarantee with 95% confidence that a calculated time that falls within the allowed deviation range will, over the calculation period, produce results within ± 5% of a 'true time value.'

Therefore, frequently occurring sub-operations, or those constrained by a shorter leveling or calculation period, have a very limited range of allowed deviation and, as a result, can absorb very little in the way of method variations in the operation. For example, the analysis for Case 2 will statistically represent the process for other assemblies, or variations of the one analyzed, that fall in a range 14.25–15.75 minutes. On the other hand, the analysis for Case 1, with its lower frequency of occurrence, will statistically represent a wider range of variation; for example, any type of similar assembly ranging from 12 to 18 minutes to prepare. This leads to the conclusion that when work has been measured by representative sub-operations of sufficient accuracy, fewer data units may be needed to establish the standard time. The data reduction that can result from this higher-level analysis will obviously save time and effort for the analyst.

ACCURACY TEST

Previous accuracy calculations were also based on the balancing time and balancing effect theories. These are graphically illustrated in Figure A.5.

As Figure A.5 shows, the desired level of accuracy of ± 5% (r_T) is required to be reached as the sum of the individual measurements (t) approaches a certain point. The total time at this point is referred to as the balancing time (T_t) which in this case is 40 hours. A balancing time of 40 hours, as in this example, allows a wide margin for variation while establishing the time of individual activities in most typical cases. Many of the predetermined motion time systems, including MOST, are typically capable of far more accuracy than this.

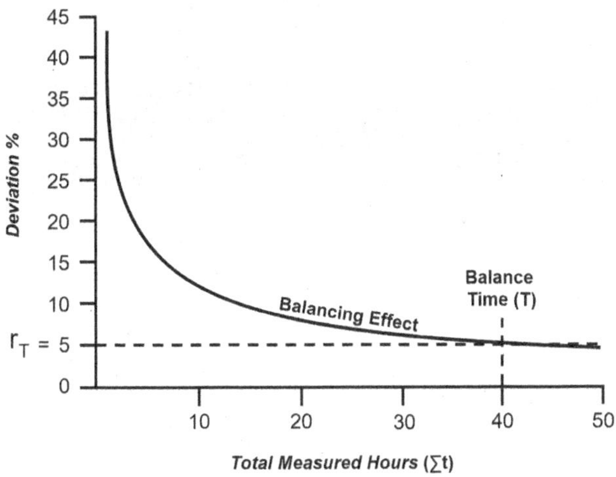

FIGURE A.5 Balancing time.

As for the balancing effect, it can be tested by evaluating the deviation between the true times for different sub-operations and the allowed times for these same sub-operations. According to theory, the desired level of accuracy should be achieved at the calculated balancing time.

Figure A.6 lists 10 time ranges covering 0.0–11.0 hours and the allowed time representing each of these ranges. In the middle columns of the table the maximum allowed deviation for each time is shown. All table values were determined from the allowed deviation formula for a 40-hour balancing period with a 95% confidence level.

To test the accuracy of this system, a series of random numbers should be used to represent the true times to perform certain sub-operations. Random numbers are usually generated by a computer program or derived from a table of random values.

Using a random number generator, the last two figures of seven-digit numbers may be used to represent the true time for the sub-operation in hours with one decimal place. For example, the number 4122375 would generate the true time of 7.5 hours.

After the true time has been established from the random number table, the next step is to select the appropriate allowed time based on the range into which the actual time falls. Using the table in Figure A.6, the time of 7.5 hours falls within the range greater than 7.2–9.0 hours; therefore, a time of 8.1 hours would be allowed since it is the midpoint of the range. In order to evaluate the balancing effect theory, enough values must be chosen so that the total of the true times is at least 40 hours (the balancing time of Figure A.5). This is necessary to ensure that the desired level of accuracy is achieved with a 95% confidence level. A complete test example is tabulated in Figure A.7.

Notice that although individual deviations were as large as 20% in one instance, the *total* deviation was less than the ±5% required. If we relate this example to an actual work measurement situation, we can see that calculated deviations can be allowed in individual measurements without losing the level of accuracy desired in

Range (Hours)	Maximum Allowed Deviation		Allowed Time Value (Hours)
	± Hours	± %	
0.0 - 0.2	0.1	100	0.1
> 0.2 - 0.6	0.2	50	0.4
> 0.6 - 1.2	0.3	33	0.9
> 1.2 - 2.0	0.4	25	1.6
> 2.0 - 3.0	0.5	20	2.5
> 3.0 - 4.2	0.6	17	3.6
> 4.2 - 5.6	0.7	14	4.9
> 5.6 - 7.2	0.8	13	6.4
> 7.2 - 9.0	0.9	11	8.1
> 9.0 - 11.0	1.0	10	10.0

FIGURE A.6 Allowed time values for a 40-hour balancing period.

True Time in Hours *(from random number generator)*	Allowed Time in Hours *(from Figure A.6)*	Actual Deviation	
		± Hours	± %
5.3	4.9	- 0.4	- 7.5
3.9	3.6	- 0.3	- 7.7
8.0	8.1	+ 0.1	+ 1.3
2.0	1.6	- 0.4	- 20.0
7.8	8.1	+ 0.3	+ 3.8
1.4	1.6	+ 0.2	+ 14.3
5.5	4.9	- 0.6	- 10.9
3.3	3.6	+ 0.3	+ 9.1
4.8	4.9	+ 0.1	+ 2.1
42.0	***41.3***	***- 0.7***	***- 1.7***

FIGURE A.7 Example of balancing effect.

the final result for a calculation period (one day, one week, etc.). This balancing principle plays an important role in the conceptual design of BasicMOST and the calculation of standards based on MOST. However, a single simulation is rather meaningless to prove the general significance of the balancing effect. Therefore, a computer program for a random simulation was written. Two runs of that program, each with a sample size of 100 simulations, showed that the average percentage error was well within ± 5%: 2.69% and 2.63%, respectively.

In each of the previous examples, the desired level of accuracy was specified to be ± 5%, which is the generally accepted standard for industry. But what about the balancing time? The 40-hour balancing period may be sufficient for calculating incentive standards based on a 40-hour pay period, but hardly acceptable for a line balancing calculation with cycle times in minutes. The use of the standard time is therefore a very important factor when considering the balancing time of a work measurement system. That is why BasicMOST was designed to have a consistent and theoretical balancing time of approximately 2 minutes.

RELATIONSHIP OF BALANCING TIME TO THE ACCURACY OF WORK MEASUREMENT

The concept of balancing time is generally regarded as a means for comparing systems to determine their applicability. For this reason, the following information is provided to ensure that this criterion may be used properly in the selection of MOST Work Measurement Systems.

Every predetermined motion time system has a balancing time which must be calculated from a large collection of operating data by statistical analysis. The relative balancing times for six predetermined motion time systems have been calculated as follows*:

Work Measurement System	Balancing Time (TMU)
MiniMOST	500
BasicMOST	3235
MaxiMOST	32,350 (estimate)
MTM-1	600 (average)
MTM-2	1600 (average)
MTM-3	16,000 (estimate)

At first glance, it would seem sufficient to compare the cycle time of the operation to be analyzed with these balancing times to make a system selection. That is, an operation at least 500 TMU long could be analyzed with MiniMOST, an operation at least 600 TMU long could be analyzed with MTM-1 and so forth. This general guideline would be technically valid, but a practical choice demands a closer look at several factors, especially since this guideline implies that any operation shorter than 500 TMU could not be accurately analyzed.

As explained at the beginning of this appendix, balancing time is defined as the theoretical total time required of a system for the summation of independent basic elements to attain a desired level of precision. For setting labor standards in most industries, the desired level is usually ±5% accuracy with 95% confidence. This means that when a work measurement time equals the system's balancing time, this measurement is expected to be from 0% to 5% less than or greater than the true value 95 times out of 100. Selection of the appropriate work measurement technique should be based on the most practical (economical) system meeting this criterion.

MEASURING SHORT-CYCLE OPERATIONS WITH MOST

A BasicMOST analysis of an operation equal in length to the balancing time has an accuracy of ± 5%. If this same analyzed time is repeated enough times to completely cover a 40-hour period (100,000 TMU per hour × 40=4,000,000/3235 TMU=1236.5) or an 8-hour period (100,000 × 8=800,000/3235 TMU=247.3), the overall accuracy for either period remains±5%. In general, if any operation is repeated identically over an entire period, the measurement accuracy for that period is the same as for one cycle. This also applies to shorter cycle operations. For example, if the cycle time is only 809 TMU (r=0.10), any number of cycles that fills a 2-hour period has an overall accuracy of ±10% when measured with BasicMOST.

However, because of the balancing effect discussed in this appendix, there is a way to combine short-cycle operations that ensures±5% accuracy over any balancing period (planning month, pay week, pay-out day, etc.). For example, each day a worker

* This list of balancing times plus other information provided in this section of the appendix is based in part on the work of Dr. William D. Brinckloe of the University Research Institute from 1975 to 1981.

could perform four *different* 809 TMU operations 247 times each and be assured that the BasicMOST analyses covering these operations would collectively be accurate to within ± 5%. Notice that the performance of each operation occupies 25% of the day (247 × 809 TMU=200,000 TMU=2 hours) and that 809 TMU is 25% of 3235 TMU. This is a special case, but it illustrates a general principle: If the percentage of the balancing period occupied by repetitions of a short-cycle operation is no greater than the percentage its cycle time is of the balancing time, the *combined* accuracy of the analyses of such operations is ± 5% or better, even though the accuracy of the individual analyses may be varied.

SELECTING A MOST SYSTEM TO ASSURE OVERALL ACCURACY

Figures A.8 and A.9 are based on this important principle. If the analyst knows the most likely percentage of the calculation period occupied by repetitions of the operation and the approximate length of an operation in TMU, he or she can quickly determine whether a BasicMOST analysis of the operation will be sufficiently accurate to ensure ±5% accuracy for the group of analyses covering the entire period. This provides a useful guideline for avoiding the extra work that would be required to analyze a short-cycle operation with a system more detailed than necessary. It is important to note that Figures A.8 and A.9 should be used as guidelines in selecting a MOST System. Clearly, they will help identify a viable system to use, but it is important to remember that there is a combination of factors to consider when selecting a system. The percentage the activity is performed and the accuracy desired are two factors, but the analyst needs to consider the cycle time as well as the purpose of the standard in selecting a system.

Using the table in Figure A.8 and the chart in Figure A.9, if two operations are each performed 50% of the week, they should be analyzed with BasicMOST if their cycle times are 1667 TMU (1 minute) since 1667 is greater than 1618 TMU (reference Figure A.8). The same principle is shown in Figure A.10. If the activity totaling 1667 TMU occurs 50% of the time with a 5% accuracy level, BasicMOST is the preferred system.

Referring to Figure A.11, if the typical operation in a department takes about 15 seconds (a little over 400 TMU) and is repeated only enough times to occupy about 10% of the day, BasicMOST would suffice because the intersection of 400 TMU and 10% falls in the BasicMOST region ('A' in Figure A.11). If the operation is repeated enough times to occupy about 20% of the day, however, MiniMOST may be used because the intersection of 400 TMU and 20% falls in the MiniMOST region ('B' in Figure A.11). Figure A.11 is based on a 5% accuracy level. If the accuracy level required is at 10%, Figure A.12 shows that BasicMOST could be used to analyze the operation if it occurs 10% ('A' in Figure A.12) or 20% ('B' in Figure A.12) of the day.

The graph presented in Figure A.9 can also apply to analyses of different lengths as shown in Figure A.13. If a day is comprised of 5% of a 162 TMU operation ('A' in Figure A.13), 25% of an 809 TMU operation ('B' in Figure A.13) and 70% of a 2265 TMU operation ('C' in Figure A.13), the overall accuracy of the BasicMOST analyses will be ± 5%.

Percentage of Period	Operation Time (TMU)	Accuracy of Analysis (±r)
100	3235	.050
95	3073	.051
90	2912	.053
85	2750	.054
80	2588	.056
75	2426	.058
70	2265	.060
65	2103	.062
60	1941	.065
55	1779	.067
50	1618	.071
45	1456	.075
40	1294	.079
35	1132	.085
30	971	.091
25	809	.100
20	647	.112
15	485	.129
10	324	.158
5	162	.224
4	129	.250
3	97	.289
2	65	.354
1	32	.500

FIGURE A.8 Limits to ensure ±5% combined accuracy for the BasicMOST analyses of short-cycle operations.

In each example presented, the frequency of occurrence happens to be the same for all operations, but the table and graph also apply to relative frequencies other than 1:1. For example, during each 4000 TMU of a balancing period, a 2000 TMU operation is performed once and a 1000 TMU operation twice. Since each operation occupies 50% of the period, the 2000 TMU operation can be analyzed with BasicMOST, but the 1000 TMU operation may need to be analyzed with MiniMOST. However, during each 8000 TMU of another balancing period, three different 2000 TMU operations are performed once and a 1000 TMU operation twice. Since each operation occupies only 25% of the period, all the operations can be analyzed with BasicMOST.

To summarize, when all analyses of the operations that fill the balancing period fall within the charted limits, overall accuracy within ± 5% is assured. This holds

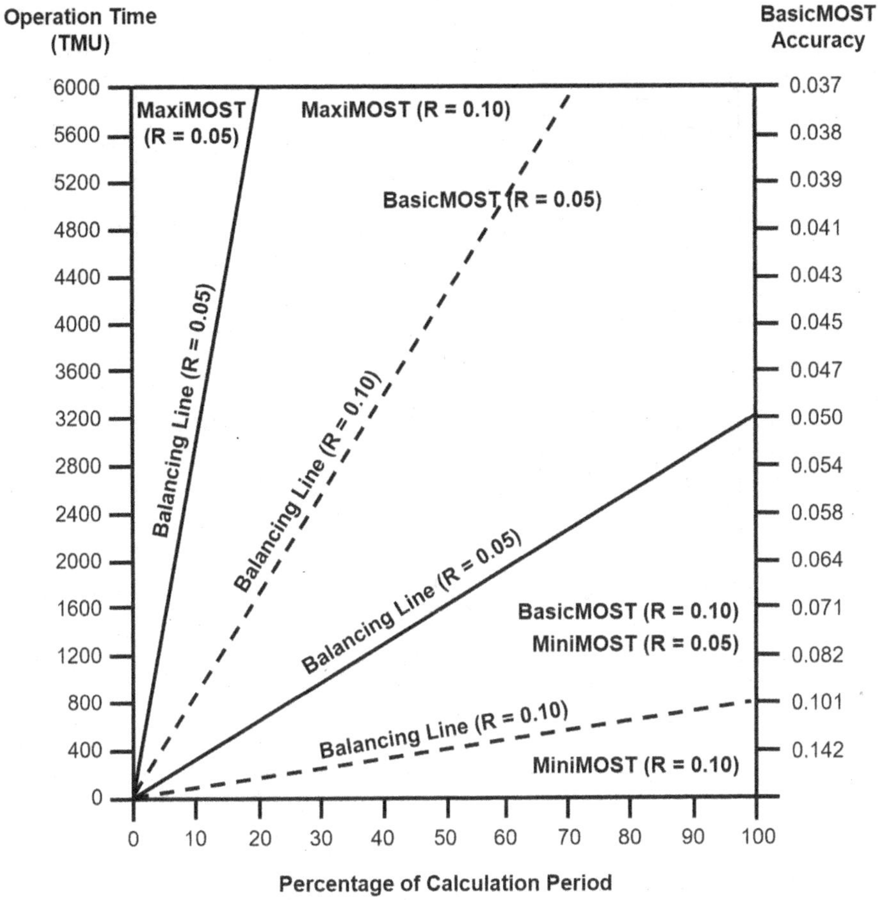

FIGURE A.9 Selection of a MOST version for the analysis of short-cycle operations where R = required accuracy.

true even for an analysis as short as 32 TMU (± 50% accuracy), although it is unlikely that BasicMOST would be used to analyze an operation this short. These charts illustrate the principle that the smaller the portion of the balancing period devoted to a particular operation, the less accurate its analysis can be without sacrificing overall accuracy beyond accepted limits.*

EFFECT OF VARIATIONS WITHIN AN OPERATION CYCLE

In a typical operation cycle, many sub-activities occur. These sub-activities are analyzed as Action Distances, Body Motions, Placements, etc. Each sub-activity differs

* Figure A.9 is also useful in another way: If the accuracy of any unit of standard data is known (regardless of its cycle length), its allowable percentage of the balancing period can be determined and vice versa. Start with the error ratio on the right of the graph, go left to balancing line and then down to find the percentage.

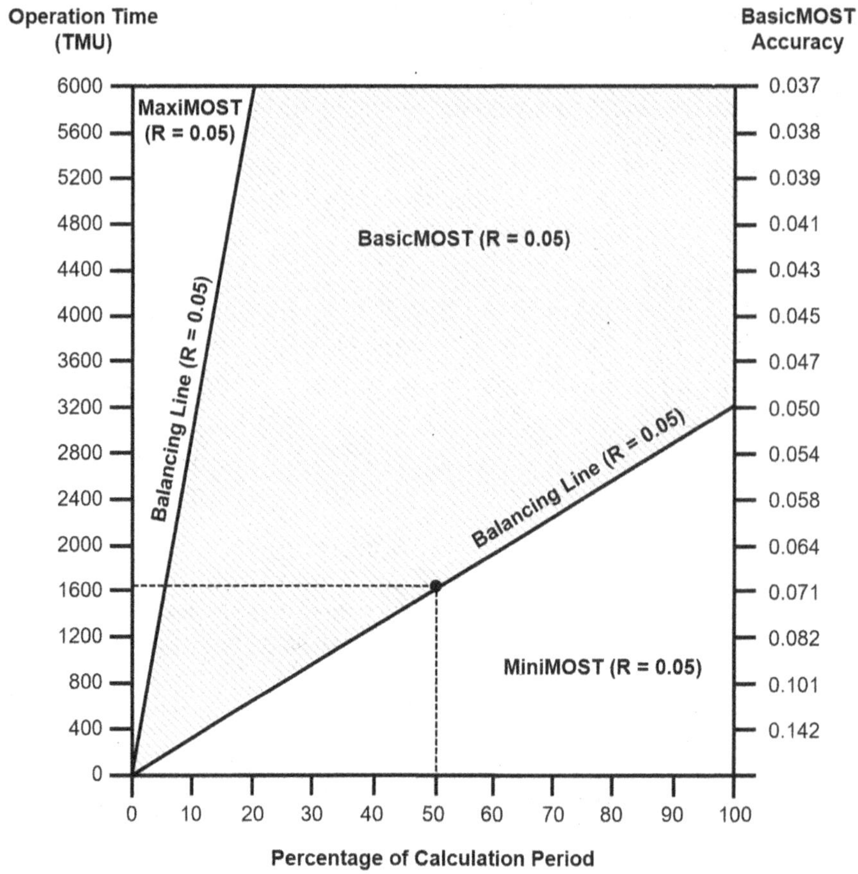

FIGURE A.10 Selection of a MOST System for an activity occurring 50% of the calculation time.

from the others to some extent, even though the same index values may be assigned. For example, an $\mathbf{A_1}$ in BasicMOST may be assigned for a 10 inch reach, a 17 inch move, a 26 inch reach or a 5 inch move. The time allowed for each of these Action Distances is 10 TMU which usually differs from the actual time for the sub-activity. In accordance with statistical principles, the algebraic sum of these differences approaches zero as more of them are included in the sum. This is equivalent to saying that the measured time approaches the true time for the operation as more sub-activities are included in the analysis. This illustrates the balancing effect, as defined above, which is the basis for the accuracy of MOST. It also illustrates the need for these variations to occur within the operation cycle so that a balancing effect will occur.

Balancing time is the measured time that ensures the precision of the measurement is within acceptable limits. The determination of a system's balancing time is based on the assumption that only independent elements enter into the calculation.

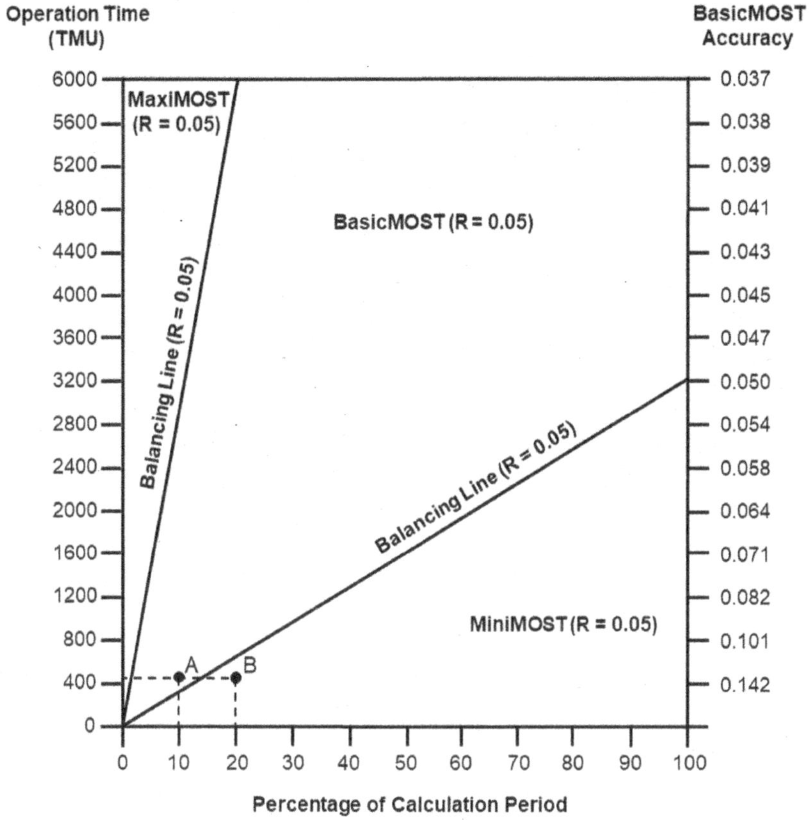

FIGURE A.11 Selection of a MOST System based on a 5% accuracy level.

This means that each element randomly differs from the others to contribute to the balancing effect. Therefore, any sequence of steps repeated identically makes no contribution to balancing. For this reason, repeated steps should be disregarded when using the system selection charts in Figures 2.3, 2.4, A.8 and A.9, which are based on the balancing time. Figures 2.3 and 2.4 can be found in Chapter 2 and can also be used to assist in system selection. For example, if an operation takes 1000 TMU but 20% of this time is for identical repetitions, use 800 TMU when using the charts. A rough estimate of the repetitive portion is sufficient to preserve accuracy. (Note that the percentage of the balancing period occupied by repetitions of the operation should still be based on its total time, 1000 TMU.)

EFFECT OF CYCLE-TO-CYCLE VARIATIONS

Assuming the system has been applied properly in accordance with the specified guidelines, any MOST version always provides a combined precision within

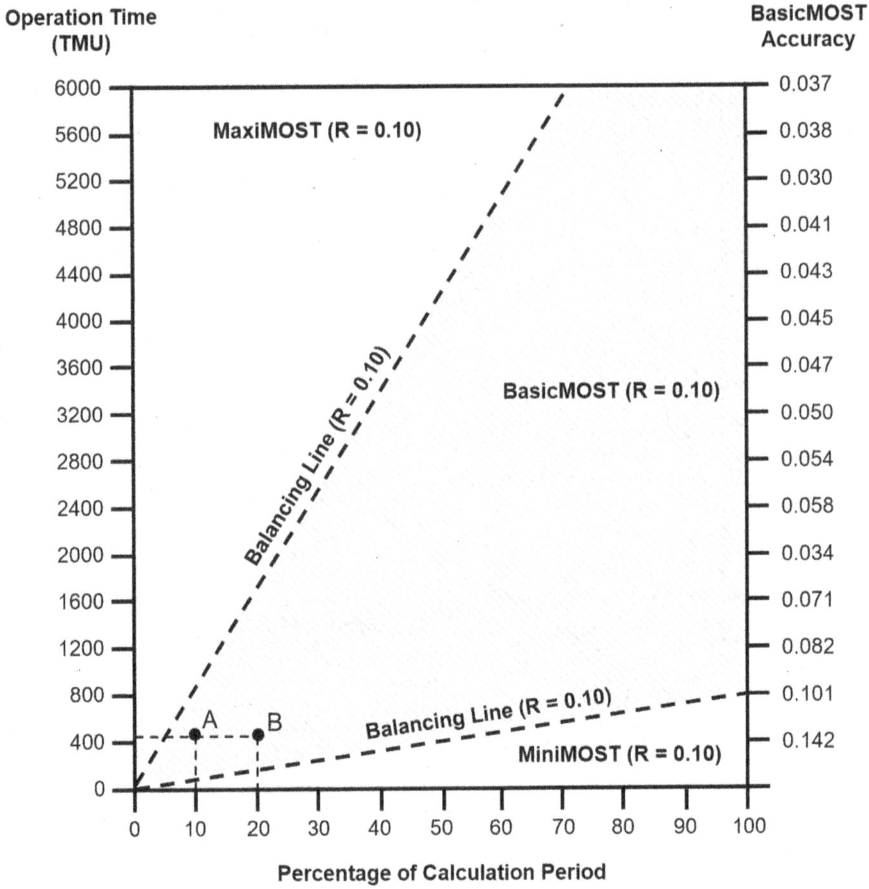

FIGURE A.12 Selection of a MOST System based on a 10% accuracy level.

±5% with 95% confidence if all the work being analyzed is performed exactly the same from cycle-to-cycle. However, because of the lack of operator training or practice, differences in parts or their orientations and numerous other factors, cycle-to-cycle variations are common. How does this affect the accuracy of a MOST analysis?

The answer lies in the fact that when cycle-to-cycle variations fall within the range of the index assigned, there is no significant effect on system precision. In a MOST analysis, an index value is assigned for each sub-activity (Action Distance, Gain Control, etc.). Each index value represents a range of possible times (variations) for the sub-activity being analyzed. One of the great benefits in the design of MOST is the optimal selection of index ranges to accommodate most of the cycle-to-cycle variations likely to occur for each sub-activity. For example, the BasicMOST value of $\mathbf{A}_1$ very conveniently includes every spatial reach and move distance from greater than 2 inches (5 cm) to the full extent of the operator's reach. Furthermore, since only

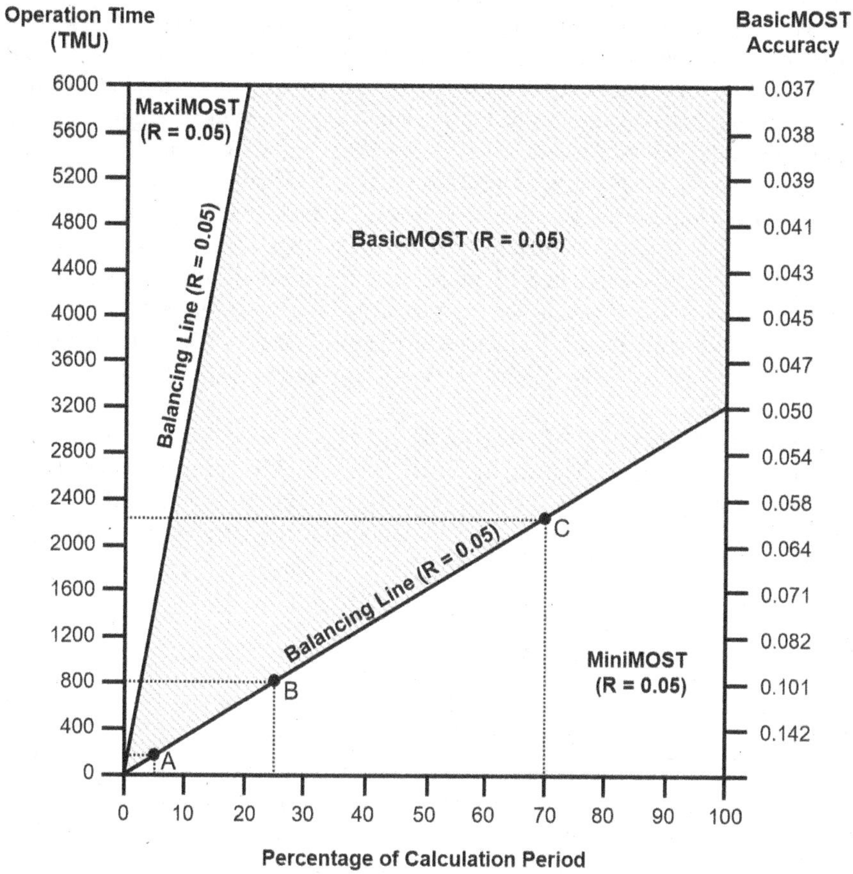

FIGURE A.13 Example of selecting a MOST System with different operation times.

the motions actually required to accomplish the work are analyzed, the application of MOST effectively eliminates almost all variations from consideration during the analysis.

However, especially in the use of a lower-level system, the expected range of variations still exceeds the index range. This problem is most likely to arise in the application of a detailed system, such as MTM-1, which has many narrow ranges for the data card values. Even though an effort has been made to improve the design, layouts and method for an operation, variations spanning two or more ranges may occur in some sub-activities because of the nature of the work. In these cases, the analyst may separately analyze each variation weighted according to its frequency of occurrence or simply choose an average value on which to base the analysis. The first approach is usually tedious; the second is usually inaccurate. If either approach is needed more than a relatively small number of times in the analysis, a higher-level system should be used.

This recommendation is based not only on practical considerations (which are illustrated in the following section), but also on research that indicates that the balancing time of a system is achieved when an analyst's average samples include a substantial number of values outside the ranges of the assigned indexes. This is because the calculation of a balancing time is based on the variances *within* index ranges. Therefore, when averaged variations exceed an index range, the variances *between* ranges must be added, which effectively increases the balancing time. If the collective averaging while using a lower-level MOST System includes a total range of variations that equals about 50% of the analyzed time for the entire operation, the balancing time is reduced to the point at which it actually equals the balancing time of a higher-level system. Under these circumstances, a MiniMOST analysis although more detailed, would be no more accurate than a BasicMOST analysis of the same operation.

Fortunately, users of MOST will rarely be concerned with these findings for two reasons: (1) cycle-to-cycle variations are least likely to occur in the most often repeated operations primarily because operator movements evolve into consistent patterns, and (2) when the system selection guidelines (Chapter 2) are followed, the MOST System used will almost always be the best choice.

AVERAGING CYCLE-TO-CYCLE VARIATIONS

It is well-established that we must deal with averages in most work measurement situations; that is, average distances, average weights, average types of motion, etc. Consider the following situation: seated at a punch press, an operator gets a part that has just been formed and moves it to one of 12 spaces in a parts tray within reach to the right. The distance from the press to the spaces on the tray varies from 8 to 32 inches (20–80 cm). If that were the entire operation to be measured, which work measurement technique would you choose—MTM-1, MTM-2 or BasicMOST? Which technique *could* you choose? The cycle time is obviously very short. To achieve an accurate analysis, your choice should be a very detailed system. Or should it?

An MTM-1 analysis of the punch press operation (Figure A.14) might be:

MTM-1	TMU
R24A	14.9
G1A	2.0
M24B	20.6
RL1	2.0
Total	39.5

Notice that in the MTM-1 analysis, the analyst assumed an average distance of 24 inches (60 cm), represented by the 'R24A' for the operation. In actuality the distance varied from a minimum of 8 inches (20 cm) to a maximum of 32 inches (80 cm). Analyzing the same operation using the minimum and maximum distances would result in the analysis shown as:

MTM-1 Minimum (8 inches, 20 cm)	TMU	MTM-1 Maximum (32 inches, 80 cm)	TMU
R8A	7.9	R32A	18.3
G1A	2.0	G1A	2.0
M8B	10.6	M32B	25.5
RL1	2.0	RL1	2.0
Total	22.5	Total	47.8

Deviations: Total deviation (8–32 inches, 20–80 cm), 64%
Minimum to average (8–24 inches, 20–60 cm), 43%
Average to maximum (24–32 inches, 60–80 cm), 21%

Since variation did exist, the MTM-1 analyst chose to average the distance rather than separately analyzing each of the 12 variations.

An MTM-2 analysis of the same operation will give:

MTM-2	TMU
GB32	23
PA32	20
Total	43

With BasicMOST and MiniMOST, using 24 inches (60 cm) as the Action Distance, one General Move Sequence Model can be used as shown on the next page:

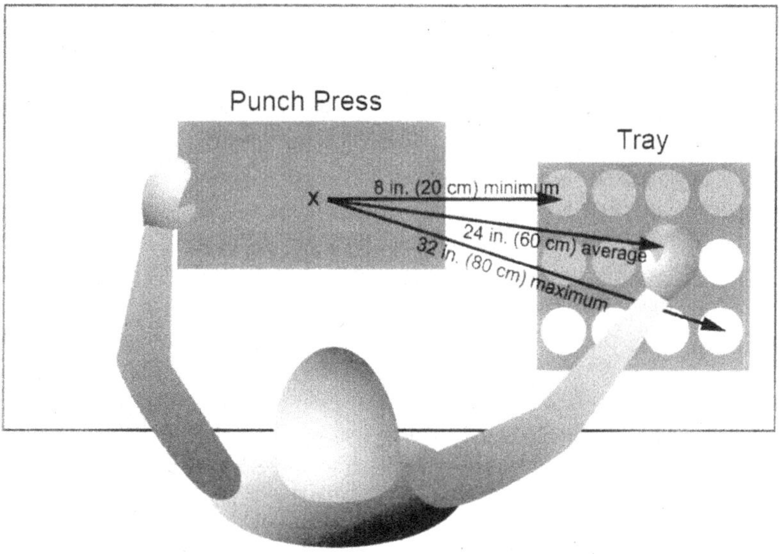

FIGURE A.14 Punch press operation.

BasicMOST	TMU
A_1 B_0 G_1 A_1 B_0 P_1 A_0	40

MiniMOST	TMU
A_{16} B_0 G_6 A_{16} B_0 P_6 A_0	44

The four total TMU values in the above analysis (39.5, 43, 40 and 44 TMU) are based on averages.

With MTM-1, the analyst selected an average distance for the reach and move motions based on a subjective judgment. MTM-2 time values are determined from the weighted average of different MTM-1 motion patterns. Index values in MOST are based on statistically calculated averages. The question is no longer "which is the correct analysis?" but "which is the most acceptable average?" No one can say with certainty which average is better. Therefore, when dealing with situations in which variations in the operation occur from cycle-to-cycle, BasicMOST gives results that are as accurate as the more detailed systems. In the analysis of an operation that contains substantial variations, MTM-1, MTM-2, BasicMOST and MiniMOST all produce an acceptable time value from an accuracy standpoint.*

In order to use a detailed work measurement system like MTM-1, made up of a large number of more or less independent elements, considerable subjectivity is required in making decisions for an analysis. 'Subjective averaging' can be good or bad. One thing is certain: this approach for work measurement does not use a consistent method, and the results are likely to be greatly influenced by the individual's experience and performance. BasicMOST is a more objective tool and consistent in this respect because the averages have been statistically established and can be consistently applied.

CONCLUSION

When analyzing an operation that varies from cycle-to-cycle, even the most detailed systems concede accuracy to the analytic technique of averaging. The question is then a subjective one of choosing the average that appears to best fit the situation.

Through the use of MOST sequence models, analysts are aided in making the correct decisions. The result is smaller deviations among analysts compared to other predetermined motion time systems. The use of a statistically derived index scale further assures the consistency of MOST.

These and other factors discussed in this appendix play an important role in selecting the choice of a work measurement technique to use in analyzing an operation. However, the selection of which MOST System to use is as straightforward as the guidelines provided in Chapter 2, which assure both *accuracy and consistency* in analyzing any operation and *economy* in the time and effort required.

* The practical accuracy of MTM-based standards is discussed in *The Impact of Variation in Method or Workplace on the System Precision of MTM Based Standards*. University Research Institute, March 1979.

Index

The index consists of two parts:

1. Subject categories.
2. Subject index.

Subject categories are in bold in the subject index.

The same subject may appear under more than one subject category to facilitate the search for information.

The same subject may appear in more than one chapter.

To use the index:

1. Select a subject category.
2. Go to the selected subject category in the subject index (alphabetical order).
3. Select a subject from the subject list.
4. Go to the indicated chapter number and/or page number.

Subject Categories

Action Distance
Alignment
Assemble or Disassemble (*see also* Fasten or Loosen or Tighten or Loosen)

BasicMOST
Body Motion

Controlled Move Sequence Model
Crane
Cut

Data Cards

Equipment Use Sequence Model

Fasten or Loosen (*see also* Assemble or Disassemble or Tighten or Loosen)

Gain Control
General Move Sequence Model
General Tools I (MaxiMOST)
General Tools II (MaxiMOST)

Keyboard
Keypad

Letter/Paper Handling

Machine Handling Sequence Model
Managing Performance
Manual Crane Sequence Model
Material Handling Equipment
MaxiMOST
Measure
Measuring Tools
Method Descriptions
Method Levels
MiniMOST
MOST Work Measurement Systems
Move Controlled

Parameter Frequencies
Part Handling Sequence Model
Placement
Powered Crane Sequence Model
Powered Truck Sequence Model
Process Time
Productivity

Record

Simultaneous Motions
Standard Time
Surface Treat

Terms
Think
Tighten or Loosen (*see also* Assemble or Disassemble or Fasten or Loosen)
Tools
Tool Use Sequence Model

Validation of Process Time

Workforce Performance
Work Measurement

Subject Index